W9-APK-111

Historical Dictionary of Data Processing

Historical Dictionary of Data Processing

BIOGRAPHIES

James W. Cortada

Greenwood Press
NEW YORK
WESTPORT, CONNECTICUT
LONDON

Library of Congress Cataloging-in-Publication Data

Cortada, James W.
Historical dictionary of data processing—biographies.

Includes index.
1. Electronic data processing—Dictionaries.
2. Electronic data processing—Biography. I. Title.
QA76.15.C66 1987 004'.092'2 [B] 86-31805
ISBN 0-313-25651-9 (lib. bdg. : alk. paper)

Library of Congress Catalog Card Number: 86-31805
ISBN: 0-313-25651-9

First published in 1987

Greenwood Press, Inc.
88 Post Road West, Westport, Connecticut 06881

Printed in the United States of America

The paper used in this book complies with the Permanent Paper Standard issued by the National Information Standards Organization (Z39.48-1984).

10 9 8 7 6 5 4 3 2 1

Contents

Preface

This historical dictionary is one of three companion volumes being published by Greenwood Press to provide basic research tools for the study of data processing. The first volume covers institutions such as companies, societies, and laboratories (*Historical Dictionary of Data Processing: Organizations*); the second reviews a range of software and hardware achievements (*Historical Dictionary of Data Processing: Technology*); and this volume presents biographies of important figures in the development of data processing: *Historical Dictionary of Data Processing: Biographies*. Taken together, the three volumes provide over 400 entries on all aspects of data processing, from the earliest to contemporary times. The size of the entries ranges from 300 words to many pages. Basic factual material is presented, and where appropriate, historical interpretation is provided. For those who need additional material, the entries conclude with bibliographic references.

When an entry mentions a topic that has its own entry in one of the three volumes, a cross-reference is indicated by one of three symbols. The * indicates that the entry may be found in the technology dictionary; the ** points to this dictionary; and † indicates that the topic is in the volume on organizations.

The indexes in each volume provide access to material within entries. Some readers may feel that additional topics could have been included but should keep in mind that space limitations, economics, and the unavailability of good hard data made it necessary to limit the entries somewhat. As the field of data processing history matures, a more complete publication can be considered.

This dictionary offers a ready reference to the biographies of over 150 individuals prominent in the history of data processing, including scientists, engineers, government officials, executives of major computer manufacturing companies, and users of data processing technology. Biographees range from ancient Europeans concerned about astronomy down to software entrepreneurs of the 1980s. For each entry the historical significance of an individual's contribution is defined and that person's place in the history of data processing is indicated. Background data, such as birth and death dates, education, employment history, and other activities, are fully covered, and, in addition, key issues as reflected in current historical literature are suggested. The reader will thus find, for example, discussions about who invented the modern electronic digital computer in several entries. The book is not intended, however, either to provide a definitive history of data processing or to be the ultimate guide for the specialist.

This work provides a balance between the needs of the general reader and the specialist. For the general user, the volume reviews subjects with the assumption that the reader's knowledge of data processing technology and its history is limited. For the specialist, it is a starting point and a reference tool.

There is a general lack of published biographical data on many of the scientists who built computers or wrote software. For this reason, of the three dictionaries, this one was the most difficult to compile. Historians have done little work in this area, even on major figures. Most of the biographees lived in the twentieth century and are still living. Many have been interviewed as part of various oral history projects.[1] In some cases, I turned to the subjects themselves for basic information, enriching American Telephone and Telegraph, one of the major firms in the data processing industry, in the process! Important biographies have been written for individuals such as Herman Hollerith,[2] the inventor of modern punch card equipment, and Thomas Watson, Sr.,[3] the founder of the International Business Machines Corporation (IBM). But no serious biographies exist, for example, on John von Neumann, a key architect of the modern computer.[4] Recently, Alan Turing, a British scientist, was the subject of a major biography by Andrew Hodges which can serve as a model of the science biography.[5] Collections of smaller chapter-length biographies have begun to appear, serving as stop-gap measures.[6] Fortunately, some published memoirs have proven to be extremely useful to historians.[7]

In the 1920s and 1930s, Thomas Watson, Sr., enjoyed visiting sales offices and manufacturing plants to meet with IBM's employees. Invariably, he would speak about the company's basic beliefs—respect for the individual, customer service, and excellence. But nothing captivated him more than people and their individual accomplishments. He had profound respect for the "power of one," for what a single person could accomplish.[8] Two decades later his son still hammered away on the same theme; the "wild ducks" in IBM, he said, made a difference and a positive contribution. Salesmen who met or exceeded their quotas were made members of the One Hundred Percent Club and engineers and scientists who made important contributions were made IBM Fellows. The Watsons, particularly Thomas Watson, Sr., recognized that data processing was an enterprise that depended as much on individual accomplishments as on the structure provided by a company, a university, a government, or a market.

The data processing industry was built on technologies that did not exist until individuals stumbled across them. This was certainly the case, for example, at Bell Labs with the transistor and a few years later with the computer chip. These people often worked alone and sometimes even without vision. Others functioned as teams pursuing common goals that were only partially defined at the beginning of their project and much changed by the time it ended. The WHIRLWIND, which Jay W. Forrester at the Massachusetts Institute of Technology built in the early 1950s, was very different from the system he had envisioned in the late 1940s. A few years earlier John Mauchly and J. Presper Eckert had a similar experience while building the ENIAC, the world's first electronic digital computer. In each case, however, engineers were bent on solving specific problems—usually in finding faster ways to compute mathematics—and along the way built calculators. Obvious examples include Gottfried Wilhelm von Leibniz with his calculator in 1673, Charles Babbage and his analytical engines built in the early 1800s, and Steven P. Jobs with the Apple microcomputer in the 1970s.

While I was preparing these biographies, several observations occurred to me about the group as a whole which gave credence to Watson's faith in the power of one. First, many of the technological advances and products came from very young people. For example, Leibniz was twenty-seven years old when he constructed a calculator, Babbage in his early thirties when he designed his difference engine, and Hollerith twenty-two when he began developing punch card equipment for the U.S. Bureau of the Census. That pattern has continued in our own day. Mauchly was thirty-five years old when he began to build the ENIAC, and, more recently, Bill Gates, the founder of Microsoft, the industry's largest manufacturer of software for microcomputers, sold his first microcomputer product while he was still a teenager. Steve Jobs at Apple was barely twenty-one years old when he introduced his first machine. And even at IBM, Don Estridge, the father of the Personal Computer (PC), was only forty-two when the company introduced the PC—making him very old by industry standards.

Second, most of the results described in the three dictionaries were the products of individuals operating alone or in very small groups. Estridge developed the IBM PC in the early 1980s beginning with a dozen employees; Hollerith built his first machines alone and then hired an assistant; and nearly all the scientists of the sixteenth and seventeenth centuries operated alone. The largest of the early twentieth-century projects, involving the construction of the ENIAC in the mid-1940s, involved only a handful of people. Forrester's WHIRLWIND made use of the largest group—several hundred by the time the system was completed. Not until IBM decided to develop the S/360 family of computers was a major and complex effort of vast proportions undertaken. This project involved hundreds of millions of dollars and almost every IBM scientist and engineer for over five years. Yet even by today's standards and in light of major projects of the preceding twenty years, the S/360 project was exceptional. Throughout the history of the data processing industry, the most important advances have usually come from small groups or individuals. One recent study on excellence in American corporations repeatedly cites people and companies in the data processing industry as examples of how progress can be made precisely through the use of small groups.[9]

Brilliance along with solid academic achievements is conspicuous in the biographies presented in this book. Nearly half of the individuals in this dictionary earned their master's or Ph.D. degrees. Several won the Nobel Peace Prize (one even twice), and as a group they authored thousands of articles and books. Child prodigies are generously scattered among them. Almost all the engineers and scientists (and many executives as well) included became interested in radio, electronics, physics, science, and, later, computing as children or as teenagers and made considerable strides in mastering these subjects before reaching their majority. The computer hacks of the 1980s are simply variants of the ham radio operators of the 1930s. Many were uninterested in their formal education until they reached college or university, and others were socially awkward. These people frequently pushed back the boundaries of technology without at first knowing the potential of their work. The pages that follow describe people who developed machines in garages (the Apple, for example), in guest bedrooms in their parents' homes (at least three major software vendors began that way), or in college dorms (Bill Gates at Harvard, for instance).

After these people recognized the commercial possibilities of their work, their companies were often capitalized with less than $1 million for hardware firms, and frequently under $1,000 (especially for software vendors). Not uncommonly, their firms grew 50, 300, or even 500 percent each year for several years when their products arrived on the market. Most engineers made poor business managers, and so such rapid rates of growth too often turned into major declines when either their companies became too big (Osborne Computer Corporation, for example) or their products were superseded by new technologies (UNIVAC machines).

Invariably, the professional manager came to dominate many companies that had been originally formed by scientists and engineers peddling one product. When this happened, they often survived and grew. Consolidations of highly specialized companies led to larger organizations and survival as well. The business managers ranged from ex-scientists (as, for example, almost all the top executives at Northern Telecom who began as engineers) to people with only business backgrounds (such as W. Michael Blumenthal at Burroughs Corporation). The challenge they all faced was to allow their companies to grow evenly and profitably in the face of rapid technological change without giving up proven business practices. The ability to manage a business in this industry first as a business and then as a purveyor of technology has been only rarely achieved. For most firms, it has been almost impossible to sustain that balance over any appreciable period of time. The one major exception has been IBM. Its executives were businessmen first, and enthusiasts of data processing technology second. At IBM science was made to serve business.

Studies have shown that the data processing industry has been one of the most competitive in the world's economy because so many individuals have entered the market with new products to compete against existing mainstays.[10] This circumstance has been largely caused by the leapfrogging of new technology into the market. A common pattern emerged in which a scientist would pursue a notion successfully and then form a company to market a product based on that work. Because many such scientists became rich and successful, ruthless competition was inevitable. It was a very volatile industry, with the successful firms of one year going out of business in the second or third year. The courage of people to formulate ideas into products characterized many industry activities.

Over half of the major contributors to data processing in the twentieth century included in this book are still alive and productive. But almost all of them experienced a distinct period in their lives when they were most clearly productive. For scientists the period usually lasted between ten and fifteen years, but for businessmen it frequently was closer to only five years. I do not have a good explanation for this pattern and hope that future historians will study the phenomenon. Examples abound. Maurice V. Wilkes built important digital computers in the late 1940s and early 1950s in Great Britain. Although he worked diligently throughout the 1960s and 1970s, and even into the 1980s, he never again matched the impact of his earlier years. The same pattern emerges with William Shockley, who developed the chip in the 1950s; Robert Noyce, who created Intel; and Eckert and Mauchly, whose work peaked with their UNIVAC in the 1950s.

Perhaps one explanation of this pattern lies in the phenomenon of rapidly developing technology bypassing individuals. Clearly, changes have come very fast in each of the past four decades. Another explanation may be that a shift in duties from research to administration, as happened to Wilkes over a period of years, may have curtailed productivity. Or perhaps turning to other areas of interest may be responsible, as happened with Jay Forrester when he gave up building machines. Or we may not as yet have recognized that the periods we now characterize as less fruitful were actually of great historical importance.

For executives the shorter period of productivity might be explained by the nature of authority. Presidents or chairmen of companies generally serve in such positions for less than five years. Yet it is precisely in such jobs that executives can have the most influence on events. There have been some dramatic exceptions, especially Hollerith and each of the Watsons and Noyce, all of whom were chief executive officers for decades. The explanation remains to be written. At least for events in the United States and Great Britain beginning in the 1930s, research directly enhanced work done by earlier scientists. Most engineers and scientists put existing technology to new uses and configured technology differently by merging various lines of scientific endeavor, including the use of electricity, a greater appreciation for quantum mechanics and physics in general, and the application of chemistry. Digital computers in the 1940s were built in large part out of "off-the-shelf" technologies—vacuum tubes, cabling, electricity, air conditioning, card punch equipment, and typewriters. The ENIAC was built that way in the 1940s, as was the IBM PC in the 1980s. Software usually required more creativity, but even by the 1960s, it was beginning to take on a structure of its own, allowing researchers to build on the work of their predecessors.[11]

Many of the companies and individuals involved were supported by the U.S. government. In fact, almost every major project from the 1930s to the 1960s was completed for the military community. ENIAC, WHIRLWIND, and SAGE are only three examples. There is a long precedent for government sponsorship of technological projects. Much earlier, in the years following the Napoleonic Wars, the British government put Charles Babbage under contract to build a machine that could calculate navigational tables. During the 1880s and 1890s, Hollerith constructed equipment for the census takers of the United States and Europe—all government agencies. The Office of Naval Research (ONR) was perhaps the single largest supporter of scientists in the United States in the years immediately following the end of World War II. ONR's Mina Rees supported many mathematicians and critically important computer-related projects. Whole companies began as suppliers of computers, primarily to the military. Two early and typical examples included Engineering Research Associates (ERA), which was established in 1946 to supply the U.S. Navy, and Computer Research Corporation (CRC) which came into existence in 1950. Obviously, then, in future years historians of data processing will have to study the role of major European and American governments in the development of the industry and its technology. The same issues to be studied are evident with the governments of France, Great Britain, both Germanies, the Soviet Union, and possibly now China. Canada and Australia have also given generous support to their scientists. Each one of these nations, as part of national policy, have supported a local data processing industry. Japan is the most dramatic example of this policy.

The biographies presented in this book are a celebration of the human spirit, testifying to the good that intelligence, ambition, hard work, and even frustrations can bring to society. These individuals transformed what was to have been the age of nuclear energy into the age of information. Technology in the 1930s was condemned as a dangerous and sterile intellectual exercise; today, we have come to the realization that it produced a renaissance such as the world has never seen before.

The lives of those who made these giant contributions to our existence in the late twentieth century remain little known. This first edition therefore represents an initial attempt to gather data relevant to historians. More remains to be learned about them.

NOTES

1. The four major collections in the United States are located at the Computer Museum (100 taped lectures), the National Museum of American History at the Smithsonian Institution (several hundred interviews), MIT Archives (several dozen interviews), and the Charles Babbage Institute (with over 70 and still growing). For details on these collections, see William Aspray and Bruce Bruemmer, "Oral Histories of Information Processing," *Oral History Association Newsletter* 18, No. 4 (1984): 2. Also useful is

''AFIPS/Smithsonian Project on Computer History,'' *Communications of the ACM* 14, No. 7 (1971): 494.

2. Geoffrey D. Austrian, *Herman Hollerith: Forgotten Giant of Information Processing* (New York: Columbia University Press, 1982).

3. Thomas Belden and Marva Belden, *The Lengthening Shadow: The Life of Thomas J. Watson* (Boston: Little, Brown any Co., 1962); William Rodgers, *Think: A Biography of the Watsons and IBM* (New York: Stein and Day, 1969).

4. However, see Nancy Stern, ''John von Neumann's Influence on Electronic Digital Computing, 1944–1946,'' *Annals of the History of Computing* 2, No. 4 (October 1980): 349–362. Although almost totally avoiding his work with computers, biographical data are available in S. J. Heims, *John von Neumann and Norbert Wierner: From Mathematics to the Technologies of Life and Death* (Cambridge, Mass.: MIT Press, 1980).

5. Andrew Hodges, *Alan Turing: The Enigma* (New York: Simon and Schuster, 1983); and for a short treatment, B. E. Carpenter and R. W. Doran, eds., *A. M. Turing's ACE Report of 1946 and Other Papers* (Cambridge, Mass.: MIT Press, 1986).

6. Recent examples include Paul E. Ceruzzi, *Reckoners: The Prehistory of the Digital Computer, from Relays to the Stored Program Concept, 1935–1945* (Westport, Conn.: Greenwood Press, 1983); David Ritchie, *The Computer Pioneers* (New York: Simon and Schuster, 1986): Joel Shurkin, *Engines of the Mind: A History of the Computer* (New York: W. W. Norton and Co., 1984).

7. Vannevar Bush, *Pieces of the Action* (New York: William Morrow and Co., 1970); Herman H. Goldstine, *The Computer from Pascal to von Neumann* (Princeton, N.J.: Princeton University Press, 1972); Emerson W. Pugh, *Memories That Shaped an Industry* (Cambridge, Mass.: MIT Press, 1984); Maurice Wilkes, *Memories of a Computer Pioneer* (Cambridge, Mass.: MIT Press, 1985). A useful compendium of article-length memories can be found in N. Metropolis et al., eds., *A History of Computing in the Twentieth Century* (New York: Academic Press, 1980). Almost every issue of the *Annals of the History of Computing* is dominated by articles written by scientists on their early computer projects.

8. See, for example, Thomas J. Watson, *As a Man Thinks* (New York: IBM Corporation, 1936), *Human Relations* (New York: IBM Corporation, 1949), and *Men-Minutes-Money* (New York: IBM Corporation, 1934).

9. Thomas J. Peters and Robert H. Waterman, Jr., *In Search of Excellence: Lessons from America's Best-run Companies* (New York: Harper and Row, 1982).

10. Franklin M. Fisher et al., *Folded, Spindled, and Mutilated: Economic Analysis and U.S. v. IBM* (Cambridge, Mass.: MIT Press, 1983) and fisher et al., *IBM and the U.S. Data Processing Industry: An Economic History* (New York: Praeger Publishers, 1983); Katharine Davis Fishman, *The Computer Establishment* (New York: Harper and Row, 1981); Stephen T. McClellan, *The Coming Computer Industry Shakeout* (New York: John Wiley and Sons, 1984).

11. Jean E. Sammet, *Programming Language: History and Fundamentals* (Englewood Cliffs, N.J.: Prentice-Hall, 1969): 30–128, 722–736; Richard L. Wexelblat, ed., *History of Programming Languages* (New York: Academic Press, 1981), passim.

Acknowledgments

Completion of this project would not have been possible without the encouragement, suggestions, and improvements of many people. Librarians at Vassar College, the U.S. Library of Congress, the Smithsonian Institution, the University of Virginia, Vanderbilt University, and the International Business Machine Corporation in Poughkeepsie, New York, all made materials available to me. To each of them I owe an enormous debt of gratitude.

So many individuals helped that it would, of course, be difficult to mention them all. A few deserve special recognition, however. Hank Tropp suggested many entries and gave me leads to information. Nancy Stern edited my original list of proposed entries, and although she may have wondered if they would ever be written, she supported my plan. Paul Ceruzzi was also supportive, as was George Ledin, Jr., who provided enormous quantities of biographical data on people associated with ALGOL. In the course of writing these volumes, I had the enormous pleasure and honor of talking to living "pioneers" such as Mina Rees and Emerson W. Pugh. Finally the editor of the *Annals,* Bernard A. Galler, cheered me on in the beginning.

At Greenwood Press, I want to make special acknowledgment of Cynthia Harris, editor at Greenwood, who served as an outstanding manager of the project as well as an intellectual asset. I thank all of the Greenwood staff for their efficient and pleasant supervision.

BIOGRAPHIES

A

ADAMS, HENRY (1838–1918). This great American historian, the author of many political histories of the early days of the Republic, frequently commented on American scientific and social events.

Adams was born on February 16, 1838, and died on March 28, 1918. He graduated from Harvard College in 1858 and did additional study at Western Reserve. He taught history at Harvard from 1870 to 1877 and subsequently edited journals and published books. In 1900, while attending the Trocadéro Exposition in Paris, he concluded that machines were a symbol of his times, much as cathedrals had been for the Middle Ages. In *The Education of Henry Adams* (1906), he wrote that "The new American, like the new European, was the servant of the powerhouse." He gave approval and respectability to an age of technology; for example, he commented that "if man during his period of activity is alive, so is a machine when in operation. As an inventor and constructor of machines I know that to be true." Together with many of his contemporaries, Adams argued that not only were new technologies inevitable but they should be used for the benefit of all. His views encouraged the acceptance of technology and the development of machines, including aids to mathematical calculations in science and business.

For further information, see: Charles Eames and Ray Eames, *A Computer Perspective* (Cambridge, Mass.: Harvard University Press, 1973).

AIKEN, HOWARD HATHAWAY (1900–1973). Aiken built the Mark I,* one of the first automatic computers in the United States, during the early 1940s. It was the first computer made known to the public (in 1944) and signalled the dawn of the computer age. Although not the only computer being built at the time, the publicity it generated revealed that International Business Machines Corporation (IBM)† had played a role in its development, and in turn the computer gave IBM engineers experience with sophisticated computational equipment.

Aiken was born in 1900 and grew up in Indianapolis, Indiana. He completed his B.S. and M.S. degrees at the University of Wisconsin. Between 1923 and 1935, he worked as a power engineer with the Madison (Wisconsin) Gas and Electric Company and at Westinghouse Electric Company. After 1935, he went back to graduate school at the University of Chicago, studied at Harvard, and completed his work for the Ph.D. in 1939. While preparing his doctoral thesis, Aiken became concerned about the tedium associated with manual calculation of numbers in physics and engineering; as a result, he began studying the possibility of using calculating machinery to perform these functions. His only tool for computing sequences of numbers was a desk calculator which proved too slow and cumbersome. He was particularly frustrated by his inability to calculate differential equations quickly. His work at Harvard in resolving this problem resulted in the construction of the Mark I.

Aiken was an instructor at Harvard from 1939 to 1941, at which point he was drafted into the U.S. Navy. He served out the majority of the war commuting back and forth between Harvard and the IBM plant at Endicott, New York, building and consulting on his computer. After the war, he rejoined the faculty at Harvard, and from 1946 until his retirement in 1961, he was director of Harvard's computational laboratory. In that capacity, he hosted a number of important seminars and conferences for other scientists interested in computers during the late 1940s and early 1950s. Available information on computers was disseminated at these sessions. After his retirement in 1961, Aiken served as professor emeritus. His major contributions to the history of data processing, or at least to the development of the modern computer, had come much earlier, however—in the late 1930s and early 1940s.

Aiken's concern for doing calculations rapidly led him to define what a computational device should be able to do. His concern lay with nonlinear differential equations which were solved through numerical methods. As any mathematician knows, solutions come from an evaluation of integrals, that is, in identifying the area that lies under the curve of a particular equation, assuming one plots such data on paper. Analytical solutions can be determined by developing formulas to identify precisely the area under a curve desired. Yet for very complicated problems this method does not work well unless the region can be broken up into smaller rectangular areas that have known heights and widths. Defining the width in advance and then calculating height by use of the original equation is a straightforward task because the calculations are performed for each rectangular slice. The problem, however, is that in order to increase accuracy these rectangular areas have to be divided into increasingly smaller ones which then are all tabulated and totaled. The calculations, although understood, are too numerous to be conducted by hand. Aiken concluded that if a machine could perform this function, the accuracy of calculations would improve and the amount of time needed to perform them would decline.

Aiken envisioned lashing together currently available card punch equipment of the sort then available from IBM into a system to do these calculations. The sheer act of writing down subtotals by machines on cards would be faster and could reduce errors of transcription. Aiken thought that a card reader could provide the initial introduction of values (numbers) for the function to be integrated and that a multiplying punch could calculate each area and generate output in the form of more cards. An accumulating punch machine could total these up and punch out the results on cards. When fed into a printer, the output from these cards could be printed on paper. Upon further thought, Aiken concluded that the whole process could be refined to eliminate any punching of cards during the calculations by simply connecting all the machines electronically, allowing them to pass data from one to another until the entire set of transactions was completed and the output printed on paper. This was the seed of the idea for a computer which he called the Automatic Sequence Control Calculator, first known as the ASCC and later as the Mark I.

In 1937, while still a graduate student, Aiken proposed the construction of such a device. He suggested that existing card punch equipment be wired together rather than be used independently, thereby sequentially carrying out transactions that earlier were executed by one machine and fed manually to another. He also specified how such a machine could handle both positive and negative numbers, use preset tables of sines, cosines, logarithms, probabilities, and so on, many of which were commercially available in prepunched card form. He even worked out how values could be incremented as functions were executed sequentially. Thus, the entire system would permit numerical integrations. Aiken proposed that the equipment be repackaged so that the functions currently available in tabulating equipment (motors, trays, etc.) could be mounted on racks and all interconnected through electrical wiring and wire boards.

Punched cards could initially serve as the means of controlling operations. When Aiken built the machine, paper tape performed this function. He postulated that cards should not be used only to store data but should also control sequential operations of machines. For that matter, punched paper tape could do the same. He also suggested using registers and fixed decimal points but with greater capability of handling larger amounts of data than was possible with existing technologies, thereby making his device more usable for scientific calculations. Aiken's ideas were largely his own, yet, as he himself acknowledged in 1937 and again later, he also borrowed from Charles Babbage's** work in the mid-nineteenth century on calculating engines.

Aiken first approached the Monroe Calculating Machine Company for help in constructing his machine but had no success. Harlow Shapley, an astronomer at Harvard, urged the young scientist to approach IBM through a Harvard Business School professor, Theodore H. Brown,** a friend of IBM's president Thomas Watson.** IBM's agreement to help marked the beginning of its serious involvement with computers. After quibbling over whether the machine should be built at Harvard (Aiken's wish) or at Endicott (Watson's position so that he could control the project), they finally agreed to construct it at the IBM plant.

Work on the machine began in 1939 at Endicott, where a group of engineers worked under the leadership of James W. Bryce,** a distinguished and highly experienced inventor of card punch equipment. Aiken designed the machine as a consultant, and IBM's engineers built it. During 1940, 1941, and 1942, Aiken divided his time between work on this machine and duties as a commander in the U.S. Navy. Watson managed to have him freed up sufficiently from naval duties to spend two summers at Endicott. By January 1943, the IBM engineers at Endicott had completed construction of the ASCC. After initial tests were performed to insure that it would work, they moved the device to Harvard. In May 1944, calculations were begun on the machine for U.S. Navy projects. Its racks of machinery and wiring were packaged in steel and glass covers, making it more attractive and less like mysterious scientific laboratory machinery. It now had the appearance of the most attractive of the earliest computers. It was also a general-purpose calculating machine.

IBM's role and the cost of the machine were clearly defined. During the 1930s, Watson had expressed a desire for more advanced card punch technology, and indeed he had sponsored a number of projects toward that end at Columbia University on computation and within his own firm for products. He was also interested in Aiken's project because of its value as a military program. In 1939, there was apparently little interest in developing such a machine into a commercially viable computer product; such interest would not surface until after World War II. The U.S. Navy paid nearly $300,000 for work on the project in exchange for the right to use it for calculations after its construction. IBM poured additional sums and resources into it, primarily for supporting and maintaining the equipment; the total cost of the project was just over a million dollars—a sizable amount for that time.

The final product had seventy-two counters to store data (numbers); each number could be as large as twenty-three digits. Punch tape gave the machine its instructions and punched cards its data. All instructions were expressed as twenty-four binary digits. Human operators ran sixty registers. Wheels on rotating shafts supported the registers, while clutches served as the means for setting them at predetermined positions. Electrical contacts sensed all numbers preset on these wheels. The clutches, in concert with the electrical sensors, passed numbers (data) from one wheel to another for processing in the next step of a calculation. It could do multiplication and division and such common mathematical functions as logarithms, exponentials, and trigonometry. It was relatively quick. For example, two twenty-three digit numbers could either be added or subtracted in about one-third of a second, faster than a person could. Results could either be passed on for other calculations or be printed out on cards or teletype paper. When operating all its functions, the noise it made was likened to that in "a room of ladies knitting."

The machine was large. It had over 750,000 parts, and its electromechanical relays were bulky and noisy. The device looked like a bank of sandwich vending machines. It weighed five tons, and was 51 feet long, 8 feet high, and 2 feet wide. Attached to it were one card punch, one tape punch, and two automatic typewriters. Also known as the IBM ASCC or the Harvard Mark I, it consumed considerable amounts of electricity and floor space. It was impressive in appearance, however.

The Mark I was used in the months before its public announcement to calculate firing tables for naval ballistics. Soon after the construction of the Mark I, the ENIAC* came on stream to perform the same application. Calculations related to the development of the atomic bomb were performed on the ASCC, and in all probability (the files are still classified and closed to historians) on the ENIAC as well. Scientists at Harvard also used the Mark I for considerable periods of time to prepare tables of mathematical functions, many of which were Bessel functions. So many of these were done that the machine acquired the nickname "Bessie."

The joint venture between Aiken and Watson had a negative end on the personal level. Earlier, they had clashed over where the machine should be built, and the animosity between them grew so intense that they avoided ever being together. Both were temperamental and often blunt, and both have been accused of being arrogant and demanding. Nevertheless, their strong wills made it possible for the project to go forward. Their unfriendly relations, however, resulted in a split when the machine was publicly unveiled.

Aiken was proud of his project and welcomed publicity for it, along with credit for its construction. Watson, always in search for any public opportunity to promote IBM's image and pleased at the contribution he had made to the war effort through the construction of this naval project, also wanted recognition for the effort. In May 1944, a public demonstration and press conference were held at Harvard to show off the machine. With Watson in attendance at the event, Aiken took the lion's share of the credit for the machine, practically ignoring IBM's important role. Watson, whose company had actually built the machine, was infuriated.

Watson informed officials at Harvard that he would no longer support such research projects there, and he leased the machine to the U.S. Navy. Aiken cared little for Watson's feelings on the matter; he had taken credit for the machine without much thought of the consequences. Watson did not completely lose interest in computational devices, however. Throughout the rest of the decade he quietly supported other efforts on a small scale and kept abreast of various projects in the United States. It would not be until the early 1950s, however, that significant interest would swell within IBM for the development of computers as viable commercial products. Watson ordered his engineers to have few dealings with Aiken and instead to focus on constructing their own machines as developmental and experimental projects. IBM's engineers had learned enough on the Aiken machine and through consultations with the engineers building the ENIAC at the Moore School of Electrical Engineering to carry on with their own work. Thus while Aiken's actions denied Harvard and himself possible additional funding from IBM for computer research in the immediate post-war period, they did not stop Watson's firm from moving ahead.

Aiken built other machines, called the Mark II, III, and IV, during the late 1940s. They were, however, anachronisms. At a time when other devices (e.g., the ENIAC) relied increasingly on vacuum tubes, his still used older relays. Meanwhile electronics came to dominate computer technology but were absent from his machines. Furthermore Aiken relied on decimal numbering systems when, by 1950, the trend clearly favored the use of binary systems. He resisted designing stored program capabilities when all other major projects outside of Harvard incorporated such functions by the late 1940s and early 1950s.

During the late 1940s IBM constructed machines while Aiken worked on the Mark. In 1948 IBM announced the Selective Sequence Electronic Calculator (SSEC),* which relied primarily on the experience and design efforts gleaned from the construction of the Mark I. The following year IBM announced the Card Programmed Calculator (CPC), the first commercially available computer-like device to be manufactured in a standard fashion, as opposed to the SSEC, which had to be built one copy at a time. The CPC, an automatic calculator of modest proportions, was accurate and commercially viable. The SSEC, like the Mark I, was a relay calculator rather than a general-purpose computer, but it gave the company valuable experience both with problems of construction and manufacture and with the commercial demand for computational equipment in general. These experiences would prove crucial when IBM finally entered the computer business in earnest.

Aiken's significance as a designer of computers was over by the late 1940s. His technology was primitive, running counter to current trends in the use of stored-program capabilities, vacuum tubes, more electronics, and finally transistors. During the period 1939 to 1943/1944, he made his greatest contributions to computer technology, and for years afterwards, he served as a conduit of information for other scientists working on computers. Although he thus indirectly continued to aid the growth of this branch of technology, his personal research had little, if any, effect after about 1948. On the other hand, Aiken's laboratory did train many students in computer science. Some historians have even argued that his laboratory at Harvard, rather than the Mark I, was his most significant contribution to data processing. Among his students were Frederick P. Brooks,** who helped design the IBM System 360,* and Ken Iverson, the designer of APL,* one of the most useful computer languages to emerge in the 1970s.

Professor Aiken died in 1973.

For further information, see: Howard H. Aiken, "Proposed Automatic Calculating Machine," *IEEE Spectrum* (August 1964): 62–69 (his original proposal of 1937); Paul E. Ceruzzi, *Reckoners: The Prehistory of the Digital Computer, from Relays to the Stored Program Concept, 1935–1945* (Westport, Conn.: Greenwood Press, 1983).

AMDAHL, GENE MYRON (1922–). Amdahl helped to design the International Business Machines Corporation (IBM)† S/360* family of computers in the early 1960s; and, during the 1970s, established and ran Amdahl Corporation,† the largest manufacturer of IBM-compatible computers.

Amdahl was born in Flandreau, South Dakota, on November 16, 1922. He completed his B.S. in electrical engineering at South Dakota State University in 1948 and a Ph.D. at the University of Wisconsin in 1952. Between 1952 and 1955, he was a project manager at IBM's research laboratory at Poughkeepsie, New York, home of the company's computer manufacturing facilities as well. While there, he was the chief designer on the IBM 704* computer.

In the late 1950s Amdahl worked outside of IBM in California on various computer-related development projects, first at Ramo-Wooldridge Corporation and then at Aeronutronics. He rejoined IBM in 1960 as manager of system design, advanced data processing systems, holding the position until 1970. During this decade, he participated in the creation of the S/360, a family of computers that ushered in a third generation of computers with new technologies and greater levels of reliability and price/performance than ever before. It also became the most successful product in the history of American business. The S/360 was announced in April 1964, with additional products introduced over the next several years. In 1970, when Amdahl became increasingly frustrated in his attempts to convince IBM to replace the high end of the S/360 with a supercomputer, he left the company and formed his own.

In 1975, the Amdahl Corporation began to introduce computers to displace IBM's largest processors. They were relatively compatible, that is, software* that ran on an IBM computer could usually do the same on one of his, and all of Amdahl's machines relied on IBM's operating systems.* His company did very well for several years until it ran into cash flow problems in the late 1970s, causing him to sell stock to remain liquid. In the process he relinquished control of his company in 1979. In 1980, he formed Trilogy, another venture firm to build large computers.

Amdahl served in the U.S. Navy during World War II (1942–1944). In 1965, he was made an IBM Fellow, which meant that he could work on any research project he wanted with minimal direction from the company. He was recognized for his important contributions to the development of the modern digital computer* throughout the data processing industry. In 1976, he was named Data Processing Man of the Year by the Data Processing Management Association (DPMA),† and in 1983, he received the Harry Goode Memorial Award from the American Federation of Information Processing Societies (AFIPS).† He had already been named an Institute of Electrical and Electronics Engineers Fellow in 1969 as a result of his work on the S/360.

For further information, see: Thomas O'Donnell, "Gene Amdahl's White Whale," *Forbes*, June 18, 1984, pp. 46, 50; Emerson W. Pugh, *Memories That Shaped an Industry: Decisions Leading to IBM System/360* (Cambridge, Mass.: MIT Press, 1984).

ANDREWS, ERNEST GALEN (1898–1980). Andrews was one of the leading developers of relay computers at Bell Laboratories† during the 1930s and 1940s. His work represented some of the most important design work in the formative years of the computer industry.

Andrews was born on January 10, 1898, in Topeka, Kansas, and spent his childhood there. He served in the U.S. Navy during World War I. Subsequently, he attended William Jewell College in Liberty, Missouri, graduating in 1922 with a major in mathematics. Andrews then joined the Western Electric Company in Kansas City and, in 1925, Bell Laboratories, which Western Electric established that year in order to conduct research into telephonic technology. During the 1920s and 1930s, Andrews worked on switching equipment for the telephone system. He took charge of designing radar trainers in 1941, creating training manuals for users of radar equipment. He also worked on some radar development while at Bell, perhaps as early as 1937.

Beginning in 1943, Bell Labs began to construct what would turn out to be five large-scale electromechanical computers, and Andrews worked with each of them. In the late 1940s, he focused considerable attention on automatic message accounting systems for telephones, but throughout the 1940s, he remained heavily involved in the development of military systems based on computer-like technologies. His most significant contribution to the development of computers revolved around his work with electromechanical relay computers. Andrews played a crucial role as a designer of the BTL Models II and III and influenced subsequent models of devices used to transfer telephone calls. The National Aeronautics and Space Agency's (NASA's) Langley Field used the Model IV, and he helped install another copy at the Aberdeen Proving Ground for the U.S. Army. He was also a founding member of the Association for Computing Machinery† in the 1950s.

Andrews' interest in computers as well as that of both Western Electric and Bell Labs, grew out of the need to have an automatic telephone switching system that transmitted digital data from telephone calls. Relay devices allowed the telephone company to switch calls automatically without human intervention, thereby driving down the cost of operation. Numbers dialed at one location could be transmitted to a central point as electronic impulses, trunked to the correct city or circuit to complete a telephone call at another receiver, in addition to transmitting electronically translations of human voices over phone wires. Relays made possible simplified circuit logics along with the technology associated with relay equipment. Work on these important projects continued throughout the 1920s and 1930s; these programs gave Andrews considerable experience and knowledge about electronics, mechanisms, and mathematics.

Beginning in 1937 at Bell Labs, George R. Stibitz,** a fellow scientist and the leading computer specialist at the laboratories, concluded that a computer could be constructed to do addition, multiplication, subtraction, and division. Further more, dial switching equipment made during the 1930s could be made to execute simple mathematical functions. Consequently, he diverted resources to the study of how to use binary codes in switching equipment to perform mathematical functions. Named the Complex Computer, the project to build the device was headed up by Samuel B. Williams, with Andrews heavily involved. The machine went into operation in January 1940 as an eight-place decimal device for the purpose of handling four complex number arithmetic operations. On September 11, 1940, Stibitz performed the first remote job entry (RJE) task in data processing history when he transmitted work from Hanover, New Hampshire, to New York City over telephone lines for the computer to execute. The results were also transmitted back to New Hampshire.

With the onset of World War II, Andrews devoted his time to perfecting subsequent models of the Complex Computer for use by the military. The names of these machines changed to Model I, then II, with subsequent versions appearing throughout the 1940s. The first four models were special-purpose military machines. The Models V and VI represented a major thrust forward in technology, moving from fixed point decimals to floating point, to a broad range of mathematical functions, and faster performance. Models V and VI were also used for civilian applications, particularly in solving complex mathematical problems.

By 1940s' standards, the relay devices built at Bell Labs were reliable and could handle ordinary telephone circuits well. Although slow, these machines were very accurate, having been designed to check their own calculations constantly; thus, they were well suited to accurately switching telephone calls. The work done by Andrews and other scientists at Bell Labs was eclipsed by 1950 as others developed faster machines that were designed for more generalized calculations in scientific and commercial circles. These newer machines relied less on electromechanical technologies, which now faded into the past, and increasingly on vacuum tubes. During the 1950s, the young computer industry also began using transistor technology more frequently.

Yet the work which scientists such as Andrews performed in the design of the early computers proved important. They worked out many of the details necessary for binary-coded-decimal systems for coding numbers, processes still in use in the 1980s. Their efforts illustrated how information could be described as numbers using binary codes—the basis for much future computer technology. Finally, Andrews in particular applied redundancy of function in computational devices to increase error-free operations, a design feature which today is basic to all computers.

Andrews, a pioneer in the development of binary computers, retired from Bell Labs in 1959 and became a consultant for Sanders Associates of Nashua, New Hampshire. He died on October 13, 1980.

For further information, see: E. G. Andrews, "Telephone Switching and the Early Bell Laboratories Computers," *Annals of the History of Computing* 4, No. 1 (January 1982): 13–19 and his "Use of Relay Digital Computer," ibid.: 5–13; W.H.C. Higgins et al., "Defense Research at Bell Labs," ibid. 4, No. 3 (July 1982): 218–236; B. D. Holbrook and W. S. Brown, *A History of Computing Research at Bell Laboratories*

(1937–1975) (Murray Hill, N.J.: Bell Telephone Laboratories, 1982); D. La Porte and G. R. Stibitz, "Elogue: E. G. Andrews, 1898–1980," *Annals of the History of Computing* 4, No. 1 (January 1982): 4–5.

ARTYBASHEFF, BORIS (1899–1965). This leading American illustrator of the mid-twentieth century provided some of the earliest comic illustrations of data processing equipment to appear in widely read American magazines.

Artybasheff was born on May 25, 1899, in Kharkov, Russia, and moved to the United States in 1919 after the Russian Revolution, becoming a naturalized citizen of the United States in 1926. He illustrated books and drew for such magazines as *Time, Life,* and *Fortune*. His best known illustrations of data processing equipment were two pictures that appeared in a booklet describing the subscription services of Time, Inc. These two pictures depict sorters as cartoon characters selecting computer cards and afterwards, sorting them. The equipment items portrayed are cartoon characterizations of IBM card readers and sorters of the 1930s. Over the course of his later years he drew some of the most widely seen cartoons on computers. Artybasheff died on July 16, 1965.

For further information, see: Charles Eames and Ray Eames, *A Computer Perspective* (Cambridge, Mass.: Harvard University Press, 1973).

ATANASOFF, JOHN VINCENT (1903–). Atanasoff, a physicist, was one of the first scientists to work on digital computers* in the 1930s, partially constructing one called the Atanasoff-Berry Computer (ABC). He became the subject of considerable controversy during the 1970s when the courts debated who had actually invented the first electronic digital computer. Some students of early computers argued that Atanasoff was one of the first to do so, and a U.S. federal judge ruled that he was indeed the first. Although historians now generally agree that the first electronic digital computer was the ENIAC,* designed and built by a team of engineers managed by J. Presper Eckert, Jr.,** and John W. Mauchly,** they still recognize Atanasoff as an important figure in the early history of the computer.

Atanasoff was born on October 4, 1903, in Hamilton, New York, the son of a Bulgarian immigrant electrical engineer. He was raised in Florida and early in life displayed considerable interest in mathematics and physics. By the age of ten he understood slide rules* and had a good working knowledge of algebra and calculus. He studied electrical engineering at the University of Florida, graduating in 1925, and pursued graduate studies at Iowa State College (now Iowa State University) in mathematics and physics, completing the master's degree there in 1926. Atanasoff taught at the school until he began his work on his doctorate at the University of Wisconsin in 1929. He finished his work on his Ph.D. in physics in July 1930, with a dissertation entitled "The Dialectric Constant of Helium." As part of the effort required to complete it, Atanasoff had to spend long hours doing calculations with the help of only a Monroe

Calculator. He soon began to question whether there were not faster ways to do calculations, particularly large systems of equations.

After returning to Iowa State College Atanasoff began to tinker with tabulating equipment from International Business Machines Corporation (IBM)† and with other available calculators on campus to see if they could be made to process mathematics electronically. As early as 1936, he published his first paper showing how some problems in theoretical physics could be solved by linking various tabulating equipment. During the late 1930s, he surveyed all existing computational devices, dividing them into two categories: digital and analog. Decades later, he claimed to have been the first to use the word *analog* (spelled *analogue* the first time he used the word in 1940) to describe a type of computer, such as the slide rule and the Bush Differential Analyzer.

In the late 1930s, Atanasoff decided to build an electronic digital computer. In 1938, he developed several ideas regarding an ideal computer; he decided that it should be electronic and not mechanical, be binary based, employ condensers for memory* of a regenerative type, and compute by direct logical action instead of by enumeration. All four concepts influenced the design of the ABC. He also acquired a graduate assistant who would play a key role in the design of the computer, Clifford E. Berry. Unlike the theoretical physicist Atanasoff, Berry had a background in mechanics and electronics. He was brilliant and soon became fascinated with the idea of a computer.

In the fall of 1939, Atanasoff and Berry began serious work on a machine. The device they constructed did mathematics electronically. Although not a complete computer, it could do differential equations using simple arithmetic. Electric capacitors stored numbers on two Bakelite drums, each of which had the capability of housing thirty binary numbers (fifty digits each). Numbers were read when the drums rotated. Each drum was 11 inches long and 8 inches in diameter. Atanasoff attached to these drums thirty units to do additions and subtractions. For input he relied on punched cards, each of which was capable of holding up to five numbers. Therefore, either through cards or a keypunch, numbers were entered into the device, each number equal to fifteen digits. These were then converted to binary numbers whereupon arithmetic could take place. An operator would manually trigger each step in the process.

Atanasoff and Berry had built portions of a computer system. The ABC had memory function, arithmetic units, and input facilities; yet its functions were very limited. The fact that Atanasoff built components of what would become the standard architecture of a computer is historically important. Furthermore, because of the onset of World War II and the need to work on other things, he never completed the project. Like Charles Babbage** a century earlier in Great Britain, Atanasoff and Berry had ideas which, if carried to completion, would have resulted in a working computer.

Work on the project progressed from 1939 to 1942, important dates in the history of the controversy that would later erupt over who first invented electronic digital computers. In the midst of this early period, in December 1940, Atanasoff attended the annual meeting of the American Association for the Advancement of Science in Philadelphia where he met John Mauchly, then a young professor at Ursinus College. They discussed their mutual interest in designing both analog and digital electronic computers, and Atanasoff invited Mauchly to visit Iowa to see the work being done on the ABC.

Mauchly visited Atanasoff in June 1941 and saw a demonstration of those portions of the ABC which had been built. Mauchly later argued that he had not been impressed with the slow use of vacuum tubes or the heavy reliance on manual intervention with the machine which also contributed to its slowness. Mauchly was already envisioning the elements of a digital computer, features that would later appear in the design of the ENIAC. After the visit, Mauchly returned to Philadelphia to attend a class in defense electronics at the Moore School of Electrical Engineering.† Immediately after this class' he accepted a job at the school.

Mauchly's trip proved critical in the controversy over the ENIAC because it could be cited as proof that he acquired ideas from Atanasoff which appeared in the computers he built. Mauchly's work on computers came *after* the trip to Iowa. His great contributions began during World War II when he and Eckert built what today is recognized as the first fully functioning electronic digital computer. Eckert and Mauchly subsequently applied for and received patents on this equipment, and they went on to build other machines as enhancements to the ENIAC in the late 1940s (EDVAC* and BINAC* and, in the early 1950s, the UNIVAC*). When Remington Rand† bought out Eckert and Mauchly's company, it acquired all rights to patents related to the UNIVAC. In the 1950s, Remington Rand passed these rights to Sperry Rand† as a result of a merger. In the late 1960s, Honeywell's† decision to challenge Sperry's patents on computers resulted in a court case which Honeywell won in 1973. As a byproduct of that trial, the court ruled that "Eckert and Mauchly did not themselves first invent the automatic electronic digital computer, but instead derived that subject from Dr. John Vincent Atanasoff, and the ENIAC patent is thereby invalid." The judge referred to Mauchly's trip to Ames, Iowa, in 1940, discounting the originality of the work he had done at Ursinus College either prior to that time or afterwards.

Many of the key figures in the development of computers in the 1930s, 1940s, and early 1950s testified in the case or commented on it. For example, Arthur W. Burks,** an engineer and mathematician at the Moore School who worked on the ENIAC, for years afterwards defended Atanasoff's claim. Herman H. Goldstine,** also associated with the ENIAC, was yet another Atanasoff supporter. Eckert and Mauchly, of course, felt they had built the first working digital computer and had not stolen ideas from Atanasoff. Mauchly argued that he had technically superior ideas on the use of computers, particularly regarding vacuum tubes, before his visit to Ames, Iowa. Historians in general view Eckert and Mauchly as the real developers of the electronic digital computer because they managed to make machines that actually were used. They acknowledge that Atanasoff built components of a possible computer, but not a whole one. One authority called the ABC a "near-miss."

Mauchly intended to design and build a general-purpose computer, whereas Atanasoff's documented intent was to design an electronic device to perform the functions of a differential analyzer.* Letters dated 1940 reveal that, although Mauchly was impressed with what he saw, despite the inadequacies which he felt were inherent in Atanasoff's machine, he had different ideas about what a computer should be like. By 1940 Mauchly had already built several components on his own, held distinct ideas about the use of vacuum tubes, and had considerable experience with flip-flop technology and scaling circuits. Mauchly consistently denied stealing any of Atanasoff's ideas. If anything, historians generally agree that Atanasoff merely confirmed in Mauchly's mind that vacuum tubes and other electronics could be used to build a computer.

Atanasoff knew of the ENIAC, which certainly received considerable press attention in 1946 and later years. Atanasoff never believed in these years that the ENIAC had been derived from his own work. That thought, he claimed, was planted by a patent lawyer in 1954 and reinforced by another in 1967. When Atanasoff finally realized that his work was more important than he had understood, he began to defend his contributions to the new industry. Amongst the list of his technical innovations, he claimed to have "conceived of an electronic digital differential analyzer in early 1941." Although this concept was not incorporated in the ABC, Atanasoff believed it contributed to the ENIAC's design. The implication that he left in 1984 was that he had developed many of the concepts and had built important components that would later be part of computers (such as small capacitors for memory, direct logic circuits, the notion that calculations should be conducted serially, and the nonrestoring method of division).

In 1942, Atanasoff stopped work on the ABC and undertook a project to develop fire control for antiaircraft batteries. In September 1942, he joined the U.S. Naval Ordnance Laboratories where he remained until 1948. His first project there involved acoustics and naval mines. In September 1945, the government put him in charge of building a computer for the Naval Ordnance Laboratories. Atanasoff continued to perform duties with the Acoustics Division, preparing for studying the sound effects of atomic blasts at Bikini Atoll. The project to build a computer was canceled before it could be fully activated. In 1949, he took the position of chief scientist, U.S. Army Field Forces at Fort Monroe, Virginia, but soon after returned to the Naval Ordnance Laboratory as director of the Fuze Program. In 1952, he left government service and established the Ordnance Engineering Corporation to do developmental work for the government. He built a computer called the Firing Error Trajectory Recorder and Computer (FIRETRAC) that tracked the course of projectiles. In 1956, Aerojet General Corporation (AGC) acquired the firm, and in 1959, Atanasoff was made vice-president of the Atlantic Division, his old company. Under his tutelage it developed computer applications for handling and sorting parcels. He resigned from AGC in 1961 and began consulting on these applications instead. Beginning in April 1967, he became enmeshed in the legal action that would involve Honeywell and Sperry Rand for the next half dozen years. During the period following the trial (1973 to the present), he has been active in debates concerning the origins of computers and his role.

Only late in his life was Atanasoff recognized as a pioneer in the development of computers. The lawsuit and the resulting historical debate increased his awareness of ABC's impact. Recognition of his work has increased. For example, Iowa State University hosted a celebration of his work on October 21, 1983. In 1984, he published a major memoir article in the *Annals of the History of Computing* in which he continued the battle with Mauchly (who had died in 1979) over who said what and when in Ames, contending that the ABC had been an important contribution to the development of computer science. Atanasoff still strongly contends that the ENIAC originated from ideas incorporated in his ABC.

For further information, see: John Vincent Atanasoff, ''Advent of Electronic Digital Computing,'' *Annals of the History of Computing* 6, No. 3 (July 1984): 229–282; A. W. Burks and A. R. Burks, ''The ENIAC: First General-Purpose Electronic Computer,'' *ibid.,* 3, No. 4 (October 1981): 310–399; John W. Mauchly, The ENIAC,'' in N. Metropolis et al., eds., *A History of Computing in the Twentieth Century* (New York: Academic Press, 1980): 541–550; Brian Randell, *The Origins of Digital Computers* (New York: Springer-Verlag, 1973); Joel Shurkin, *Engines of the Mind: A History of the Computer* (New York: W. W. Norton and Co., 1984); Nancy Stern, *From ENIAC to UNIVAC: An Appraisal of the Eckert-Mauchly Computers* (Bedford, Mass.: Digital Press, 1981).

AUERBACH, ISAAC LEVIN (1921–). This computer scientist was active in important projects in the 1940s and 1950s, became a well-known consultant by the 1960s, and published studies on the data processing industry. He also helped to establish organizations that represented that industry.

Auerbach was born in Philadelphia on October 9, 1921. He completed his B.S. in electrical engineering at Drexel University in 1943 and his M.S. in Applied Physics at Harvard University in 1947. He then joined other engineers who were working on the BINAC* and then on UNIVAC,* an early electronic digital computer* for the commercial market. He remained with the group until 1948. When he joined the Special Products Division at Burroughs Corporation† where he remained until 1957. At that time he established the Auerbach Corporation for Science and Technology in Philadelphia, serving as its chairman and president, positions he held as of 1986. He also became president of Auerbach Associates, Inc. (1957–1976), Auerbach Consultants (since 1976), and chairman of Auerbach Publications, Inc. (since 1960). He has been the publisher of Auerbach Computer Technical Reports throughout the same period and has served as both a director and treasurer of the Baupost Group, Inc., since 1982.

Auerbach also found time to serve as a director on the board running the Software Group, Inc., since 1983.

Very early in his career Auerbach became a spokesman for the data processing industry. In addition to the consulting firms he ran, he used other avenues to speak about data processing. In 1957–1960, he was a U.S. consultant on automation and information processing at UNESCO. In 1959, he chaired the U.S. delegation to the first International Conference on Information Processing. He was founder and first president of the International Federation for Information Processing (IFIP)† which rapidly became a major voice for the entire data processing industry. Auerbach served as its president from 1960 to 1965, and in 1969, he was made an honorary life member of the organization. He was editor of *The Auerbach Annual-Best Computer Papers* (in 1971–1974 and again in 1979–1980).

Auerbach was named an IEEE (Institute of Electrical and Electronics Engineers, Inc.) Fellow in the early 1960s as well as a distinguished Fellow of the British Computer Society. His influence within the industry was also reflected in other associations and honors: member of the National Academy of Engineering, the Franklin Institute, the Japan Computer Society, and the National Academy of Science, to mention a few.

Auerbach made his greatest technical contributions in the 1940s and 1950s. In addition to his work on the UNIVAC, he also developed technologies to support communications in computers, and worked on the development of the first ICBM guidance computer, and other U.S. defense projects during the 1950s.

For further information, see: Isaac C. Auerbach, "The Start of IFIP—Personal Recollections," *Annals of the History of Computing* 8, no. 2 (April 1986): 180–92; Joel Shurkin, *Engines of the Mind: A History of the Computer* (New York: W. W. Norton and Co., 1984).

B

BABBAGE, CHARLES (1791–1871). Babbage is best remembered as the developer of calculating machines that were the precursors of the modern computer. Nearly equal in importance, he was an early British advocate of the systematic use of science and technology in industry. He was both a student of pure mathematics and the builder of difference engines,* devices that could compute tables by using the method of finite differences. He also designed the analytical engine,* which many historians consider the forerunner of today's computer.

Although Babbage failed somewhat to convince his generation to apply science and technology effectively in industry, his work on computing devices—a byproduct of his concern for using technology to solve practical problems—remains a legacy of monumental intellectual proportions. Even the exercise of trying to construct such devices caused him to advance the state of the art in precision mechanical engineering in the 1820s.

Babbage was personally eccentric and socially popular. He enjoyed his family and had an extensive network of friends and colleagues which included many of the leading scientists of Europe during the mid-nineteenth century. He had enormous energies which he expended in many directions: he was an inventor, a writer of articles and books, he formed scientific societies, he struggled with the British government for aid to scientific research, and he was active in political groups.

Babbage was born on December 26, 1791, at Walworth, Surrey, into a middle-class family. His father worked in banking and merchandising. He was raised and educated on the outskirts of London where, as a child, he showed an early interest and skill in mathematics. In 1810, he went to Trinity College, Cambridge, where he continued his study of mathematics, a field of study that was still important at this university. While at university, he formed a circle of friends active in sailing, politics, and the study of mathematics.

In 1814, Babbage moved back to London where he soon after married. He began lecturing on various scientific topics, including astronomy, at the Royal Institution, and in 1816, he was elected a member of the Royal Society—a key institution in the encouragement of scientific research in England. Between 1815 and 1820, he pursued his study of mathematics extensively, learning more about algebra and writing on the theory of functions. He also became very familiar with scientific thinking in Europe where his thoughts on computing devices would later receive considerable attention. In these years of intellectual growth, the young Babbage also sought employment as a lecturer or professor to supplement his meager income. A political liberal in an age when Tories controlled most academic positions, he failed to land an appointment. Nonetheless, he continued his research and reading.

The 1820s were years of major significance to Babbage's life's work and the history of computing. In this decade work on the difference engine took place; the analytical engine became the primary project of the 1830s and beyond. During the 1820s, Babbage devised plans for a difference engine for which he obtained financial support from the British government. Up to that point, calculating devices had been relatively simple. Dating from the 1640s when Blaise Pascal** developed machines that could be marketed, these digital calculators had not fundamentally changed or improved since then. They all had one fundamental problem: the user had to be continually involved in the calculations, thereby insuring that errors would abound and that the process would be tedious. One of Babbage's objectives was to automate the process as much as possible.

Babbage's main contribution to computing lies in the work he did on two types of calculators. The first, known as the difference engine, was designed to calculate tables of numbers using the method of finite differences. This device was intended to have a printer to generate the results of computations as they were calculated. The difference engine represented a significant advance in the technology of calculators. Babbage's second device, the analytical engine, called for logical systems capable of a greater variety of functions than could be had with the difference engine. Although he never completed its construction, the analytical engine rather than the difference engine is the basis for calling him an early pioneer in the development of the modern computer, for in design at least this device approached the concept of a versatile and programmable machine.

Babbage invented the difference engine in order to prepare tables accurately, for example, for navigation and insurance. The need for accurate navigational tables was an urgent one, for England was the leading maritime power of the world. Existing tables were extremely prone to error, causing a large number of shipwrecks. Babbage's application provided an incentive for the British government which purchased these navigational tables for its maritime and naval fleets. As early as 1820 or early 1821, Babbage began to develop his final plans for the difference engine. First, he sought to store numbers on strips of metal, borrowing from a widely used mechanism to control the striking of hours by clocks. Second, he studied the use of toothed wheels for storing numbers much as was the case with the earlier calculators of Pascal, Gottfried Wilhelm von Leibniz,** and others. Babbage favored the toothed wheel approach because it required fewer parts. Each wheel could hold single digits from 9 to 0, with carrying going to another wheel, thus counting 9, 0, 1, 2, 3, and so on, with the wheels continuously turning in the same direction. By 1822, he had a working model of the calculating portion of his machine that could prove it was possible to have an engine calculate per his design.

Babbage also required printing capability which would eliminate another human step, thus reducing yet another possibility of generating errors in tables. He proposed that the Royal Society now support his research. His six-figure wheel model drew much attention and approval from the Royal Society, making it possible for the British government to agree to underwrite his work. Workmen were employed at his shop to machine parts while he continued to design other components. During this same period, he visited Europe, established contacts within the scientific community, and wrote about the political economy of England. Based on extensive trips to various British factories, he wrote *On the Economy of Machinery and Manufactures* (1832), an important book that contained a great deal of information on British factories and detailed his suggestions on how technology could be employed in industry. His work in political economy was substantial and, to Babbage, it was an important personal accomplishment.

In addition to these activities, Babbage continued his work on the difference engine through out the 1820s and into the 1830s. In the early stages, the government supported his work; after it withdrew support because of the lack of a completed machine, he used his own resources. While working on the difference engine (which like the analytical engine was never completed) he began to think about a more complex device. By the end of 1834, he had turned his attention away from the difference engine and to the newer, more complex challenge.

In the 1830s, years of great scientific productivity and invention for Babbage, his growing reputation and personal contacts allowed him to comment more frequently on the use of technology in industry and to become part of the mainstream of British and European scientific activities. But it was the analytical engine which would command most of his energies for the remainder of his long life, although his vast experience with the difference engine would deeply influence his work. He was now a well-established mathematician, financially independent, and only in his early forties.

From 1834 to 1836, Babbage outlined the main components of a new machine. With his new machine he sought to separate the functions of (1) storing information, (2) holding numbers within the calculator, (3) creating a processor (that portion of the machine that performed calculations such as the simple mathematics of adding, subtracting, dividing, and multiplying, and (4) producing output. In many respects, this set of concepts paralleled the modern view of computers as reflected in John von Neumann's** work and that of others in which data are stored in one part of the computer or in storage devices and are brought into the computer itself for processing and subsequent expulsion to storage units or to printers for publication. In short, Babbage envisioned a computing system made up of various devices.

In the early 1830s, Babbage generated about 300 engineering drawings for his machine; he continued to produce design work to the end of his life. For the store function he worked out a scheme of toothed wheels, each of which rotated independently of others on the same axis. Each axis was to store a number, and each wheel a single digit. Relying on punched cards which had earlier been developed to instruct looms on weaving patterns, Babbage used cards to introduce the numerical values of a constant to the machine which he called the number-card. A variable-card was used to identify the axis on which a number would be housed. Such a card could also order a number out of storage or back in. Operation-cards were used to control the operations of the processor, ordering addition, subtraction, multiplication, division, or other functions. He worked out various sequences of cards to instruct the analytical engine. The operations and order in which cards were used could be altered by calculations previously performed by the machine. Hence, by the late 1830s, the basic design of a powerful calculator existed.

Probably during 1836, Babbage also gave some thought to designing a machine that could calculate algebra. It was about this time that his ideas led him to the concept of analytical engines, devices which he thought would require even more capability for calculations. Earlier, he had thought in terms of enhancing his ideas for difference engines. He now had to put the older machine aside in his thinking, even though a significant part of it already existed and worked. Although the need for such a machine still existed, a redesign was almost necessary since tooling methods had improved immensely from the early 1820s, making it less expensive to redesign than to complete the old one. Babbage decided to focus on a more sophisticated device that would be richer in function—hence, the analytical engine. Another reason why he abandoned the difference engine was the government's decreasing interest in it—he needed the government's financial support to complete the project.

The most important years for his work on the analytical engines were the 1830s through the end of 1847 at which time Babbage felt the majority of the design work had been completed. Although the broad scheme and its principles of operation had been documented as far back as the end of 1836, the additional ten years of work proved necessary to simplify the design and work out details.

Babbage began designing yet another generation of devices without having even built the analytical engine, other than some design models of certain parts. In the late 1840s, he worked on a second and third difference engine that would be simpler to build and be smaller and lighter in design. He took advantage of his earlier experiences when possible. Thus, during the 1830s he had worked out the details of the punched card input and the use of wheels to store data, and the actual processing of calculations. This collection of design work was incorporated in newer models. He finally determined that basic functions (such as adding and subtracting) could be performed in seconds, and multiplication and division in minutes, on the newer models. He even developed a variety of arithmetic procedures to handle the carrying of numbers.

During the 1840s, Babbage focused his attention on concurrent processing, more advanced calculations, and the design of peripheral equipment for card readers and punches, printers, and plotters. He created designs for concurrent processing whereby, for example, numbers could be coming from storage to what he called the mill (processor) while the computer was performing calculations on other numbers. He harnessed his knowledge of mathematics and mechanical engineering to a logical mind. This combination of skills made it possible for him to develop detailed plans for various machines. His leading biographer, Anthony Hyman, has concluded that, given the technology of the 1840s, his analytical devices could have performed as designed if they had been built. Because he could not afford to construct them and the British government no longer had any interest in his work, the machines were not made. But by the early 1850s Babbage's greatest productivity had ended, and his personal situation had changed. Life had become less serene. First he lost his wife, and then Countess Lovelace,** a close friend and companion who worked with him on mathematics and so clearly understood his work on calculating devices, died of cancer. Many other of his relatives were now dead or gone. For the rest of his life, he would tinker with his designs, but he would never again achieve important results.

In these years Babbage began work on his memoirs, later published as *Passages from the Life of a Philosopher,* a draft of which he had completed by 1864. He intended to write a history of calculating engines, portions of which were written and subsequently published after his death. Although old friends and visitors came to see him in these years and were given tours of his partially built devices and saw his designs, few appreciated the meaning of his work, let alone understood it. On October 18, 1871, Babbage died. Henry Babbage, his son who had served the British government in India for many years, took up his father's work with the engines and published some of his father's writings, as well as his own recollections. He published the portion of his father's papers which dealt with the history of engines as *Babbage's Calculating Engines* in 1889. Much of Babbage's work came to an end with his death, not to be appreciated until the middle of the next century. He is considered a giant in the history of the computer as well an important figure in the scientific community in nineteenth-century England. Perhaps one of the tragedies of his life was his inability to complete the manufacture of any device. While he had many great ideas, his complex designs and incomplete devices made him an undervalued figure in his own time.

For further information, see: Charles Babbage, *The Exposition of 1851* (London: John Murray, 1851), *The Ninth Bridgewater Treatise: A Fragment* (London: John Murray, 1837), *Passages from the Life of a Philosopher* (London: Longman, Green, 1864); Henry P. Babbage, ed., *Babbage's Calculating Engines: Being a Collection of Papers Relating to Them, Their History, and Construction* (London: E. and F. N. Spoon, 1889), *Memoirs and Correspondence of Major-General H. P. Babbage* (London: Privately Printed, 1910). For biographical treatments, see: V. V. Bowden, ed., *Faster Than Thought* (London: Sir Isaac Pitman and Sons, 1953); J. M. Dubbey, *The Mathematical Work of Charles Babbage* (Cambridge: Cambridge University Press, 1978); D. Halacy, *Charles Babbage, Father of the Computer* (New York: Crowell-Collier, 1970); Anthony Hyman, *Charles Babbage: Pioneer of the Computer* (Princeton, N.J.: Princeton University Press, 1982); M. G. Losano, *Babbage: La Macchina Analitica, un secolo di calcolo automatico* (Milano: ETAS Kompass Libri, 1973); D. L. Moore, *Ada, Countess of Lovelace: Byron's Legitimate Daughter* (New York: Harper and Row, 1979); P. Morrison and E. Morrison, eds., *Charles Babbage and His Calculating Engines: Selected Writings by Charles Babbage and Others* (New York: Dover, 1961); M. Moseley, *Irascible Genius: A Life of Charles Babbage, Inventor* (London: Hutchinson, 1964); D. Nudds, "Charles Babbage (1791–1871)," in J. North, ed., *Mid-Nineteenth Century Scientists* (Oxford: Pergamon Press, 1969), pp. 1–34.

BACKUS, JOHN (1924–). Backus was the project leader of the group that developed FORTRAN,* the first widely used high-level language. He also co-designed with Gene M. Amdahl** the IBM 704* computer. His contributions represent two major events in the history of data processing during the 1950s.

Backus was born in Philadelphia on December 3, 1924. He graduated from Columbia University in 1949 with a B.S. degree and completed his A.M. the following year. In 1950, he joined International Business Machines Corporation (IBM)† in New York City as a programmer, serving in that capacity until 1953 when he was promoted to manager of programming research. He held that position until 1959. While manager of programming research, his staff developed FORTRAN. Originally proposed in 1953, work began on the language in 1954, and in the spring of 1957 it was announced. Besides being the first high-level language in data processing, it also became the most continuously used high-level language in the industry. A generation of scientists and engineers who programmed did so with Backus's FORTRAN. During the 1950s he also helped to develop the 704 computer which IBM sold in 1960. He also prepared the FORTRAN I and II compilers for the 704 computer.

Backus joined IBM's T. J. Watson Research Center in Yorktown Heights, New York, as a staff member in 1959 where he continued to do work on programming languages* until named an IBM Fellow in 1963. He continued to work at Yorktown Heights and later at San Jose, California. Backus served on the design committee working on ALGOL 58 and ALGOL 60.* In the early 1980s, he was doing research on function-level languages and their associated mathematics.

In 1975, Backus was given the National Medal of Science and in 1983, the

Harold Pender award from the Moore School of Electrical Engineering,† University of Pennsylvania—the original site of the ENIAC* in the 1940s, one of the earliest digital computers.* He has also received the ACM Turing award. His contributions to data processing, especially to programming languages, were among the most important in the area of software,* particularly during the 1950s.

For further information, see: Backus' memoirs of designing FORTRAN in "The History of FORTRAN I, II and III," *Annals of the History of Computing* 1, No. 1 (July 1979): 21–37 and his "Programming in America in the 1950s—Some Personal Impressions," in N. Metropolis et al., eds., *A History of Computing in the Twentieth Century* (New York: Academic Press, 1980), pp. 125–136; Jean E. Sammet, *Programming Languages: History and Fundamentals* (Englewood Cliffs, N.J.: Prentice-Hall, 1969).

BARTH, CARL GEORGE LANGE (1860–1939). Barth refined earlier mathematical slide rules* to the point where, during the nineteenth century, they became popular and widely used aids to mathematical calculations.

Barth was born in Norway on February 28, 1860, and immigrated to the United States in 1881, serving as a mechanical draftsman with William Sellers and Company of Philadelphia between 1881 and 1890. Between 1882 and 1888, he also taught mechanical drawing at the Franklin Institute at night. From 1890 to 1901, he worked as chief draftsman for various companies, including the Arthur Falkenau Company, and again for the William Sellers and Company drafting department. From 1899 to 1901, he worked for Bethlehem Steel in Pennsylvania. It was at this time that he met Frederick W. Taylor (generally considered to be the father of scientific management) and worked with him until Taylor's death in 1915.

From 1901 to 1923 Barth helped introduce Taylor's scientific management concepts to companies all over the United States. As part of that effort, he developed advanced models of slide rules that sped up mathematical calculations. His earliest model of such a device was copyrighted in 1902 and consisted of a round wheel designed to provide solutions to "speed and feed" problems. His work reflected a growing need within the industrialized world for mechanical aids to calculation and thus was a remote predecessor of the calculator of the 1970s. His hand-held device was also a competitor of the more traditional straight-line slide rule that was sold in the late nineteenth century by the Keuffel and Esser Company. K & E's slide rule survived Barth's round one, remaining in production until the early 1970s.

During 1909–1918, Barth worked for the U.S. Army, lectured at Harvard on scientific management, and did a similar teaching stint at the University of Chicago. He died on October 28, 1939.

For further information, see: Charles Eames and Ray Eames, *A Computer Perspective* (Cambridge, Mass.: Harvard University Press, 1973).

BAUM, LYMAN FRANK (1856–1919). Baum is best remembered as the author of *The Wizard of Oz,* stories about robotic creatures, and children's tales. He reflected turn-of-the-century acceptance of machines performing anthropoomorphic functions, thus presaging the arrival of robots while forming definite images of such devices in the public's mind.

Baum was born in Chittenango, New York, on May 15, 1856, and began writing for newspapers in 1880. Between 1888 and 1890, he edited the *Dakota Pioneer* (Aberdeen, South Dakota) and later (1897–1902) the *Show Window* (Chicago). It was during this later period that he first began publishing children's stories: *Mother Goose in Prose* (1897, 1902), *Father Goose—His Book* (1899), and the first of several tales about Oz: *The Wonderful Wizard of Oz* (1900). Other fairy-like tales of Mother Goose and various characters followed until his death. His Oz stories included *The Marvelous Land of Oz* (1904), *Ozma of Oz* (1908), *The Emerald City of Oz* (1909), *Little Wizard Series* of six stories in 1913, *Patch-Work Girl of Oz* (1913), *Tik-tok Man of Oz* (1914), *Scarecrow of Oz* (1915), *The Lost Princes of Oz* (1917), *The Tin Woodman of Oz* (1918), and the play *Wizard of Oz,* which was first performed in Chicago (1902) and which became the basis of the movie of the same title in 1939. Other stories about Oz were also made into plays during the early 1900s.

The stern-looking author of children's books, with dark hair parted down the middle and thick mustache, believed machines were good for society. This optimism is reflected in his character Tik-tok in the Oz stories. Tik-tok, a copper man with clockwork characteristics, ''was sure to do exactly what he was wound up to do, at least all times and in all circumstances'' (*The Road to Oz,* 1909). Tik-tok had arms and legs, a head with mustache, a metal hat that looked similar to a World War I U.S. Army helmet, and a round ball-like body. He wore spats as was the fashion of the times and, although his countenance did not appear warm and friendly, he was always admired and respected for ''his'' efficiency.

When Baum died on May 6, 1919, he was considered one of the most widely known advocates of machines and automation. Through his Oz series of stories, which had been made into movies as early as 1908 and 1909 and had become the basis of an early theme amusement park called Fairyland of Oz, Baum probably did more to create images of robotic automata in the United States in the early 1900s than anyone else.

For further information, see: J. Cohen, *Human Robots in Myth and Science* (London: Allen and Unwin, 1966); Charles Eames and Ray Eames, *A Computer Perspective* (Cambridge, Mass.: Harvard University Press, 1973); Allen Eyles, *The World of Oz* (Tucson, Ariz.: HP Books, 1985).

BECH, NIELS IVAR (1920–1975). This Danish scientist introduced the computer age to Denmark and served as the first managing director of Regnecentralen, Denmark's first designer of electronic digital computers.*

Bech was born in Lemvig, Denmark, in 1920 and only graduated from high school. During World War II, he was a teacher and was involved in anti-Nazi activities. Between 1949 and 1957, he served as a calculator for the Copenhagen Telephone Company (Kobenhavns Telefon Aktieselskab [KTA]). Bech's company assigned him to the Danish Academy of Technical Sciences to work on a computing project in the late 1940s where he participated in the construction of DASK, a computer patterned on the design of the Binary Electronisk Sekevens Kalkylator (BESK) under construction in Stockholm at the Swedish Mathematical Center. The group made DASK operational in 1957. Upon completion of the project, Bech became director of Regnecentralen, serving in that capacity until 1971.

During his tenure, the agency designed software* and hardware and trained computer specialists. Following DASK, his staff designed and built eighteen copies of another computer called GIER (1959–1962). Bech convinced his government to install a GIER computer at each Danish university. This computer specialist also saw an opportunity to sell the technology being developed in his country to other nations. Thus, for instance, in 1964 he placed a GIER at the University of Warsaw and later exported his agency's goods and services to Czechoslovakia, Hungary, Bulgaria, Rumania, the Democratic German Republic (East Germany), and Yugoslavia.

Bech attempted to form a consortium of computer developers throughout the Scandinavian countries but was unsuccessful. However, he did help establish a journal dedicated to computer science called *BIT* which has continued publication since its first issue in 1961. Bech also became one of Europe's leaders in the effort to have ALGOL* accepted and understood as an important computer language. He used his agency's facilities, for example, to publish bulletins on the language and the results of various international conferences during the 1960s.

During the 1960s Bech held various international positions within the data processing community. He was active in the International Federation of Information Processing (IFIP)† and served as an advisor to UNESCO. He also hosted a series of European conferences on computing, including those of NordSAM and NordDATA which have been held each year since 1959.

Regnecentralen evolved from a nonprofit organization into a manufacturing company, producing both hardware (input/output devices and computers) and software by the second half of the 1960s. After a battle with its board of directors in 1971 over how the company was run, Bech was dismissed from his position as head of the company's corporate management. The company went bankrupt in 1979. He died of a heart attack on July 25, 1975.

For further information, see: Isaac L. Auerbach, "Eloge: Niels Ivar Bech, 1920–1975," *Annals of the History of Computing* 6, No. 4 (October 1984): 332–334; Poul Sveistrup et al., eds., *Niels Ivar Bech—en epoke i edb-udviklingen i Danmark* (Copenhagen: DATA, 1976).

BELL, CHESTER GORDON (1934–). Bell helped build the PDP* series of minicomputers for Digital Equipment Corporation (DEC)† in the 1960s. The PDP–8, which he engineered, was the first minicomputer introduced commercially in the data processing industry. He also built the PDP–11 and the

VAX–11 processor. In the early 1980s he played a crucial role in the establishment and operation of the largest museum in the world devoted to computers.

Bell was born in Kirksville, Missouri, on August 19, 1934. He completed his B.S. in electrical engineering at the Massachusetts Institute of Technology (MIT) in 1956 and his M.S. there the following year. Bell worked in the Engineering Speech Communications Laboratory at MIT during the academic year 1959–1960, after which he joined DEC as an engineer at Maynard, Massachusetts, where he worked on the early series of PDP computers and engineered the PDP–5, 8, and 11. In the process he did considerable developmental work with interactive computing and became a leading expert on minicomputers. He left DEC in 1966 to take up duties as professor of computer science at Carnegie-Mellon University which had already become an important center for research on hardware and artificial intelligence*. While there he worked on minicomputer-related research and on interactive computing. In 1972, he returned to DEC as vice-president of engineering with the mandate to produce the VAX (an enhancement over the current PDP series) and to move the company more forcefully into the world of semiconductor (chip*) technology. He remained at DEC until his retirement in 1983. It was during the latter years of his second stay at DEC that he helped found the Computer Museum (located in Boston) with strong backing from his company. In July 1983, Bell established a new firm called Encore Computer Corporation, for the purpose of building a new generation of small computers. His co-founders were Kenneth Fisher, previously of Prime Computer, Inc.,† and Henry Burkhardt, originally from Data General Corporation.† At Encore, Bell assumed the role of chief technical officer.

Bell published several books important in the field of computer technology. He also received a variety of awards from the engineering and data processing community in the United States. In 1982, for example, he received the Eckert-Mauchly award, named after the two developers of the world's first electronic digital computer,* the ENIAC,* and later of the UNIVAC*.

In addition to contributing to the establishment and operation of the Computer Museum, Bell engineered the PDP–8, the first minicomputer built for commercial use in the 1960s. It was small enough to fit into a cabinet and thus become a subset of other systems. It had a full complement of normal functions evident in computers. Over the next twenty years it eventually evolved (by 1978) as one chip. In one form or another, the computer broke new ground in the design of compact systems beginning in the early 1960s. Although Bell acknowledged in the 1980s that the VAX–11 was intended simply to be an extension of his earlier machine, it proved to be another major contribution to the development of powerful and efficient minicomputers. The results of his experiences with PDPs and the VAX appeared in a series of books, the most important of which is *Computer Engineering* (1978). It was intended to be a text for practicing computer designers, but it included considerable amounts of information about the history of the PDP series, including Bell's work on related machines while at Carnegie-Mellon University.

For further information, see: C. Gordon Bell et al., *Computer Engineering: A DEC View of Hardware Systems Design* (Bedford, Mass.: Digital Press, 1978), with A. Newell, *Computer Structures: Readings and Examples* (New York: McGraw-Hill, 1971), and with J. Grason, *Designing Computers and Digital Systems Using PDP–16 Register Transfer Modules* (Maynard, Mass.: Digital Press, 1972).

BEMER, ROBERT WILLIAM (1920–). This member of the data processing industry was one of the major developers of ASCII, one of the industry's standard seven-bit code descriptors for data. He played an active role in the use of ALGOL* and held a variety of jobs that reflected new patterns of employment within the industry during the 1960s and 1970s.

Bemer was raised in Sault Ste. Marie, Michigan, and completed his A.B. degree in mathematics at Albion College in 1941. In 1949 he first became involved in data processing when he joined RAND Corporation as a programmer. In June 1951, he left to join Lockheed Aircraft in Burbank, California, as a group leader of mathematical analysis. In November 1952, he left Lockheed to work for Marquardt Aircraft, only to return to his old employer in December 1955 to establish a computer department for the Missiles and Space Center. Soon after, he joined International Business Machines Corporation (IBM)† as Director of Programming Standards. In April 1962, he took on the same duties at the UNIVAC Division of Sperry Rand,† a job he held until February 1965 when he moved yet again, this time to Bull G.E. and from there directly into General Electric† itself. He remained with the organization in Phoenix, Arizona, after Honeywell Information Systems acquired that GE operation, working for the new firm as of 1986. Bemer was an earlier version of the new class of industrial gypsy which was to became so common in the data processing industry, particularly in Silicon Valley, an area south of San Francisco, during the 1970s and 1980s.

As a developer of ASCII, Bemer was heavily involved in industrywide attempts to develop standard character sets since about 1960. In the mid-1980s, he became chairman of the International Standards Organization Subcommittee on Programming Languages.

For further information, see: R. W. Bemer, "A Politico-Social History of ALGOL," in Mark I. Halpern and Christopher J. Shaw, eds., *Annual Review in Automatic Programming* (New York: Pergamon Press, 1969), 5: 151–237.

BERRY, CLIFFORD E. See ATANASOFF, JOHN VINCENT

BIGELOW, JULIAN (1913–). Bigelow helped to develop John von Neumann's** computer at the Institute for Advanced Studies, (IAS) at Princeton, New Jersey, following World War II. During the war, he was an engineer in the Fire Control Division of the National Defense Research Committee (NDRC) which used science to support the U.S. war effort. Bigelow left the agency in mid-1945 to join von Neumann and in time became his chief engineer.

Bigelow had already developed a profound interest in automatic computing before coming to Princeton. He had supported related projects while at NDRC, as well as through his attachment to the Applied Mathematics Panel during the war. Bigelow's responsibilities at the IAS were to do all the engineering developmental work necessary to build a stored-program digital computer,* later known as the IAS Computer.* His contribution was in the design of circuitry that made possible the use of electrical components in the system. He assumed overall responsibility for the design concept of the machine, along with developing the detailed arithmetic and control organs of the system. In the late 1940s, Bigelow believed that a computer should perform asynchronously. This meant that a computer had to allow each operation to run as necessary to completion and then have such a task signal the mainframe that it was finished, causing the next to begin. He believed that this approach would result in a faster computer than the more traditional approach of slicing off a period of time for a task to complete before computer power was assigned to another operation. With his concept in mind, his staff designed a means for transferring data reliably, developed diagnostic code, and built the arithmetic unit which was the part of the computer that did the actual processing of applications.

During the summer of 1948, Bigdow visited Great Britain to survey various developments in the field of computing. He brought back information on the Williams tube, which became part of the technical basis for the memory* system used on the IAS Computer. It consisted of Cathode ray tubes (CRTs), instead of the more traditional mercury-based systems, and proved far more reliable. Bigelow remained at IAS throughout the late 1940s.

For further information, see: Herman H. Goldstine, *The Computer from Pascal to von Neumann* (Princeton, N.J.: Princeton University Press, 1972).

BILLINGS, JOHN SHAW (1839–1913). Billings, a medical doctor, became a leading American statistician and originally encouraged Herman Hollerith** to apply the technology of punched cards common to the Jacquard loom to tabulating statistics. As a result, Hollerith developed a series of card punch devices used to tabulate the U.S. Census of 1890. Billings' interest in card punch technology for gathering and tabulating vital statistics in general and for census taking made possible the birth of modern information processing. Hollerith's work at the Census Bureau eventually led to his creation of a company that finally became known as International Business Machines Corporation (IBM).†

Billings was born on April 12, 1839, in Switzerland County, Indiana. He earned his A.B. degree from Miami University in 1857, and he completed his formal education at the Medical College of Ohio in 1860. By the end of the century, he had also been granted honorary degrees from such universities as Edinburgh, Yale, Johns Hopkins, Harvard, Dublin, and Oxford. During the American Civil War, he served as a surgeon in the Union Army, remaining in the service until his retirement in 1895. Following the Civil War, however, he became a medical inspector, ran military hospitals, and gathered vital statistics for the Army. He also built up the Surgeon-General's medical library from 600 to 50,000 volumes and published a sixteen-volume *Index Catalogue of the Library of the Surgeon-General's Office USA,* published between 1880 and 1894. This publication instantly became a major reference work for the medical profession. He also wrote a *National Medical Dictionary* in two volumes, published in 1889, and still found time to help organize and establish the New York Public Library.

With his vast experience as a statistician, Billings took charge of vital statistics for the tenth U.S. Census (1890) and for vital and social statistics in the Census Bureau for the eleventh U.S. Census (1900). Years later, Hollerith acknowledged that Dr. Billings had encouraged him to develop machines to speed up and improve the government's ability to take the census, using cards to handle the mechanical aspects of tabulating data. After Hollerith joined the staff of the U.S. Census Bureau to study the problem of using cards, Billings continued to encourage the younger man's work. In 1890, Billings used machines Hollerith had developed.

Billings' interest in Hollerith had begun as early as 1880 with his original suggestion. In 1887, he prepared a paper on the benefits of using mechanical aids to gather statistical data and exercised his widespread influence to encourage health departments in New York City and New Jersey to use Hollerith's devices. Billings' own implementation of these units during the 1890s allowed him to tabulate the Census of 1890 faster than it had been done in recent years, even though the amount of data gathered had increased dramatically over 1870 and 1880. Thus, his prestige, support, and friendship made possible the wide acceptance of Hollerith's devices. This energetic doctor died on March 11, 1913.

For further information, see: Geoffrey D. Austrian, *Herman Hollerith: Forgotten Giant of Information Processing* (New York: Columbia University Press, 1982); F. H. Garrison, *John Shaw Billings: A Memoir* (New York: Putnam, 1915); L. E. Truesdell, *The Development of Punched Card Tabulation in the Bureau of the Census, 1890–1940* (Washington, D.C.: U.S. Government Printing Office, 1965); C. Wright, *The History and Growth of the United States Census* (Washington, D.C.: U.S. Government Printing Office, 1900).

BJERKNES, VILHELM (1862–1951). Bjerknes played a significant role in the development of modern weather forecasting techniques. He concluded that weather could be forecasted correctly only by relying on mathematical techniques that tracked weather patterns. Thus, his work focused on the use of mechanics and physics rather than on older methods which relied on charts. His concentration on mathematics created the need for tools to perform large numbers of calculations rapidly. It was not until World War I, however, that other scientists forcused on the need for calculators, and only after World War II were his suggestions implemented with computers.

Bjerknes was born in Norway where he studied hydrodynamics, a course of study he continued under Henri Poincaré. Bjerknes is best remembered for his work on electric resonance which led to the establishment of the electromagnetic theory of radiation. His first publication on the use of mathematics to guide weather forecasting appeared in an article in an Austrian metereological publication in 1904. Six years later, he published a fuller treatment, *Dynamic Meterology and Hydrography*. Bjerknes' ideas were finally implemented on the ENIAC* computer in 1950; he had lived long enough to see his ideas implemented before his death in 1951.

For further information, see: Charles Eames and Ray Eames, *A Computer Perspective* (Cambridge, Mass.: Harvard University Press, 1973).

BLUMENTHAL, WILLIAM MICHAEL (1926–). Blumenthal was Secretary of the U.S. Treasury in the Carter administration and chairman of the board, first of Bendix Corporation and later of the Burroughs Corporation.† At Burroughs he presided over the reorganization of the firm, a few acquisitions, and, in 1986, the merger of Sperry Corporation† with his company. The combined Sperry-Burroughs, named Unisys, organization was the second largest manufacturing firm in the data processing industry.

Blumenthal was born in Germany on January 3, 1926, moved to the United States in 1947, and became a naturalized citizen in 1952. He completed his B.Sc. at the University of California at Berkeley in 1951, an M.A. and M.P.A. at Princeton University in 1953, and his Ph.D. at the same institution in 1956. Between 1954 and 1957, he served as a research associate at Princeton and during most of these years was a labor arbitrator for the State of New Jersey. He was appointed to his first important position within the business community in 1957 when he became a vice-president and director for the Crown Cork International Corporation, remaining in that capacity until 1961. Then he became deputy assistant secretary for economic affairs at the U.S. Department of State. From 1963 to 1967, he held a presidential appointment as deputy special representative for trade negotiations with the rank of ambassador. In 1967, he left government service to become president of Bendix International, holding that position until 1970. During the same period, he served on its board of directors. He was elected vice-chairman of the company in 1970 and its chief operating officer in 1971. For the next five years he was chairman, president, and chief operating officer. Between 1977 and 1979 he was Secretary of the U.S. Department of the Treasury.

Blumenthal joined Burroughs Corporation, in 1979 as a director, and between

1980 and 1981, he was its vice-chairman and chief executive officer. In 1981, he was named chairman of the board and chief executive officer of the company, positions he still held as of 1987. During the first several years at Burroughs, Blumenthal cultivated the company's existing customer base while revitalizing the firm in general. By the end of 1984, revenues had reached $4.87 billion and the company had 65,000 employees. Burroughs was recognized as an important manufacturer of mainframe computers and, in an attempt to enhance its position in the market by expanding its customer base and critical mass in research and development, Blumenthal sought to acquire other data processing firms, beginning with manufacturers of peripheral equipment. He made his first attempt to take over Sperry in 1985 but failed. In May 1986, he announced that he wanted to acquire enough of Sperry's stock to take over, a goal he achieved within a month.

Looking at the mainframe business in which both firms were important players suggests the potential impact of such a merger. In 1985, and on a worldwide basis, Sperry claimed about 6.1 percent of the total market for mainframes, Burroughs an additional 6.5 percent, while all other substantial competitors (except the International Business Machines Corporation—IBM†) each had less than 3 percent. IBM was credited with enjoying nearly 63 percent of the market. The prospect of merging Burroughs and Sperry meant that Blumenthal would command an immediate market share of 12.6 percent. In May 1986, at the time he announced his intent to take over Sperry for the second time, Blumenthal was quoted by the American press as saying that the merger would "significantly raise the level of competition in the computer industry," offering an alternative to IBM. Many questioned the validity of that logic inasmuch as Burroughs would now have two lines of computers incompatible with each other which in turn would still remain (as before) incompatible with IBM's. The lack of compatibility thus remained a serious barrier to IBM's world.

In mid-June 1986, when Blumenthal announced that his firm had acquired enough stock to take over Sperry, the combined two firms were approximately one-fifth the size of IBM. Their combined assets were slightly over $10 billion, and they had some 130,000 employees. If the U.S. government permitted the merger to go through, Burroughs would emerge as the second largest manufacturing company in the data processing industry. By late 1986, the merger had been finalized under the new name of Unisys. Credit for expanding Burroughs' base of operations through a strategy of acquisitions along with the introduction of new products belonged to Blumenthal. It was a game plan with a proven history of success within the data processing community.

For further information, see: Stephen T. McClellan, *The Coming Computer Industry Shakeout* (New York: John Wiley and Sons, 1984).

BOLLÉE, LÉON (1870–1913). This French inventor built a calculator in 1887, the first machine that could multiply two numbers directly. Prior to his machine, the product of multiplication had to be obtained mechanically by adding numbers repeatedly together. Although he spent a great portion of his life building automobiles and running a family business, his calculator kept drawing his interest; he designed others, although none apparently was ever built. The 1887 machine was an important example of a Charles Babbage** class engine.

Bollée was born in 1870 at Le Mans, France, into a family that operated a foundry. Bollée was brought into the business and, at the age of eighteen (in 1887), designed a difference engine,* also called a direct multiplication machine, to help calculate the dimensions for bells made at the foundry. In subsequent years, he also experimented with other designs. Bollée exhibited his machine at the Paris Exposition of 1889 and won a gold medal for his creation. Among his papers found after his death were plans for a very sophisticated engine that could handle twenty-seven orders of difference; no other nineteenth-century device came close to that magnitude. Yet the multiplier was his major contribution. It looked like the wheels used on a 1930-vintage washing machine to wring water out of clothing, and it was equipped with gears. It was also very heavy and required two strong men to lift. Only several copies of the gadget were ever built, however, and as a consequence, it was more of a curiosity than an influence on data processing. One of the more important copies went to the Belgian Ministry of Railroads, and, of course, another was used at the foundry.

No historian has ever learned where Bollée got his inspiration for this device. He may have read about Charles Babbage or he may have been familiar with the work done by others prior to 1887. What is known is that this clever man was an interesting and creative inventor. At the age of thirteen, for example, he patented an unsinkable aquatic bicycle. Years later a bold Englishman successfully rode such a bicycle across the English Channel. Bollée, also invented a cash register and a railroad ticket dispenser. As he grew older, he became more interested in designing, building, and racing automobiles. He was best remembered for establishing the famous automobile racing track named after his home town: Le Mans.

For further information, see: Léon Bollée, ''Sur une Nouvelle Machine à Calculer,'' *Comptes Rendus à Academie du Sciences du Paris* 109 (1889): 737–739; Charles and Ray Eames, *A Computer Perspective* (Cambridge, Mass.: Harvard University Press, 1973).

BOOLE, GEORGE (1815–1864). Boole's work on the calculus of finite differences and his contribution to the study of formal logic paved the way for such important computer scientists as John von Neumann** and Alan M. Turing**. His work contributed significantly to information system theories, the theory of probability, the theory of lattices, and the geometry of sets. Boole made it possible to reduce logical steps (thoughts) to algebraic systems, thereby allowing computers to be programmened using binary mathematics.

Boole was born in Lincoln, England, into a lower middle-class family. He lived at a time of great industrial expansion in England. In a period when children of his class did not receive a thorough education, let alone opportunities to attend a university, he was able to acquire an education. He learned Latin and Greek at home as a boy, and his father introduced him to mathematics. When he went to work at sixteen to help support his family, it was at a school, making it possible for him to learn French, German, and Italian on his own. This quiet young man intended to join the ministry but was encouraged to stay in education. At age twenty he opened his own school. He continued his studies but now fully concentrated on mathematics, reading all the available literature of his time. As he studied, he realized that additional work could be done to expand the field of mathematics, and by the mid-1840s, he was publishing papers on the subject. He was fortunate in meeting the editor of the *Cambridge Mathematical Journal,* D. F. Gregory, who not only encouraged his work but also published the mathematician's papers. Bertrand Russell was later to write that in these years Boole invented pure mathematics. During the 1840s and 1850s, however, other mathematicians noticed his work on the calculus of finite differences, including Charles Babbage,** designer of the analytical engine*. Herman H. Goldstine,** an important computer scientist of the 1940s and 1950s, thought Boole's work on finite differences was important because the field of study was "the basic tool of the numerical analyist."

Like most computer scientists, Goldstine thought that Boole's work on formal logic was most important. In 1848, Boole published a short book entitled *The Mathematical Analysis of Logic,* followed by a far more important and detailed work in 1854, *An Investigation of the Laws of Thought, on Which Are Founded the Mathematical Theories of Logic and Probabilities.* In these two books, Boole explained how to convert logic into mathematical forms. For the first time a mathematician presented axioms for logic. In other words, he did for logic what Euclid and others had done for geometry. And he presented the material in algebraic notations. Although monumental in its implications, his work received more attention on the eve of World War I than earlier when Russell and Alfred North Whitehead published their massive *Principia Mathematica* (1910–1913). Beginning at this time, logic became a respectable branch of mathematics.

By being reduced to algebraic terms, logic, when combined with binary numbering systems (which had been around since the 1600s), made it possible to instruct a machine to perform commands. Boole argued that $x^2 = x$ for every x in his system which was part of his law in "which the symbols of quantity are not the subject." To quote Goldstine: "Now in numerical terms this equation or law has as its only solution 0 and 1." That thought insured a marriage between the idea of mathematically expressed logical thought (commands) and binary systems in computer design. According to Goldstine again, "Their logical parts are in effect carrying out binary operations." So for the computer scientists who had discovered that the use of binary numbering systems was useful when combined with the electronic characteristics of switching systems that permitted power to run down a line or through a device (on or off), the result was the computer. For the scientist sitting at his desk working out the logic of commands, the use of the well-known computational tools of algebra was easy. Boole relied on the symbolic language of calculus modified to meet his particular needs, and he used such well-understood algebraic symbols as *x*, *y*, and *z*—notations understood by all scientists.

Although Boole's great contribution would not be fully appreciated until the advent of the computer age, he nonetheless had an impact on his peers. Babbage, for instance, studied his ideas while working on the design of the analytical engine. Boole's ideas helped him understand the concept of a mathematical operation and the role of quantities needed to drive an operation within a machine. Some scientists working on calculating machines had therefore already begun to appreciate the usefulness of Boolean algebra. It was simply left to the twentieth century to complete the translation of his ideas into tasks reduced to terms that a computer could understand.

By the late 1840s, Boole was devoting most of his time to mathematics and had won recognition for his work. In 1849, he became a professor of mathematics at the newly established Queen's College at Cork, Ireland, where he spent the rest of his life. In the course of his career he published some fifty papers, the two books already mentioned, and two textbooks, one on differential equations in 1859 and another on finite differences in 1860.

Boole married Mary Everest in 1855; she was the daughter of Sir George Everest for whom Mount Everest was named.

For further information, see: T.A.A. Broadbent, "George Boole," *Dictionary of Scientific Biography* (New York: Scribner's, 1970): 293–298; Herman H. Goldstine, *The Computer from Pascal to von Neumann* (Princeton, N.J.: Princeton University Press, 1972); H. Tropp, "George Boole," in A. Ralston and C. L. Meek, eds., *Encyclopedia of Computer Science* (New York: Petrocelli/Charter, 1976) pp. 177–178. *For Boolean algebra, see:* R. R. Korfhage, *Logic and Algorithms* (New York: John Wiley, 1966); J. D. Peatman, *The Design of Digital Systems* (New York: McGraw-Hill, 1972).

BOOTH, ANDREW DONALD (1918–). Booth was an early pioneer in the development of British computers. He developed a series of components that went into British computers in the late 1940s and early 1950s, including memory* systems. He also designed whole systems and encouraged the growth of a British computer industry.

Booth was born on February 11, 1918, in East Moslesey, Great Britain. He completed his B.Sc. at the University of London in 1940, his Ph.D. in chemistry at the University of Birmingham in 1944, and his D.Sc. in physics at the University of London in 1951. He was a research physicist for the British Rubber Producers' Research Association (1944–1946), lecturer at Birkbeck College of the University of London (1946–1949) when he did considerable work on computers, professor of theoretical physics at the University of Pittsburgh (1949–1950), director of the computer project at the University of London (1950–1954), reader on computer methods at Birkbeck (1954–1957), and Birkbeck's head of the Department of Computer Science (1957–1962). In the 1960s and 1970s, his responsibilities shifted. First, he became a professor of electrical engineering and head of his department at the University of Sask (1962–1963), and then dean of engineering there from 1963 to 1977. During this same period, Booth also served as president and honorary professor of physics at Lakehead University (1972–1977). He subsequently retired from teaching and administration.

Booth also had a role related to data processing. He was a Nuffield Fellow (1946–1947), a Rockefeller Fellow, a member of the Institute for Advanced Study (1947–1948), and a consultant to the British Rayon Research Association (1948–1955). He was a scientific advisor to the British computing industry, including International Computers and Tabulators, Ltd., from 1950 to 1964 when the company entered the world of data processing. He was also editor of the journal published by the Royal Micros Society.

Booth's most important work relating to computers came in the 1940s. After spending six months at the Institute for Advanced Study at Princeton, New Jersey, working with John von Neumann's** engineers, he returned to Great Britain with the intent of building a small computer. His machine, called the Automatic Relay Computer (ARC), was partially funded by the British Rubber Producer's Research Association. He built the arithmetic portion of the machine by early 1948, while a magnetic drum storage component capable of housing 256 20-bit words became operational later that year. When fully completed, the ARC was capable of storing 50 numbers and performing about 300 different instructions as a sequence-controlled machine. While working on this machine, Booth was assisted by one of the first female programmers in England—Kathleen Britten whom he subsequently married.

Because of a lack of research funds, Booth concentrated on low-cost components; nowhere was this more evident than in his use of thermionic valves in 1949. He thought that small, low-cost computers could be useful as scientific processors for research organizations also operating with limited budgets. He focused on the design of a new machine which eventually became a group of devices known as the All-purpose Electronic (Rayon) Computers, or simply as the APE(R)C series. His father, S. J. Booth, got involved in the project by developing the magnetic drum used in this class of machine. The two represented the only father and son team known to have built computers together, while others had married research and programming assistants. The APE(R)C had 415 thermionic valves and began limited operations in July 1952.

Booth established relations with the British Tabulating Machine Company (BTM),† a European derivative of International Business Machines Corporation (IBM).† When the firm wanted to build computers, it turned to Booth and modeled its first important product—the Hollerith Electronic Computer (HEC)—on his APE(R)C. HEC was publicly unveiled in 1953 at the Business Efficiency Exhibition in London. The following year it became a product called the BTM 1200, and five were sold promptly. The BTM 1201 was introduced soon after to handle commercial applications; its predecessor was marketed to scientific users. The 1201 sported drum storage of 1,024 words and was first shipped in

1956. Seventy were eventually sold. Therefore, for BTM Booth's inspiration had proven significant during the 1950s, and its products were some of the earliest commercially available small systems.

For further information, see: Simon Lavington, *Early British Computers* (Bedford, Mass.: Digital Press, 1980).

BOYER, JOSEPH (1848–1930). Boyer was president of the Burroughs Adding Machine Company (later known as Burroughs Corporation†) from 1905 to 1920. During his tenure as chief executive officer, the company became a leading supplier of office and information handling equipment.

Boyer was born near Toronto, Canada, in 1848. He became an apprentice in a machine shop in St. Louis, Missouri, in 1866, and during his thirty-one years in that city, he proved to be a successful creator of manufacturing enterprises. He also invented and worked with William S. Burroughs** to develop an adding machine. Boyer invented the first widely used pneumatic hammer, and to market that product he established the Boyer Machine Company. In 1901, that organization merged with the Chicago Pneumatic Tool Company. The Boyer Machine Company built the first adding machines sold by Burroughs as products of the Arithometer Company in the 1880s. Boyer was thus linked to Burroughs from the earliest years and, in fact, was an officer of the company. In 1901, he moved the firm to Detroit where he believed a larger pool of labor was available than in St. Louis or any other Midwestern location. He became president of Burroughs in 1905 and its chairman of the board in 1920. In 1920, his son-in-law, Standish Backus, became president. In the year Boyer died (1930), Burroughs logged sales of $30 million and built 150,000 machines. The plant in Detroit had some 10,000 workers. In the early 1900s, Boyer also invested heavily in automotive companies.

While Boyer was at the helm at Burroughs, the company developed a series of products that over time became smaller, more user friendly, and less expensive. These developments allowed the company to grow into a major provider of office equipment in the United States and in Europe. Innovations in products appeared constantly, a strategy that would be prevalent in the data processing companies decades later. Boyer acquired other firms to broaden the company's base, such as the Moon-Hopkins Billing Machine Company in 1921. He built manufacturing plants in Latin America and in Europe. His primary competition after World War I was International Business Machines Corporation (IBM),† with such firms as Remington Rand† and the National Cash Register Company (NCR)† also major rivals. Boyer died on October 24, 1930.

For further information, see: L. J. Comrie, *Modern Babbage Machines. Bulletin* (London: Office Machinery Users Association, Ltd., 1932); B. Morgan, *The Evolution of the Adding Machine; The Story of Burroughs* (London: Burroughs Machines, 1953).

BRAINERD, JOHN GRIST (1904–). This engineer was in charge of the project at the Moore School of Electrical Engineering† which built the ENIAC,* the first electronic digital computer.* He also participated in the construction of its sequel, the EDVAC.* In his role as overall project leader on the ENIAC in the last years of World War II, Brainerd made a significant contribution to the emergence of the modern computer, particularly toward the management of that kind of developmental effort.

Brainerd was born in Philadelphia on August 7, 1904. He completed his B.S. at the University of Pennsylvania in 1925 and a Sc.D. there in 1934 and he also studied periodically at the Massachusetts Institute of Technology (MIT). He was a reporter in Philadelphia between 1922 and 1925. In 1925, he joined the new Moore School of Electrical Engineering as a faculty member, and in 1954, he became a full professor. He retired from the Moore School in 1970 and was named University Professor of Engineering. During his career he held various posts at the University of Pennsylvania, including director (1954–1970) and, earlier, chairman of the division of physical sciences within the graduate school (1942–1948). Between 1935 and 1937, he served as acting state director for one of the U.S. government's relief efforts, the Public Works Administration (PWA), and advised the federal authorities on various scientific projects during the 1930s and 1940s. He was a member of the Scientific Advisory Committee for the National Bureau of Standards (NBS)† from 1959 to 1965, and he worked with related agencies interested in scientific developments throughout the 1960s and 1970s.

But the centerpiece of Brainerd's contributions to data processing came in the 1940s while working on the ENIAC. It was he, for example, who actually hired John W. Mauchly,** one of the co-inventors of the ENIAC, in the early 1940s. Brainerd early recognized that Mauchly's ideas for a computational device made sense and, along with Herman H. Goldstine,** then representing the U.S. Army, supported proposals for the construction of a computer at the Moore School. He worked toward that end in 1943. When a contract for its construction was signed with the U.S. Army on June 5, 1943, Brainerd was named the project's supervisor; he was already a research director at the school. The machine was built and became operational in 1946. He subsequently worked to obtain a similar contract from the U.S. Army to support the development of the EDVAC.

After the ENIAC was built, Brainerd quarreled with J. Presper Eckert** (co-inventor with Mauchly of the ENIAC) and Mauchly. The two developers wanted to file for patent rights on the ENIAC, but Brainerd thought this violated adacemic ethics. As a result of their disagreement, he resigned as project supervisor on the EDVAC, although he retained responsibility for the final efforts on the ENIAC. The controversy over patent rights ultimately destroyed the preeminent position the Moore School had enjoyed in the young field of computing since many of its key engineers resigned, including Eckert and Mauchly who went on to establish their own company.

Brainerd's most important role in the history of computing therefore ended with the decline of the Moore School as a major center for research on computers in the late 1940s. But while active with the ENIAC he had played a solid role. He had been interested in computational equipment and was a proven manager. He was able to balance the needs of the government on the one hand and the various communities within his university on the other. Ultimately, his administrative abilities allowed the ENIAC to be built and the EDVAC started. These were complex projects involving the coordinated activities of many individuals and organizations, and the politics of academic and governmental research. Science and engineering alone were no longer enough. Brainerd was thus one member of a new generation of engineers in the mid-twentieth century who had to coordinate many activities and could no longer develop inventions or conduct research alone. As a member of that new caste, Brainerd operated effectively and made a lasting contribution to the development of the digital electronic computer.

For further information, see: John G. Brainerd, "Genesis of the ENIAC," *Technology and Culture* 17, No. 3 (July 1976): 482–488; Brainerd and T. K. Sharpless, "The ENIAC," *Electrical Engineering* 67, No. 2 (February 1948): 163–172; Herman H. Goldstine, *The Computer from Pascal to von Neumann* (Princeton, N.J.: Princeton University Press, 1972); Nancy Stern, *From ENIAC to UNIVAC: An Appraisal of the Eckert-Mauchly Computers* (Bedford, Mass.: Digital Press, 1981).

BRIGGS, HENRY (1561–1630). This well-known British mathematician encouraged the use of logarithms, which in turn led to the development of the slide rule* and increased pressure for other mechanical aids to computation.

Briggs was born in February 1561 at Warley Wood, Yorkshire, and in 1585 obtained a master's degree from Cambridge University. He lectured there in 1592, and he joined the faculty at Gresham College in London to teach geometry in 1596. In 1614, he learned about John Napier's** "bones" and immediately became enthused with what became known as logarithms. Applying his prestige and influence, Briggs encouraged the use of logarithms and contributed to their development by constructing new tables, which were easier to use than Napier's. He published his own set in 1624, creating the base 10 logarithms used today. Briggs was able to extend his own tables from 1 to 20,000 and from 90,000 to 100,000, carried out to fourteen decimal places. Less than 0.04 percent of his entries had errors. His work helped make logarithms a popular aid to computation in the Seventeenth Century and led to the development of improved tables throughout the next centuries. Within only twenty years of his first published tables, they were used by mathematicians around the world. The tool was a quick method for arriving at products and quotients. Briggs' tables were the most popular of all such tools because they were easiest to use.

Briggs died at Oxford, where he was professor of astronomy, on January 26, 1630. This creator of long division, which was also a major aid to mathematics, might ultimately be best remembered for his contributions to the development of computational technologies.

For further information, see: Carl B. Boyer, *A History of Mathematics* (Princeton, N.J.: Princeton University Press, 1985, original edition, 1968); Michael R. Williams, *A History of Computing Technology* (Englewood Cliffs, N.J.: Prentice-Hall, 1985).

BROOKS, FREDERICK PHILLIPS, JR. (1931–). Brooks was one of International Business Machines Corporation's (IBM's)† most important scientists responsible for the development of early computers during the 1950s and 1960s. He worked on STRETCH* and the Harvest systems, and later managed the engineering design of the System 360* family of computers. That last project proved to be the most significant computer product ever announced and made possible the rapid expansion of the data processing industry during the 1960s and 1970s.

Brooks was born on April 19, 1931, in Durham, North Carolina. He attended Duke University, graduating in 1953. He completed an S.M. in 1955 and a Ph.D. in 1956, both at Harvard University. Upon concluding his graduate work, Brooks became an engineer for IBM at its laboratory in Poughkeepsie, New York, home of that company's major efforts in computing over the next three decades. Brooks moved to IBM's research facilities at Yorktown Heights, New York, in 1959 but returned to Poughkeepsie in 1960 to work on various projects, the most important of which would become the S/360. Between 1964 and 1975, he served as a professor at the University of North Carolina (UNC) at Chapel Hill and, after 1975, as Kenan Professor. He was also chairman of the Computer Science Department, beginning in 1964 and until the early 1980s. At UNC he played a leading role in the establishment of a data center, and he expanded the educational offerings in the general field of computer science. While professor, he also wrote one of the earliest and most important studies on the management of data processing projects still consulted today: *The Mythical Man-Month: Essays on Software Engineering* (1975).

But Brooks is best remembered for his contributions at IBM. He came to Poughkeepsie with a solid background in data processing, having earned his Ph.D. within the Harvard Computation Laboratory. This facility already had a solid history of achievement and had served as an important, early source of well-trained computer scientists. The first significant projects Brooks worked on involved Harvest and STRETCH, each of which led the company to new technologies used in its computers during the late 1950s. Along with other engineers, Brooks worked on program-controlled interrupts, control words, and various methods of harmonizing diverse data formats—all involving advances in software.* After those two projects were completed, Brooks became involved in the development of the 7030, the direct byproduct of STRETCH.

In 1960, IBM was developing yet another product line called the 8000 series, a family of computers which was intended to displace existing IBM computers. Many of the engineers involved in STRETCH worked on the 8000, and Brooks soon emerged as one of its greatest proponents within the company. At the time, there was a proposal within IBM to build a different line of computers—what ultimately became known as the S/360—proposed by Robert O. Evans.** After political infighting, management reviews, and so on, Evans won and Brooks lost. Top management made the decision in May 1961 to develop the S/360 and drop support for the 8000. In a move designed to win the support of a considerable number of engineers loyal to the 8000 series, Evans asked Brooks to lead the planning effort that would be required for the S/360. Brooks accepted, any many who had resisted S/360 within IBM's engineering community subsequently joined him.

The technical problems which Brooks and Evans now faced were considerable. They wanted to develop an integrated family of computers in which programs that ran on one machine could run on all other models. They intended for peripheral equipment to be interchangeable with all models. Both insisted on compatible operating systems* up and down the line and, just as challenging, additional functions in all software to improve the amount of work a processor could handle while enhancing the management of a system through software. They sought to take advantage of recent advances in technology, such as the introduction of the transistor and later, the chip.* Brooks focused considerable attention on software and on the basic pattern that would be used for defining the systems involved. Thus, for example, he explored the benefits of standardizing on an 8-bit byte to describe an alphanumeric character, two decimal digits, or a portion of some binary number.

Brooks' concentration on the details of software proved significant. He influenced all decisions made concerning software on systems that used in excess of 16K of memory,* which meant all of the large processors in what would become the S/360. Thus, he was in the midst of developing the operating system later called OS/360, which functioned in systems that had 32K or more of memory—huge by the standards of the day. The basic architecture of that operating system also appeared in that for small systems, ultimately called the Disk Operating System, or more commonly, DOS. He ran a development team made up of 2,000 programmers by the end of 1965 and was responsible for a budget of $60 million. He managed the single largest software development project that had ever been attempted up to that time. As of the mid-1980s, it remained one of the largest software projects in the history of data processing.

After Brooks' operating systems and related software were announced during the mid-1960s, all future operating systems reflected many of the features he introduced. While historians have tended to emphasize the advances in hardware achieved with the S/360, perhaps more profound in the long run were those advances achieved with software. What in effect Brooks and his staff did was to develop full-function, relatively easy to use, highly automated operating systems that were so good (when all the ''bugs''* were worked out) that they remained the standard for the next several decades. His software made it possible for ever larger systems to be built in the 1970s and 1980s. While IBM's operating systems grew and expanded in function and reliability during the 1970s and

1980s, their basic configuration had been defined by Brooks a quarter-century earlier.

Because the S/360 became the de facto industry standard, his work influenced the characteristics of many computers manufactured by other companies. During the 1960s, numerous firms standardized on S/360 architecture. In the 1970s and 1980s, the phrase used was "S/370* architecture"—merely an updated version of Brooks' original work. Plug-compatible computers made by Amdahl Corporation,† for example, used his operating systems. His contribution was enormous: by the end of the 1960s, over 15 percent of all installed computers in the world were S/360s and by the mid-1970s, over half were standardized on his operating systems. While IBM doubled its sales between 1964 and 1970, the data processing community had recognized the benefits of the new hardware largely because of its new software.

IBM has always been as concerned with how a project is managed as with its results. Hence, in 1965 the company was as interested in software as in the development and subsequent manufacture of equipment. The company has continued to be successful because it concentrates on improving packaging and manufacturing processes as a means of enhancing reliability and efficiency. The result has been products that can be sold competitively. Such a tradition of management had long existed at IBM. Until the S/360, neither IBM nor the industry as a whole had ever experienced software development on such a massive scale.

The process of managing the most important family of software ever achieved now had to be invented. The lessons learned made it possible for IBM's engineers to develop other operating systems involving hundreds of thousands of lines of programming (such as MVS/XA used in the 1980s), in large part because of the experiences gained with the OS/360 effort. Brooks later described many of those lessons in his book on the mythical-man month.

Brooks has been recognized for his overall contributions to computing and more specifically for his work on software and operating systems. In 1970, he received the McDowell award from the Institute of Electrical and Electronics Engineers (IEEE)† Computer Society. The same organization also gave him its Computer Pioneer award in 1982. In 1970, the Data Processing Management Association (DPMA)† named him its Man of the Year. He also became an IEEE Fellow. In the 1980s, Brooks continues to do research on computer science while teaching at UNC.

For further information, see: Charles J. Bashe et al., *IBM's Early Computers* (Cambridge, Mass.: MIT Press, 1986); Frederick P. Brooks, *The Mythical Man-Month: Essays on Software Engineering* (Reading, Mass.: Addison-Wesley, 1975), and with K. E. Iverson, *Automatic Data Processing, System/360 Edition* (New York: John Wiley, 1969); René Moreau, *The Computer Comes of Age: The People, the Hardware, and the Software* (Cambridge, Mass.: MIT Press, 1984); Emerson W. Pugh, *Memories That Shaped an Industry* (Cambridge, Mass.: MIT Press, 1984).

BROWN, GORDON S. (1907–). Brown developed the electrical engineering program at the Massachusetts Institute of Technology (MIT) into one of the most important in the United States during the mid-twentieth century. It in turn was the scene of many important computer-based research projects which included the construction of WHIRLWIND* for the U.S. military and real-time computing. Brown personally conducted research on computer-based topics during the 1930s and 1940s as well.

Brown did his university work at MIT, and in the late 1930s, while working on his Ph.D., he developed a cinema integraph within the Research Division that caught the attention of Vannevar Bush,** who at that time was working on analog computing. Brown completed his graduate work in June 1938 and remained at MIT. This electrical engineer early became involved in research involving computational equipment. When World War II began and MIT was awarded the assignment to develop radar technologies, Brown participated. That massive project trained many engineers in the use of electronics and related technologies that would appear later in computers. In the Servomechanisms Laboratory, Brown and others worked on research projects under contract, usually for the U.S. government. For example, his laboratory developed a gun control system used during World War II by the U.S. Army. In addition, since he had studied feedback mechanisms along with classical electrical engineering, he was able to bring a new perspective to the work done in the department during and after World War II.

Brown was born in Australia and moved to the United States in 1929, completing his B.S. at MIT in 1931 as an honor student. While in graduate school at MIT, he studied mathematical problems along with normal scientific courses. At this young age he was already recognized: in 1939, by the time he received his appointment as assistant professor of electrical engineering at MIT, he had been in charge of the Electrical Engineering Department's laboratory for several years. Over the next few years, Brown convinced MIT and various government agencies to expend more than $1 million worth of research on servomechanisms. These efforts advanced the understanding of electronic feedback systems, which would become so critical to numerical control and other automated computers after World War II. Such work also enhanced research on analog systems, which was then in process at MIT. In 1940–1941, Brown recruited Jay W. Forrester** into the department. Forrester would later be in charge of the effort to build WHIRLWIND. During World War II, Brown conducted research, encouraged work on servomechanisms, undertook teaching the subject to the military, and developed various gun control and radar-based technologies and equipment for the military.

Following the war, Brown continued his work on servomechanisms during a period when research on various computer-related activities (including what would later be called artificial intelligence*) flowered at MIT. As head of the Servomechanisms Laboratory, he had created an environment in which managerial controls were loose enough to encourage diverse projects to be pursued effectively, as long as they were focused and were completed in a timely manner. Rather than impose rigorous administrative structures, Brown chose highly qualified people to work in his organization. By 1944, when early discussions began which ultimately led to WHIRLWIND, the Laboratory had nearly 100 employees, thirty-five of whom were trained engineers. It had an established record for developing a control system for 40mm cannon, radar antenna for ships, airborne radar, and turret equipment, along with other specialized equipment. Because of this record of doing research and delivering completed and useful devices to the military, MIT was the logical choice for additional contracts for devices such as simulated flight training equipment, beginning in 1944. These ultimately led to computer-related projects in the late 1940s and early 1950s.

While Forrester dominated WHIRLWIND, Brown provided administrative support in his capacity as head of the laboratory and as an influential scientist at MIT. In 1948, for example, when the project was under attack and underfinanced, he supported it, gaining additional funding for the tasks ahead. WHIRLWIND's historians, Kent C. Redmond and Thomas M. Smith, argue that Brown's management style made it possible for Forrester to enhance, expand, and change the technical developmental efforts associated with the project midway through. Projects of this type, when done under contract, typically would not be changed as much as this one was without enormous legal and administrative difficulties. Yet the result of this particular situation was change and, consequently, a very large computer based on much new technology that became useful to the general data processing community during the early 1950s.

On a more personal basis, Professor Brown's career at MIT prospered. During the 1940s he rose in rank, and in 1950 he became chairman of the Electrical Engineering Department, taking over from Harold L. Hazen** who had been in charge since 1938. Three years later Brown established the Acoustics Laboratory at MIT. During the years when he headed the Electrical Engineering Department (1950–1959), the number of research projects, classes, and students increased. The result was a profoundly sophisticated program at MIT that kept it in the forefront of science centers throughout the 1950s and 1960s. During this period, for instance, Brown cooperated with or encouraged the development of real-time computing (Project MAC*) and enhancements to the technology of hardware, and saw the rise of artificial intelligence as a distinct field of study at MIT. On July 1, 1959, he became dean of the School of Engineering, serving in that capacity until November 1, 1968. At that point he decided to give up his administrative responsibilities and return to teaching and was named the first Dugald Caleb Jackson Professor of Electrical Engineering. The faculty at MIT elected him Institute Professor in 1972.

For further information, see: George S. Brown and D. P. Campbell, "Instrument Engineering, Its Growth and Promise in Process Control," *Mechanical Engineering* (February 1950): 124–127; Kent C. Redmond and Thomas M. Smith, *Project WHIRL-*

WIND: The History of a Pioneer Computer (Bedford, Mass.: Digital Press, 1980); Karl L. Wildes and Nilo A. Lindgren, *A Century of Electrical Engineering and Computer Science at MIT, 1882–1982* (Cambridge, Mass.: MIT Press, 1985).

BROWN, THEODORE HENRY (1888–1973). Brown introduced Howard H. Aiken**, of Harvard University to International Business Machines Corporation (IBM)† engineers, resulting in the development of the Automatic Sequence Controlled Calculator. That experience introduced engineers at IBM to computers during the early 1940s. Brown also served on the first Board of Managers of the Watson Computing Bureau at Columbia University in 1937, a laboratory established with considerable support from IBM. Also known as the Thomas J. Watson Astronomical Computing Bureau, its mission was to foster research in astronomy. It was supported by Columbia, IBM, and the American Astronomical Society.

Brown was born in Mystic, Connecticut, on October 5, 1888, completed his A.B. at Yale University in 1910, and the A.M. and Ph.D. degrees at Harvard in 1911 and 1913, respectively, in mathematics. He returned to Yale as an instructor in 1913 and during World War I was an instructor for the U.S. Navy. At Harvard he was a member of the research staff, Bureau of Business Research, at the graduate school of business administration (1925–1926), assistant professor of business statistics (1926–1927), associate professor (1927–1932), and, finally, full professor. During World War II he was an important scientific advisor to IBM, the government, and the military, particularly the Army Air Force which was then investing in various computational projects. He was also chief economist in the Office of Civilian Requirements, 1943–1944.

Brown did considerable research on sampling techniques, research that contributed to the field of opinion polling such as that conducted by George H. Gallup, which in later years required the use of computers. He discovered, for instance, that any prediction will be accurate to within 5 percent if it is based on a sampling of 900 opinions. His pioneer work opened up a vast new use of computers following World War II, even though the majority of his important work was completed during the 1920s and 1930s, first for market research and later in the area of political forecasting. He died in December, 1973.

For further information, see: T. H. Brown, *Use of Statistical Techniques in Certain Problems of Market Research* (Cambridge, Mass.: Harvard University, Bureau of Business Research, 1935); Charles Eames and Ray Eames, *A Computer Perspective* (Cambridge, Mass.: Harvard University Press, 1973); Herman H. Goldstine, *The Computer from Pascal to von Neumann* (Princeton, N.J.: Princeton University Press, 1972).

BRYCE, JAMES WARES (1880–1949). Bryce was International Business Machines Corporation's (IBM's)† chief engineer for a third of a century and one of the least known, yet most important, contributors to the evolution of IBM's product line and technical management in the early twentieth century.

He was a giant in the world of card punch equipment, the basis of early modern data processing technology.

Bryce was born in 1880 and studied mechanical engineering at New York City College during 1897–1900. He began his career in 1900, first as a draftsman and later as a designer in New York. He then became a consulting engineer, and, in 1917, Thomas J. Watson** (head of C-T-R before it became known as IBM) hired Bryce as a supervisory engineer. He was sent to the company's plant in Endicott, New York, home of C-T-R's recording division, a plant established to enhance the existing product line of card punch equipment. During his early years at IBM, products were improved and new ones, such as the Type 80 Sorter (introduced in 1925), came out. Bryce and Clair D. Lake** Perfected a new computer card in 1928 which would eventually become the familiar eighty-column variety. Their card allowed more data to be stored than earlier versions, and over the next several years Bryce directed the efforts required to modify all of IBM's products to use it. Within a short period of time this card accounted for 5 to 10 percent of all of IBM's revenues during any given year during the 1930s.

New products continued to appear in the 1930s as a result of Bryce's efforts. In 1931, he and others had perfected a multiplying punch later called the Type 600 which in 1933 was replaced with the Type 601. During his tenure the Type 285 tabulator appeared, as did the IBM Type 405 Alphabetetical Accounting Machine, perhaps the most popular product from IBM in the 1930s. The 405 remained in the product line until after World War II.

In 1937, it Howard H. Aiken** wrote to Bryce seeking help on what would become the Harvard Mark I* computer. It was Bryce who assigned Lake to be Aiken's technical contact with IBM. Bryce also served as a liaison between the U.S. Army and its project at the University of Pennsylvania, what later became known as the ENIAC,* the first electronic digital computer.* His role involved providing various IBM-built components and card peripheral equipment for the project.

After World War II, IBM increasingly turned its attention to the new electronics of the vacuum tube. Yet its storehouse of experience with electronic components was largely the result of Bryce and his general interest in scientific developments. In the 1920s and 1930s, he had encouraged some experimentation with electronics, including work on vacuum tubes—the building block of computers in the 1940s and early 1950s. But Bryce, ever a practical manager, did not propose that his executive management during the 1930s develop products made up of such components, for he realized the limitations of their technologies and functions. He had supported the evolution of the card punch product line from mechanical models to electrical units that could be configured as accounting systems; yet even these machines lacked all the features he wanted. Budgets were limited, and his experience indicated that product-oriented research would pay earlier dividends than basic research. That lesson is more obvious to technical managers today than it was in the 1920s and 1930s.

Given the limitations of existing products and the lack of enough details about electronics, Bryce decided that the general field of electronics should only be observed. In the depths of the Great Depression IBM could ill afford to make significant investments in that area, but the subject could not be entirely ignored either. Thus, he instructed that his patent development department join forces with the electrical laboratory in Endicott to do some modest research. In 1936, for example, A. Halsey Dickinson (who worked for Bryce) investigated relay and tube switching circuits, an exercise that would later allow IBM to contribute to the technical development of the Mark I and, later, the company's own electronic calculators in the late 1940s. Dickinson's early work ultimately led him to file patents in 1940.

Bryce also had great foresight, perhaps the most notable example being the recording of digital information on magnetic medium for use in accounting applications. The thought was more of a dream in the 1920s, but by the mid-1930s enough technical developments had taken place to make that a reality in time, shifting the problems from basic or abstract consideration to applied research. One major problem he had to overcome was that, once punched with data, a card could not be reused to house other information. Discovery of a reusable medium—and magnetic entraption of data promised that possibility—would dramatically reduce the cost of storing information in machine-readable form. As early as 1937, nearly twenty years before IBM announced the first disk drive and a decade before magnetic tape became a viable part of most computer configurations, Bryce was searching for ways to convert data to a magnetic medium. He investigated magnetic surface coatings and even laminated internal layers of magnetized materials on cards. By the time World War II had begun, he and his staff in Endicott had gathered considerable information on the subject. They had also rigged up a National Cash Register Company (NCR)† machine to record on magnetic tape and ran experiments on sensing magnetized dots. Experiments conducted at Endicott in 1941 and 1942 led to the development of a test version of a ledger-posting device in 1943 which magnetically recorded balances read from ledger paper. That experience formed the basis for future research at IBM in the 1940s and early 1950s.

Bryce operated IBM's patent department until 1948 when Dickinson took it over. Between the two of them (from 1932 through 1954) their staffs earned 243 patents, with an additional 56 pending approval. Bryce had personally accumulated 400 patents by 1936. He was an acknowledged genius within IBM and was the company's "patron saint of engineering." In 1936, during its centennial celebration, the U.S. Patent Office honored Bryce as one of the ten "greatest living inventors."

Bryce died in 1949, leaving behind a strong tradition of innovation within IBM's engineering community. He remained an unrecognized giant in the world of card punch equipment, however, being second only to Herman Hollerith* in influence. Indeed, he represented a logical extension of Hollerith.

Bryce contributed greatly to card punch technology from World War I to the late 1940s. The move from mechanical to electronic devices, along with printing card punch gear, came in his time. The expansion of card punch applications and the variety of equipment to support them were largely his accomplishments. And always there were the people, a solid collection of experienced engineers and scientists at IBM organized to build new products that were ultimately based on computer technology.

Much remains to be learned about this engineer's life, particularly about his formative years between 1900 and 1915. What little is known indicates that he was consistently a practical, creative inventor and, later, a manager who balanced the adventure of exploration and tinkering with the practical realities of costs and schedules. He was IBM's most important scientist in the first half of the twentieth century.

For further information, see: Charles J. Bashe et al., *IBM's Early Computers* (Cambridge, Mass.: MIT Press, 1986); Herman H. Goldstine, *The Computer from Pascal to von Neumann* (Princeton, N.J.: Princeton University Press, 1972); William Rodgers, *Think: A Biography of the Watsons and IBM* (New York: Stein and Day, 1969).

BURKS, ARTHUR WALTER (1915–). Burks participated in the creation of the ENIAC* and the EDVAC* computers in the 1940s, the ENIAC was the first operational large-scale electronic digital computer and the EDVAC the first digital computer* with stored-program capability incorporated into its design. That would ultimately become a universal feature of all computers.

Burks was born on October 13, 1915, at Duluth, Minnesota. He graduated from DePauw University in 1936 with an A.B. and the following year with an A.M. from the University of Michigan. He completed his Ph.D. in 1941 and subsequently went to work for the Moore School of Electrical Engineering† at the University of Pennsylvania, staying until 1946 while teaching philosophy in another department. Between 1946 and 1948, he continued his work on the design of computers at the Institute for Advanced Studies at Princeton, New Jersey, with John von Neumann.** Between 1948 and 1954, he was a consultant on computers for the Burroughs Corporation.† He also worked on the Oak Ridge Computer (ORACLE) between 1950 and 1951. After World War II, he held academic positions at several schools, but primarily at the University of Michigan: assistant professor, 1946–1948, associate professor, 1948, and full professor, 1954. Between 1967 and 1971, he served as chairman of the university's Department of Communications Sciences. In the next twenty years he served as visiting professor at various universities, including Harvard, Illinois, and Stanford. He has written a number of works on philosophy and computer science, and, since 1975, he has served as editor of the *Journal of Computer and System Sciences*.

Burks made his most important contributions to the development of the computer while at the Moore School. Unlike his peers at Moore who were engineers, all of his training had been in mathematics and logic. He first came to the Moore School as a student to take a course in electrical engineering and subsequently was given a job there. Years later he observed that part of the design effort for the ENIAC involved taking existing technologies (for example, vacuum tubes and card punch equipment) and building a machine quickly for the war effort. Yet the final device that emerged could calculate ten times faster than contemporary equipment and do complicated calculations as were required, for instance, in ballistics trajectories. The ENIAC made full use of existing vacuum tube technologies, which was a first with digital computers. He was responsible for insuring that all the electronic circuits logically were correct and could work. This task was undertaken during the design phase of the machine's development. Burks also kept detailed minutes of all meetings associated with the ENIAC and subsequently wrote many of its technical specifications. In February 1946, he performed the earliest public demonstration of the ENIAC. He did two calculations: the first, a simple addition of 5,000 numbers, and the second, a calculation of a trajectory for a cannon shell.

The Moore School undertook a more advanced model of the ENIAC during World War II; it was eventually called the EDVAC and Burks contributed to it as a senior engineer. The scientists wanted to include some use of the internal program control within the computer (a feature which today is common in all computational devices while expanding the machine's memory* capacity. Work began in the fall of 1944, and a contract was obtained from the U.S. Army to fund the project beginning January 1, 1945. During that year, while the majority of the staff continued working on the ENIAC, Burks joined von Neumann and Herman H. Goldstine* in laying out the mathematical-logical structure of the new machine. Although von Neumann eventually was credited with producing the final report describing the logic of a computer's architecture, Burks contributed to many of its components.

For further information, see: A. W. Burks and A. R. Burks, "The ENIAC: First General-Purpose Electronic Computer," *Annals of the History of Computing* 3, No. 4 (October 1981): 310–399; Nancy Stern, *From ENIAC to UNIVAC: An Appraisal of the Eckert-Mauchly Computers* (Bedford, Mass.: Digital Press, 1981).

BURROUGHS, WILLIAM SEWARD (1855–1898). Burroughs was the founder of the Burroughs Adding Machine Company (later Burroughs Corporation†), a major supplier of office equipment originally established to market adding machines but later producers of other office equipment. After World War II it also became a large supplier of computers.

Burroughs was born on January 28, 1855, in Auburn, New York. His early employment included work at a bank, in retail stores, and in lumber operations. At the age of twenty, he moved to St. Louis where he worked first for his father in a model-making shop and then for the Future Great Manufacturing Company which made woodworking equipment. In the early 1880s, he tinkered with mechanical equipment. He established the American Arithmometer Company in 1885, capitalized at $100,000, for the purpose of marketing an adding machine he had developed. By 1887, he had fifty copies of the machine which the Boyer

Machine Company of St. Louis had built for him. When the machines proved impractical, he scrapped them and worked on developing a better device. He built his second version of the adding machine in 1892 and by 1895 had sold 284 copies of it. Just as the company was finally expanding, he died on September 15, 1898.

Although Joseph Boyer** now came to dominate the company, its momentum had developed under Burroughs and it did well. In 1900, for example, it sold 1,500 adding machines, and, by 1903, the annual sales volume had grown to 4,500. In 1905, the Arithmometer Company was renamed the Burroughs Adding Machine Company.

For further information, see: Daniel Boorstin, *The Americans: The Democratic Experience* (New York: Random House, 1973); B. Morgan, *The Evolution of the Adding Machine; The Story of Burroughs* (London: Burroughs Machines, 1953).

BUSH, VANNEVAR (1890–1974). Professor Bush, one of the most influential scientists in the United States during the 1930s and 1940s, was the inventor of an important analog computer* at the Massachusetts Institute of Technology (MIT). His pioneering work on computers would be used in the development of other computers. Some of Bush's students invented computers during the 1940s and 1950s. His work on differential analyzers* and analog devices was thus extremely influential in the history of data processing.

Bush was born on March 11, 1890, at Everett, Massachusetts, and was raised in Boston. He studied at Tufts University, graduating in 1913, and he completed his doctorate at Harvard University and at MIT in 1916, after which he went back to Tufts to teach. In 1919, he joined the faculty of MIT to teach electrical engineering as an associate professor. During World War I he conducted research for the U.S. Navy on submarine detection technologies. MIT promoted him to full professor in 1923, and, between 1932 and 1938, he served as vice-president and dean of engineering. He was president of the Carnegie Institution in Washington, D.C., between 1939 and 1956 and, during World War II, director of the Office of Scientific Research and Development (1941–1946). In that capacity he was able to funnel government funds into many scientific projects, including research on the atomic bomb. His efforts led to the establishment of the Manhattan Project.

Bush was the author of a wide variety of publications, the most important of which were *Principles of Electrical Engineering* (1922), *Operational Circuit Analysis* (1929), *Endless Horizons* (1946), *Modern Arms and Free Men* (1949), *Science Is Not Enough* (1967), and his memoirs, *Pieces of the Action* (1970).

Bush's work as a scientist at MIT during the 1920s and 1930s is of greatest interest to any historian of data processing. At MIT he began work on a mechanical means of solving mathematical problems of concern to electrical engineers. One of the most common, time-consuming types of calculations performed in the early 1920s was tracking electrical currents in power grids. By late 1927, he and his students and colleagues had designed a device that could perform simple equations to help the cause. They called it a product integraph. One of his students, Harold L. Hazen,** worked with Bush to expand the machine's capabilities to perform differential equations—in short, calculus. In 1930, these two scientists constructed a machine which they called a differential analyzer,* and throughout the decade Bush made modifications and improvements to what was the first general equation solver ever built.

Bush disseminated the results of these efforts within the academic community in a number of papers and articles. During the 1930s, his were the best computational machines available to the scientific community, being made up of gears, wires, and shafts. His differential analyzer was designed much like an analog computer in that it measured movement and distances and then performed calculations using these variables in a continuous manner. The machine constructed in 1930 had six integrators, capability to store numbers, and an arithmetic control function. It was made from glass disks on tables that moved. One set of measurements would cause a table to move, another the rotation of a disk, and yet another wheel mounted on a glass disk could track a third variable by measuring its distance from a disk. The entire machine was controlled by using several devices called torque amplifiers, invented in 1927 by C. W. Niemann at Bethlehem Steel Company. Like power steering on a car, they controlled wheels and shafts, moving with minimal slippage, thereby insuring a high degree of accuracy in calculations.

This mechanical device used electricity primarily to drive printers and the shafts which were attached to servomotors and the torque amplifiers. Additional variations of this machine appeared throughout the 1930s. Thus, for instance on the second model of the differential analyzer, shafts of light were used to determine whether disks were correctly aligned. Bush built two models of the machine; the second was a major improvement over the first. The initial device was completed in 1930 and the second one in 1935.

Bush's differential analyzers were fast by the standards of the day but were already slow by the mid-1940s. Yet, it was a remarkable machine. Two to three days' work was needed to set up the first analyzer to do a particular problem. Shafts and wheels had to be set precisely to indicate specific, single digits. The x-axis reflected one combination of numbers and the y-axis yet another. To solve some problems the machine had to be operated for weeks. The longer a problem had to run, the greater the chance the machine would break down. The most common problem occurred when a part would slip out of place, forcing the staff to stop the device and make repairs and adjustments before continuing. As a consequence, the results were not always as accurate as desired, but it was effective when many solutions to a particular set of equations were needed. Throughout the 1930s, and particularly after 1935 when he was working on the second differential analyzer, Bush automated an increasing amount of human steps and introduced greater use of electrical components. Thus, for example, the positions of various shafts were monitored by electrical means and instructions were fed to the machine through paper punch tape. Eventually, gear boxes were also electrified.

The second machine proved to be a significant step forward. First, it was of larger capacity: this time eighteen rather than six integrators were used with the capability of adding twelve more. Second, it performed more accurately. It was very large: 100 tons housing some 2,000 vacuum tubes, thousands of relays, and 200 miles of wiring connected to 150 motors.

Two copies of this second machine were made for use in other laboratories. One of these machines was placed at the U.S. Army's Ballistic Research Laboratory at the Aberdeen Proving Ground, reflecting that agency's early interest in calculators. Beginning in the 1940s, other computers would also be installed for use there. The military's enduring interest in calculators grew out of its need to calculate as quickly as possible ballistics projections, speed, and flight paths, taking into account many variables such as wind direction and velocities. It was a natural application for a computational device since ballistics involved vast quantities of mathematics.

The second copy of the machine went to the Moore School of Electrical Engineering† at the University of Pennsylvania. This was the home of the most advanced development work on computers during the early 1940s. The ENIAC* and the EDVAC* computers were built at the Moore School in the mid-1940s.

Bush's role as chairman of the National Defense Research Committee (NDRC) turned out to be a less positive chapter in the history of computing. Although this committee had provided funds for some computer projects during 1941 and 1942, Bush did not want the government to support the Moore School's work on the ENIAC. First, because of his work with analog computers, Bush was not inclined to support the development of digital equipment—a rival technology to his analog approach. Other members of the committee were also from MIT (Samuel Caldwell and Harold L. Hazen), and they too were wedded to their university's technological directions, even at a time when the benefits of electronic digital computers* were slowly becoming obvious. Another member of the committee provided a second reason for resisting ENIAC: George R. Stibitz,** of Bell Laboratories.† He had constructed electromechanical relay machines and thus represented another early tradition in the construction of computational devices different from that now emerging at the Moore School. A third problem, less obvious, was that this committee was made up of prestigious scientists from the Boston area who might have felt nervous about supporting work being done by very young scientists and engineers, many without advanced degrees, and at a university outside the Cambridge area. In 1943, the NDRC's resistance was ignored, and a contract was signed on May 7 with the U.S. Army to build the ENIAC, one of the most important computers ever made.

The battle over funds during the early 1940s reflected a growing concern within the scientific community regarding the future development of computers. Within the small circle of computer scientists, the 1930s was dominated by those who favored the development of analog machines. Analog devices measure and process data in a continuous manner. A simple example of such a device is a watch which measures time continuously. Slide rules* are another example of an analog device because all data are presented in a continuous manner on a scale. Because electrical engineers were interested in measuring the flow of electricity, it was natural that the devices they developed would have analog qualities. Thus, their need to measure voltage, pressure, and temperature, and to model physical phenomena required the development of analog computers of the type Bush worked with in the 1920s and 1930s.

Other requirements for calculating devices eventually led scientists to question the value of continuing to build only analog computers. For one thing, analog devices could only present approximations of actual results since the computation of data was represented in a continuous form. These devices, however, were widely used by scientists in the 1930s and early 1940s, for there were no other types of sophisticated computers. The only other option possible was to employ the relay computers used by American Telephone and Telegraph (AT&T)† to manage telephone calls, but even these were too specialized for general scientific use.

A second major tradition in the history of computing evolved in the early 1940s, which led to the development of digital computers, such as the ENIAC. The difference was conceptually simple in that a digital computer could measure and present quantities as discrete numbers. For example, the number 1.5 on a digital output would be 1, followed by a decimal point and a 5. On an analog machine, the same data would be two numbers, 1 and 2, with an indication that the actual number lay between the two digits. The precision of data handling in digital devices meant that, in the decades to come, a wide variety of applications were possible which could not be conducted on analog devices. Thus, even in the 1980s, analog machines were almost exclusively used for scientific analyses and process control.

In the early 1940s, when some electrical engineers and other scientists were studying the possibilities of constructing digital devices, they were working contrary to the short tradition that existed in computing, that of the analog. Yet the benefits of analog technologies were declining in the face of developments with digital devices. These new machines were also constructed using materials and technologies that were far superior to those employed just a few years earlier in the building of analog machines. For example, almost from the beginning of digital computing, machines relied heavily on electromechanical means and soon moved to electronics, while analog devices evolved from their initial construction form of mechanical machines into electromechanical and eventually to electronics.

By the late 1940s, analog computers were no longer dominant within the scientific community and the slowly emerging data processing industry. Indeed, historians argue that it was overcome by digital devices before the end of World War II. In its heyday, Bush's analyzers were the supreme machines, and the work done by scientists at MIT was some of the most important. In the center of that early tradition was Vannevar Bush. Professor Bush died on June 28, 1974.

For further information, see: Vannevar Bush, "The Differential Analyzer: A New Machine for Solving Equations," *Journal of the Franklin Institute* 212 (October 1931): 447–488, *Pieces of the Action* (New York: Morrow, 1970); Herman H. Goldstine, *The Computer from Pascal to von Neumann* (Princeton, N.J.: Princeton University Press, 1972); C. Shannon, "Mathematical Theory of the Differential Analyzer," *Journal of Mathematics and Physics* 20 (1941): 337–354; Joel Shurkin, *Engines of the Mind: A History of the Computer* (New York: W. W. Norton and Co., 1984).

C

CARR, JOHN WEBER (1923–). This professor of mathematics was a president of the Association for Computing Machinery (ACM)† and helped develop ALGOL,* once a strong candidate to become a universal programming language.*

Carr was born and raised in Durham, North Carolina, in 1923 and completed his B.S. in mathematics at Duke University in 1948. He earned an M.S. in 1949 and a Ph.D. at the Massachusetts Institute of Technology (MIT) in 1951 and remained there as a research associate. In 1952, he became a research mathematician at the University of Michigan, holding that position until 1955. He then joined the faculty at the University of North Carolina as an associate professor and director of research at the school's computer center. In 1959, he joined the Moore School of Engineering† at the University of Pennsylvania, the research center that built the ENIAC.* By the early 1980s, Carr had become chairman of the Graduate Group for Computer and Information Sciences. During 1957–1958, he served as president of the ACM, one of the most important organizations within the data processing community. During his tenure there the ACM established contacts with European computer scientists, as a result of which a task force was established to study the possibility of a universal language. This language, first called IAL, became ALGOL by the early 1960s. Carr also helped establish *Computing Reviews* and served as its first editor between 1959 and 1963.

For further information, see: I. L. Auerbach, ''Association for Computing Machinery (ACM),'' in Anthony Ralston and Chester L. Meek, eds., *Encyclopedia of Computer Science* (New York: Petrocelli/Charter, 1976), pp. 128–129; Jean E. Sammet, *Programming Languages: History and Fundamentals* (Englewood Cliffs, N.J.: Prentice-Hall, 1969).

CARY, FRANK TAYLOR (1920–). This American executive was chairman of the board and chief executive officer at International Business Machines Corporation (IBM)† during the 1970s and early 1980s. IBM is today the single largest company within the U.S. data processing industry and one of the top fifteen largest in the world. During Cary's tenure the company continued to grow and was also sued for antitrust violations by the U.S. government, a legal action that dominated Cary's time in the late 1970s. The government dropped its suit after twelve years, making it one of the longest and most complicated legal actions in the history of American jurisprudence.

Frank Cary was born in Gooding, Idaho, on December 14, 1920, and completed his B.S. at the University of California at Los Angeles (UCLA) in 1943 and an M.B.A. at Stanford University in 1948. Cary also served in the U.S. Army between 1944 and 1946. He joined IBM in 1948, pursuing a career in marketing which in the late 1940s took him into the ranks of the company's salesmen on the West Coast. By the early 1950s, he had served as a branch manager and was rapidly working his way up the organization. He headed the company's first major effort to automate the administrative functions involving asset control, ordering, and billing (known as the AAS system); this system was still being used in modified form in the 1980s to manage many of the company's administrative relationships with customers. While working on this project, however, Cary acquired an enormous amount of knowledge about the company. In 1964, he was named president of the Data Processing Division, a position he held for two years and that involved running IBM's largest marketing organization in the United States. It was at this time that the company announced the S/360* family of computers, which proved to be the most successful product in the history of American business and a boon for Cary's marketing division.

Cary was named an IBM vice-president and group executive in 1966 and a senior vice-president in 1967. The following year he was elected to the company's board of directors. He became president of the company in 1971 and chairman of the board in 1973, a position he held until 1983. Subsequently, Cary served on the board of directors and on various committees it staffed; he was chairman of the Executive Committee which determined many of the company's strategic plans.

Under Cary's leadership IBM invested over $17 billion by the early 1980s in retooling all its plants and in replacing the entire product line with new offerings that were less expensive, more competitive, and could be sold in greater quantities than ever before. As a result, IBM's revenues shot up from less than $30 billion to over $40 billion.

While chief executive officer, Cary faced a number of lawsuits, the greatest being with the U.S. government. The very existence of IBM was at stake: The Justice Department Sought to break IBM up into several companies if the firm was found guilty of antitrust behavior. At the same time, Competitors filed over a dozen lawsuits. IBM won all of them or saw them dismissed without finding the company at fault. The government's suit cost IBM several hundred million dollars to fight, made public several billion documents on the history of the data processing industry, and ultimately led to the government collapsing its case in 1981.

Cary reorganized the company in the 1970s, splitting the marketing structure in the United States into the Data Processing Division (DPD) to handle large accounts and the General Systems Division (GSD) to work with smaller customers and products. Groups were established to which divisions reported in marketing and manufacturing. He hoped these series of moves would spark sufficient rivalry to generate new ideas and products.

Cary also made the basic decision to broaden the company's products, so that mainframes, which had dominated sales in the 1960s (and did again in the 1970s), would no longer represent the lion's share of the firm's revenues. As it had already become evident that mainframes would ultimately represent a declining proportion of his customers' expenditures, Cary elected to enter the world of small computers, and he distributed processing and telecommunications by the late 1970s as a method of insuring IBM's continued success. As a result of these decisions, many new products appeared by 1980 which soon fostered growth in the company near or at the same rate as various segments of the data processing industry.

Cary was considered an efficient and effective manager who worked well within the bureaucracy and culture of the company. He led a complex enterprise that had a presence in over 100 countries and a staff of several hundred thousand employees. While chairman, he also found time to serve on the Business Council, which advised the U.S. government, and on various boards, including those of J. P. Morgan and Company, Hospital Corporation of America (HCA), the Conference Board, the Brookings Institution, the Massachusetts and Institute of Technology, to mention a few. As of 1986, Cary was still on IBM's board of directors.

For further information, see: Franklin M. Fisher et al., *Folded, Spindled, and Mutilated: Economic Analysis and U.S. v. IBM* (Cambridge, Mass.: MIT Press, 1983); Katharine Davis Fishman, *The Computer Establishment* (New York: Harper and Row, 1981); Robert Sobel, *IBM: Colossus in Transition* (New York: Times Books, 1981).

CHARNEY, JULE GREGORY (1917–1981). Charney was a leading meteorologist who, with John von Neumann** by 1950, was applying computers extensively to weather forecasting.

Born on January 1, 1917, Charney grew up in California and attended the University of California at Los Angeles (UCLA) where he majored in meteorology and studied mathematics extensively; he completed his doctorate in 1946. In the late 1940s, he applied his knowledge of meteorology to von Neumann's idea that numerical weather prediction could be further advanced through the use of computerized computations. The idea that weather prediction could be an expression of those physical laws related to weather as numerical algorithms had been discussed since the 1920s, but all earlier calculations had been done by hand and were either too slow or not accurate. The idea of using computers was new. Charney joined Neumann's Institute for Advanced Study (IAS) at Princeton, New Jersey, where he began his most important early work.

At Princeton Charney conducted research and calculations using the ENIAC* and then the IAS Computer.* In 1954, he participated in the establishment of the Joint Numerical Weather Prediction Unit of the U.S. Weather Bureau, Air Force, and Navy to do routine weather forecasting, particularly atmospheric flow patterns. He continued to find new ways of applying computers to weather forecasting throughout the 1950s and 1960s. In 1956 he joined the faculty of the Massachusetts Institute of Technology (MIT) to teach meteorology. There he studied the Gulf Stream, hydrodynamic energy in the atmosphere, hurricanes, and deserts. By the mid-1960s, he was able to start describing weather conditions as a worldwide physical system, focusing primary attention on the atmosphere. He died on June 16, 1981, of cancer.

For further information, see: Jule G. Charney et al., *The Feasibility of a Global Observation and Analysis Experiment,* Report to the Committee on Atmospheric Science (Washington, D.C.: National Research Council, 1966); Norman A. Phillips, ''Eloge: Jule G. Charney, 1917–1981,'' *Annals of the History of Computing* 3, No. 4 (October 1981): 308–309; G. Plazman, ''The ENIAC Computations of 1950—Gateway to Numerical Weather Prediction,'' *Bulletin of the American Meteorological Society* 60 (1979): 302–312.

CHEVION, DOV (1917–1983). Chevion was a leading figure in data processing in Israel and served as head of the Centre for Office Mechanization (MALAM) within the government of Israel.

Chevion was born on April 16, 1917, in Lodz, Poland, and moved to Palestine in 1935. He studied at the Hebrew University in Jerusalem and after World War II held a position with the British forces as a statistician. In the late 1950s, he began working with computers and helped establish MALAM which he directed from 1964 to 1982. During these years he developed training programs that created a cadre of hundreds of data processing professionals. He was also active in establishing data processing organizations in Israel, most notably the Information Processing Association of Israel (IPA) in which he served as chairman (1966–1976) and president (1976–1982). He also participated in the establishment of the first International Jerusalem Conference on Information Technology (JCIT) which initially met in 1971 and subsequently throughout the 1970s and 1980s. He represented Israel and IPA in the International Federation for Information Processing (IFIP)† from 1964 to 1978, was an IFIP trustee (1965–1967, 1970–1973, 1973–1976), and an IFIP vice-president (1967–1970). Throughout the 1970s, he served on other international committees dedicated to the expansion of data processing and to the education of computer professionals. His most valued contribution was to the development of educational programs concerning information processing.

For further information, see: Isaac L. Auerbach, ''Elogue: Dov Chevion, 1917–1983,'' *Annals of the History of Computing* 7, No. 1 (January 1985): 4–6.

COMRIE, LESLIE JOHN (1893–1950). This British mathematician managed the British Nautical Almanac Office and established the Scientific Computing Services, Ltd. (1937), both of which relied heavily on electromechanical hand calculators and assorted punched card technology to produce mathematical tables and to do complex scientific research, each for the first time on a broad scale. With his wide knowledge of such technologies, during the 1920s and 1930s, Comrie was a unique and early consultant on the subject to the scientific and business communities. In addition, he produced astronomical tables of considerable importance.

Comrie was born in 1893 in New Zealand and studied mathematics and astronomy at Cambridge University. In about 1924 he came to the United States where he taught at Swarthmore College and at Northwestern University briefly during the 1920s. He returned to Great Britain in 1926 to assume the position of first deputy superintendent of the H.M. Nautical Almanac Office of the Royal Naval College, and between 1930 and 1936 he served as its superintendent. As a result of the nautical office's mission to produce mathematical tables, he came to learn a great deal about punched card equipment manufactured by Herman Hollerith** and Brunsviga,† as well as British products. Comrie encouraged their use in the development of nautical tables. In 1928, he employed such technology to calculate tables showing the position of the moon. His calculation of Brown's Tables of the Moon pushed the application of punched card technology into new directions. Before 1928, such equipment had been used for traditional business and statistical applications. With his new work, it was shown to be equally useful in complex scientific research.

Comrie had also done other studies that advanced the use of punched card gear. For example, in 1931, he concluded that the National Accounting Machine could be the basis of a new difference engine.* He armed the existing device, an adding machine with a twelve-column keyboard, with six registers which had been used differently in the past, and he used the entire device to capture numbers into each of the registers, transferred them around, and printed results of each calculation. Moving left to right, the carriage activated appropriate registers to cause the first difference to be accumulated (added) to the second difference from the first. The step was repeated many times across as many of the registers as necessary. In short, he made it into a difference engine.

Comrie next employed card punch machines and the National Accounting Machine to do Fourier series calculations, which made it possible for him to prepare Brown's Tables for the years 1935 to 2000. Work on this project had started as early as 1928. Having learned a great deal about punched card equipment in the process, in 1932 Comrie persuaded Hollerith's company to modify its devices so that whatever was in one register could be transferred to other registers, enabling a plugboard wired program to be set up that simulated

Charles Babbage's** original difference engine. In the Late 1930s, International Business Machines Corporation (IBM)† modified its product line to accomplish the same task. Similar efforts to use card punched equipment were going on in the United States at Columbia University under the guidance of Wallace J. Eckert,** at Harvard University with Howard H. Aiken,** and at Bell Laboratories under the direction of George R. Stibitz.**

In the late 1930s and 1940s, Comrie ran the Scientific Computing Services, which relied on punched card equipment. He also continued to encourage the use of computational technology, including electronic digital equipment when that became available. For example, in 1945, he supported the British effort to do research on computing through the establishment of a Mathematics Division within the National Physical Laboratory (NPL).† The following year, after visiting the United States, Comrie made available the results of his newfound knowledge of American computational projects to Maurice V. Wilkes** and others at Cambridge University and at Manchester University. For example, he brought back material on the design of the EDVAC* written by John von Neumann** which would directly influence the design of the British EDSAC,* a major early electronic digital computer.* His help to Wilkes was typical, and that of the EDVAC report was only one in a long series of favors to Wilkes.

Comrie was admired for his unorthodox methods. For example, when he came to the Nautical Almanac Office, he found that the staff was still using logorithms, and so he had to push the organization hard to persuade it to adopt punched card equipment.

During World War I Comrie had lost the lower half of his left leg in France while a member of the New Zealand Expeditionary Force, and in his later years he became increasingly deaf. But neither of these handicaps deterred him—although his methods sufficiently irritated government officials that he was forced to set up his own consulting firm in 1937. Comrie died in 1950.

For further information, see: Leslie J. Comrie, "The Application of Commercial Calculating Machines to Scientific Computing," *Mathematical Tables and Other Aids to Computation* 2 (1946): 149–159, "The Application of the Hollerith Tabulating Machine to Brown's Tables of the Moon," *Monthly Notices, Royal Astronomical Society* 92, No. 7 (1932): 694–707, "Calculating Machines, Appendix III," in L. R. Connor, ed., *Statistics in Theory and Practice* (London: Pitman, 1938): 349–371, "Computing by Calculating Machines," *Accountants Journal* 45 (1927): 42–51, *The Hollerith and Powers Tabulating Machines* (London: Privately printed, 1933), "Inverse Interpolation and Scientific Application of the National Accounting Machine," *Supplement to the Journal of the Royal Statistical Society* 3, No. 2 (1936): 87–114, "Mechanical Computing," in *Plane and Geodetic Surveying* (London: Constable, 1934), 9: 284–294, "Modern Babbage Machines," *Bulletin, Office Machinery Users Association, Ltd.* (London) (1932): 1–29, "On the Application of the Brunsviga-Dupla Calculating Machine to Double Summation with Finite Differences," *Monthly Notices, Royal Astronomical Society* 5 (March 1928): 447–459, "Recent Progress in Scientific Computing," *Journal of Scientific Instruments* 21 (August 1944): 129–135, Leslie J. Comrie et al., "The Application of Hollerith

Equipment to an Agricultural Investigation,'' *Supplement to the Journal of the Royal Statistical Society* 4, No. 2 (1937): 210–224; Herman H. Goldstine, *The Computer from Pascal to von Neumann* (Princeton, N.J.: Princeton University Press, 1972).

CORBATÓ, FERNANDO JOSÉ (1926–). Corbató, a professor at the Massachusetts Institute of Technology (MIT), played an important role in the development of online computing at MIT in the late 1950s and 1960s. His work with real-time systems represented some of the earliest and most impressive efforts in what by the mid-1970s became common throughout most U.S. universities and companies.

Corbató was born on July 1, 1926, in Oakland, California. He attended the University of California at Los Angeles (UCLA) between 1943 and 1944, completing his undergraduate studies in physics at the California Institute of Technology in 1950 and a Ph.D. at MIT in 1956. He then went to work at the Computation Center at MIT, remaining until 1966. During this period he worked to create a campus-wide network for students and faculty to access computers from terminals. Between 1963 and 1966, he served as deputy director of the Computational Center, and from 1963 to 1972, he was head of the computer systems research group first associated with Project MAC.* During these years he also headed an automatic programming division. Yet from 1962 forward, he taught classes in computer science and engineering while managing various projects. He was named Cecil H. Green Professor of Computer Science and Engineering in 1983, and from 1980 to 1983, he was also director of computing and telecommunication resources. He found time to write articles and books on real-time computing and on data processing at MIT. He was an Institute of Electrical and Electronics Engineers (IEEE) Fellow, and received both the W. W. McDowell award (1966) and the Computer Pioneer award (1982). In 1980, the American Federation of Information Processing Societies (AFIPS)† recognized his contributions with the Harry Goode Memorial award. As of 1986, he was still doing research on computer operating systems* and their architecture and was participating in related projects in end user computing.

Corbató's first experiences with data processing at MIT involved WHIRLWIND,* and thus he was at the center of major computational projects there from his earliest days on campus. He spent a great deal of his time in the mid-1950s working with programming and helping graduate students become familiar with that technology. In 1961, he reported on the results of his efforts and on those of others on the staff in creating the Compatible Time-Sharing System (CTSS),* one of the first such systems in the United States. It was rapidly and widely used at MIT. In 1962 the system was fully implemented and expanded under his direction. In June 1963, a new release of CTSS made it possible for twenty-one users to be on the system—a huge number at that time.

Yet CTSS and the Computation Center ran essentially batch jobs; hence, further work was done to expand the capabilities of time-sharing. This led to Project MAC, a major effort of the early to mid-1960s to provide new computational tools. Corbató participated in that project and with its follow-on, the Multiplexed Information and Computing Services (Multics), which was also created during the 1960s. Both efforts led to major expansions in real-time computing at MIT.

For further information, see: Fernando Corbató et al., "An Experimental Time-Sharing System," *Proceedings, SJCC* 21 (1962): 335–344 and *The Compatible Time-Sharing System: A Programmer's Guide* (Cambridge, Mass.: MIT Press, 1963); R. M. Fano, "Project MAC," in *Encyclopedia of Computer Science and Technology* (New York: Marcel Dekker, 1979), 12: 339–360; Karl L. Wildes and Nilo A. Lindgren, *A Century of Electrical Engineering and Computer Science at MIT, 1882–1982* (Cambridge, Mass.: MIT Press, 1985).

CURTISS, JOHN HAMILTON (1909–1977). Curtiss, a scientist, played a critical role in the acquisition of computers by the U.S. government in the 1940s and early 1950s. While serving as chief of the Applied Mathematics Division (AMD) at the National Bureau of Standards (NBS)† (1946–1953), he fostered construction of the SEAC* and the SWAC* and installed the first UNIVAC* computers within the government, thereby helping computers to win acceptance as useful machines.

Curtiss was born on December 23, 1909, and graduated from Northwestern University in 1930 with a degree in mathematics. He continued his studies in statistics at the University of Iowa for a master's degree and completed his Ph.D. at Harvard University in 1935. He taught mathematics at Johns Hopkins University (1935–1936) before moving to Cornell University where he was a member of the Mathematics Department until January 1943, when he enlisted in the U.S. Navy, serving as an officer in Washington, D.C. After World War II, Curtiss took a job at the National Bureau of Standards as an assistant to the director, E. U. Condon. On July 1, 1947, he became division chief within NBS for the National Applied Mathematics Laboratories, later renamed the Applied Mathematics Division (AMD). Curtiss continued at the NBS until July 1953. Between 1954 and 1959, he first worked for the Courant Institute of New York University for one year and then became executive director of the American Mathematical Society (AMS). He joined the faculty of the University of Miami, Coral Gables, in 1959.

Curtiss' role in introducing computers specifically to the NBS and to the U.S. government at large was most significant for the early history of modern data processing. The NBS had used tabulating equipment and other computational devices during the 1930s and worked on war-related projects in the early to mid-1940s. In 1946 Curtiss was asked to conduct a survey to determine the government's computer needs and the role NBS should play. Curtiss issued his report in February 1947, recommending in part that the AMD be established and divided into functional areas that could perform numerical analysis, run a computer laboratory (today we would call this a data center), do statistical engineering, and research on computers. The recommendations were accepted, and his recommended organization remained the same throughout the 1950s and 1960s.

By late 1946, Curtiss had obtained a budget for the acquisition of two computers. He was instrumental in the government's negotiation with J. Presper Eckert, Jr.,** and John W. Mauchly** for one UNIVAC (1946) and with Raytheon Company† (1947) for another. One UNIVAC went to the Census Bureau in 1951, and another went to the Air Comptroller in February 1952. In April the U.S. Army took delivery of its own. Raytheon never delivered its computer, although a variation of the device originally contracted for was built for the Navy in 1952.

Under Curtiss' guidance the NBS built a computer called the Standards Eastern Automatic Computer (SEAC)* for the Air Comptroller (between fall of 1948 and April 1950). The NBS also took on the project of making a second computer called the Standards Western Automatic Computer (SWAC) (between January 1949 and April 1950) as a scientific application machine and the first to use Williams tubes technology. SEAC continued to be used until 1964 and the SWAC until 1967. As a consequence of these projects and his encouragement of others, the U.S. government became a world leader in the use of computers in the 1950s in defense, statistical analysis, and scientific applications.

In addition to his work within the government, Curtiss also had research interests. After joining the University of Miami, he studied approximation theory in the complex domain. As his father had done in 1926, he wrote a book on complex variable theory, this one entitled *Introduction to Functions of a Complex Variable* (New York: Marcel Dekker, 1978). He had continuously studied numerical analysis since the 1940s, publishing a series of papers on related topics beginning in the early 1940s. His last technical paper was published in 1972.

Curtiss' activities within the data processing profession included participation in the creation of the Eastern Association for Computing Machinery, soon renamed the Association for Computing Machinery (ACM)† which he served as its first vice-president in 1947. He also supported publication of computer and mathematics-related journals such as the *Mathematical Tables and Other Aids to Computation* and the *Pacific Journal of Mathematics*. Curtiss died on August 13, 1977, of heart failure.

For further information, see: Gertrude Blanch and Ida Rhodes, "Table-Making at the National Bureau of Standards," in B.K.P. Scaife, ed., *Studies in Numerical Analysis, Papers in Honor of Cornelius Lanczos* (New York: Academic Press, 1974), pp. 1–6; John H. Curtiss, "A Review of Government Requirements and Activities in the Field of Automatic Digital Computing Machinery," in C. C. Chambers, ed., *Theory and Techniques for Design of Electronic Digital Computers. Lectures Delivered 8 July 1946–31 August 1946* (Philadelphia: Moore School of Electrical Engineering, University of Pennsylvania, 1948): 29.1–29.32; Mina S. Rees, "Mathematics and the Government: The Post-War Years as Augury of the Future," in D. Tarwater, ed., *The Bicentennial Tribute to American Mathematics, 1776–1976* (Washington, D.C.: Mathematics Association of America, 1977), pp. 101–116; John Todd, "John Hamilton Curtiss, 1909–1977," *Annals of the History of Computing* 2, No. 2 (April 1980): 104–109.

D

DAHL, OLE-JOHAN (1931–). Dahl was a co-developer of SIMULA* with Kristen Nygaard,** a fellow Norwegian computer scientist who worked with him, primarily with programming languages,* from the 1950s through and 1980s. SIMULA languages were important European computer-based simulation tools of the 1960s and early 1970s.

Dahl was born in 1931 in Mandal, Norway, and was educated at the University of Oslo. In 1957, he completed his master's thesis on ''Numerical Mathematics'' which concerned the representation and manipulation of multidimensional arrays on a two-level store computer and was thus an early European thesis dedicated to computer science. He worked for the Norwegian Defense Research Establishment (NDRE) from 1952 to 1963. The bulk of Dahl's early work was in the field of programming, but beginning in 1956 he became increasingly involved in the development of software.* This shift reflected a more widely visible phenomenon in Europe in which the scientific and engineering communities began paying more attention to software generally and to programming languages and operating systems* specifically. While at the NDRE Dahl developed a programming language called MAC. It was first described in 1957, later modified to reflect many of the features of COBOL,* and then implemented for the first time in the early 1960s.

In 1963, Dahl went to work for the Norwegian Computing Center (NCC), which had been established in 1958, and devoted all of his attention to the development of SIMULA. He became professor of computer science (a new offering) at the University of Oslo in 1968. Throughout the 1970s and into the 1980s, he devoted his research to program architectures, specification tools, and the development of verification techniques in programming.

Dahl also played an active role in the newly emerging computer science community, serving as Norway's delegate to the International Federation for Information Processing (IFIP)† from 1964 to 1976 in its Technical Committee

2, which was devoted to programming languages. From 1970 to 1977, he was also a member of the IFIP Working Group 2.2 which focused on language definition. Since its founding in 1969, he has been a member of the IFIP Working Group 2.3 which looks at programming methodologies.

For further information, see: Kristen Nygaard and Ole-Johan Dahl, "The Development of the SIMULA Languages," in Richard L. Wexelblat, ed., *History of Programming Languages* (New York: Academic Press, 1981), pp. 439–493.

DANTZIG, GEORGE BERNARD (1914–). Dantzig is considered the father of linear programming (LP). In operations research linear programming involves procedures for locating the minimum or maximum of a linear function of variables that are controlled or subject to linear constraints. Put in nontechnical terms, Dantzig developed mathematical techniques that were used extensively to derive decisions in the study of problems in economics, military issues, transportation, and business management. These methods relied on mathematical (algebraic) notation and first came into being during the late 1940s. Linear programming provided much impetus for the use of computers in the 1950s and 1960s to support decision-making.

Dantzig was born in Portland, Oregon, on November 8, 1914, the son of a well-known mathematician, Tobias Dantzig. George Dantzig earned an A.B. in mathematics at the University of Maryland in 1936, an M.A. at the University of Michigan in 1937, and a Ph.D. at the University of California at Berkeley in 1946. He obtained his first job at the U.S. Bureau of Labor Statistics in Washington, D.C., in 1937 and remained there until 1939 as a junior statistician. His next position was as chief of combat analysis statistics for what became the U.S. Air Force in the War Department (later Department of Defense) between 1941 and 1946. From 1946 to 1952, he served as an advisor in mathematics to the Pentagon, and between 1952 and 1960, he was a research mathematician for the RAND Corporation in Santa Monica, California. Subsequently, he returned to Berkeley as professor and head of its program in operations research (1960–1966). In 1966 he moved to Stanford University in the shadow of what would become known as Silicon Valley in the 1970s and home for hundreds of computer and data processing companies. He remained at Stanford as professor of operations research and computer science.

Dantzig's contributions to the study of operations research using linear programming began in the late 1940s. He combined his training in mathematics with experience in government and applied that knowledge to practical problems in statistics. Following World War II, the U.S. Air Force began what was called Project SCOOP* (Scientific Computation of Optimum Programs) in order to mechanize and increase the speed with which planning and deployment took place, using mathematical and, later, computer-based methods. Dantzig participated in the project from its beginning. In 1947, he worked on a model of how planning was conducted. As a consequence of these efforts, Dantzig developed the "simplex method" which soon became the most widely used process for computing "best" strategies in the mid-twentieth century.

The simplex method for solving problems begins with a basic solution that can work and then searches through sets of solutions so that the value of the linear form never increases. The exercise involves finding nonnegative values for variables

$$x_1, x_2, \ldots x_n$$

that satisfy constraints expressed as

$$a_{i1} x_1 + a_{i2} x_2 + \ldots + a_{in} x_n = b_i, i = 1, 2, \ldots, m$$

while at the same time minimizing the linear form

$$c_1 x_1 + c_2 x_2 + \ldots + c_n x_n.$$

This process allows one to maximize problems expressed quantitatively and those that have unrestricted variables. An optimum solution can thus be worked on. In operations research, Dantzig called an optimum solution a "basic feasible solution." Such a solution lives within the constraints defined and has at most m positive x_i values. Thus, Dantzig's simplex method began with a basic feasible solution as a base and then defined solutions within constraints. In nontechnical terms, it was used to solve, for example, for the best shipping routes to use to carry a certain amount of cargo across the Atlantic Ocean within a given period of time.

Such problems were common during World War II and immediately after and became the backdrop against which Dantzig did his research. Following the war, it became very obvious that fundamental problems in an economy could also be studied using his technique. He refined and applied his methods to various problems throughout the 1950s, and at RAND (which had become an early user of computers), Dantzig was able to employ this technology in operations research. By the time he began teaching at Berkeley, the use of linear programming for problem-solving had been established as one of the more important uses of computers. The use of LP was further cemented, and following the publication of his book on the subject in 1963, his work became the classic in the field.

Dantzig's techniques were refined further during the 1960s and 1970s, these methods are still widely used in the 1980s. During the 1950s and 1960s LP was used for specific projects, many of which were related to U.S. government interests or to University research. By the early 1970s, decision-making software programs began to appear which were based on his principles of linear programming. These programs expanded the capability of many nontechnical users of computers who were searching for optimal solutions to problems that could be expressed mathematically.

Dantzig's contributions were recognized at an early date. In addition to a large number of honorary degrees, he received the Exceptional Civilian Service medal from the War Department in 1944 and the U.S. National Medal of Science in 1975. In 1977, he received an award from the National Academy of Science for his work in applied mathematics and numerical analysis. He continues to do research on mathematical models for planning and scheduling and is still regarded as one of the major researchers in the field.

For further information, see: George Dantzig, *Linear Programming and Extensions* (Princeton, N.J.: Princeton University Press, 1963); William S. Dorn and Herbert J. Greenberg, *Mathematics and Computing: With FORTRAN Programming* (New York: John Wiley and Sons, 1967); Philip M. Morris, *Methods of Operations Research* (Cambridge, Mass.: MIT Press, 1951).

DEEDS, EDWARD ANDREW (1874–1960). This American manufacturing executive developed automotive devices and founded Delco, an important supplier of parts for cars and trucks. Deeds was also an executive at the National Cash Register Company (NCR)† and ultimately its president and chairman of the board. During his tenure as its chief executive officer in the 1930s, NCR remained one of the largest and most important suppliers of office equipment and information handling products in the world.

Deeds was born on March 12, 1874, attended the Granville (Ohio) Academy, and graduated from Denison University. While in college, he gained considerable experience with electricity and operated the electric light and water plants for both his town and its university. He next studied electrical engineering at Cornell University and in 1898 joined the Tresher Electrical Company in Dayton, Ohio, as a draftsman. Tresher built electrical motors, and within one year Deeds had become its superintendent and chief engineer. In 1900, he joined NCR, which was already a well-established supplier of office equipment. At that time, the future founder of International Business Machines Corporation (IBM),† Thomas J. Watson, Sr.,** was an employee of the company. Deeds performed his first important service for NCR as a cost and maintenance engineer by installing electrical power in the plant in Dayton. He left NCR in 1901 and also electrified the Shredded Wheat Company of Niagara Falls, New York. In 1903 he returned to NCR as an assistant general manager but soon after was promoted to vice-president of engineering and production. In the decade before World War I he expanded the company's manufacturing capability by establishing plants in Canada, Great Britain, Germany, France, and Italy. He left NCR again in 1914 to pursue other business ventures.

Deeds returned to NCR as chairman of the board in 1931. Between 1931 and 1935 sales at NCR doubled. He became president of NCR in 1937, replacing Frederick B. Patterson, and he held on to the chairmanship. During the 1930s, despite a world Depression, Deeds expanded the company's manufacturing operations, particularly in Canada. He merged NCR's functions with those of the Krupp Cash Register Company in Germany and the Fujiyama Company in

Japan. He retired in 1940, having made NCR a major supplier of information handling equipment. By that time, however, IBM had became a giant in the office products market and NCR's chief rival.

Deeds' interest in NCR's product line and in electrical matters in general also influenced events in the automotive industry. In 1904, for example, he hired Charles F. Kettering for NCR. Kettering was the inventor of the first truly automatically operated cash register. He worked with Deeds to develop a torque motor which made possible electrical business machines. These two men next developed another version of the motor as an ignition system for automobiles, replacing the hand crank of the early cars. They named their device Delco after the Dayton Engineering Laboratories Company which they had formed to build and market their invention. Deeds served as president of this company until 1916 when it was sold to General Motors. The two inventors also developed the Delco-light, a portable power plant that could be used on farms, and in schools, hotels, churches, and so on. This device contributed to the enormous success of their company, sending total sales up from $2.5 million in 1916 to $20 million in 1919. During the 1920s Delco was acquired by General Motors.

Meanwhile, Deeds had become president of the Smith Engineering Company, a firm that made gas products. He was also chief executive officer of the Domestic Building Company, which had been established to develop sites in Moraine City, just outside of Dayton, Ohio. Deeds participated in the consolidation of over 100 sugar firms in Cuba into a single enterprise called the General Sugar Company in 1922. Later, this concern became known as the Vertientes-Camaguey Sugar Company. Deeds found time to be its president and chairman of the board until 1946. He was also president of the Niles-Bement-Pond Company (later Pratt and Whitney Company) until 1943. Deeds died on June 1, 1960.

For further information, see: Isaac F. Marcossen, *Wherever Men Trade* (New York: Dodd, Mead, 1948) and his earlier book, *Colonel Deeds: Industrial Builder* (New York: Dodd, Mead, 1947).

DICK, ALFRED BLAKE (1856–1934). This American businessman invented the mimeograph and founded A. B. Dick and Company, a major supplier of printing and duplicating technology during the twentieth century. Mimeographic products reflected yet another facet of a general exploration of the late nineteenth century into mechanical ways of gathering and moving ever increasing amounts of data. While mimeographic machines were used heavily until about 1970, they were eclipsed, though not entirely eliminated, by the introduction of xereographic technology and word processing in general. Thus, mimeographic technology was a distant relative of data processing because it created a demand for the use of information handling equipment in offices along with card punch equipment. By the mid-1980s, the handling and printing of data almost became one very large application. The evolution of that process can be traced in part to the work of Alfred Dick.

Dick was born on April 16, 1856, in Bureau County, Illinois. In 1863, his family moved to Galesburg, Illinois, where he was raised and attended public schools. Between 1872 and 1879, he worked for the George W. Brown and Company which manufactured agricultural equipment in his home town. Dick went to Moline, Illinois, in 1878 to work for the Deere and Manson Company, and he also became a partner in the Moline Lumber Company. In 1883, he established his own lumber company in Chicago called A. B. Dick and Company. He incorporated it with himself as both president and treasurer in 1884.

While managing the firm in Chicago, Dick realized that he needed a way of sending inquiry sheets to mills and yards on a daily basis in order to track inventories and to know where supplies could be drawn should his own yard not have stock to meet an order. However, hand-writing such sheets was too labor-intensive, and so he sought a means of duplicating them. He eventually stumbled across the idea of developing a sheet of paper with a wax coating which he called an ''automatic stencil.'' Using a stylus, one could, in effect, write a form letter. These sheets could then have ink applied to their surfaces with a hand-held roller on a flatbed press, allowing hundreds of duplications. Dick carried the process further by obtaining from Thomas A. Edison an electric pen to use in the reproduction of multiple copies of original writings. Edison granted him permission to use a device that coated the waxed stencil sheets. Edison also invested in the enterprise. The two men now had a system called the Edison-Dick Mimeograph.

It soon became clear to Dick that he could make more profit from mimeograph than from lumber, and, therefore, in 1887, he sold off the lumber portion of the business and turned full time to the manufacture and sale of mimeograph equipment and supplies. Improvements were made into the next century. The system was adapted to typewriters, then a rotary mimeograph machine appeared, while stencil fabric improved. Like many devices of the late 1800s and early 1900s, existing technology was infused with electrical power. By the 1930s, his machines were common in most offices and schools. By 1934, the company sold its products around the world. In the United States alone, it employed 1,700 people and had sales offices in almost every major city. By then the firm was known simply as A. B. Dick Company. It had its own manufacturing and sales force. Growth had come quickly as was typical of firms with new technologies successfully marketed. By 1900, the company was selling its products throughout the United States, and by the 1920s, it dominated the mimeograph/duplication market. It maintained that lead until the 1970s when xereographic technology made it possible to compete against mimegraphics on a cost-effective basis.

For further information, see: Daniel J. Boorstin, *The Americans: The Democratic Experience* (New York: Random House, 1973).

DICK, ALFRED BLAKE, JR. (1894–1954). Dick was the son of Alfred Blake Dick,** founder of the A. B. Dick Company, the world's most important manufactuer of mimeographic equipment and supplies. Dick, Jr., inherited the top management position of his father's company and led it into various areas of the duplication market. After his own death in 1954, the firm remained an important provider of duplication equipment and eventually entered the world of word processing.

Dick was born in Chicago on February 11, 1894, and graduated from Yale University in 1915. He joined his father's company first as a clerk and then as a salesman. During World War I he served in the U.S. Navy. He returned to the company as vice-president and treasurer. In 1934, upon the death of his father, the board of directors made him president and treasurer, positions he held until 1947 when he became chairman of the board. During World War II, like many other executives in the United States, Dick converted his manufacturing operations to the production of war-related goods. In the case of A. B. Dick Company, nearly 50 percent of productive capacities were devoted to the war. Along with mimeograph equipment the firm now manufactured bomb sighting heads. Following the war, the company introduced a new line of products called the 400 series of A. B. Dick Mimeographs.

Dick moved all manufacturing to one location in Miles, Illinois, in 1949, and in the next few years the entire product line expanded. He introduced a folding machine and also brought out a variety of products for the lithograph market. Impression paper and spirit duplicating fluid also appeared during the 1950s. These products made A. B. Dick the only firm that could provide a complete set of offerings covering mimeo, spirit, and offset—the three types of duplication used by offices in the 1950s. Although the younger Dick died on October 24, 1954, his company used its well-established foothold in the office market to continue introducing other products in the 1960s and 1970s, including word processing in the 1970s. His son, Alfred Blake Dick III, was born in 1918 and joined the company in 1939. After various promotions, in 1947 he became president of the firm, serving in that capacity until 1961 when he became chairman of the board.

For further information, see: John N. Ingham, *Biographical Dictionary of American Business Leaders, A-G* (Westport, Conn.: Greenwood Press, 1983).

DIEBOLD, JOHN (1926–). Diebold was one of the first consultants in the American data processing industry. He was preeminent from the late 1950s through the 1970s. Before 1970, there were few experts in the industry, and by the 1970s, he was a widely known consultant within American business in general. The data processing industry had thousands of consultants by then, representing a subclass within that community, while the consulting industry was already nearly a $5 billion operation. Diebold, as one of the first and very successful members of that community, became a recognized leader. With nearly three decades of experience as a consultant in the general field of data processing, he is considered the archtype of the technical consultant.

Diebold was born in Weehawken, New Jersey, on June 8, 1926. He earned a B.S. degree from the U.S. Merchant Marine Academy in 1946 and a B.A. in economics from Swarthmore College in 1949. He was awarded the M.B.A. with distinction from Harvard University in 1951, after which he joined Griffenhagen and Associates—a management consulting firm—becoming an owner of the company in 1957. In 1960, he was named chairman of the board of Griffenhagen-Kroeger, Inc. Earlier, he had established Diebold Group, Inc., as a management consulting firm in 1954 in New York City, serving as president and chairman of the board. In 1958, he expanded outside the United States by establishing Diebold Europe, S.A. He created John Diebold, Inc., in 1967 to serve as both a management and investment firm, holding the position of chairman of the board. In 1967, Diebold also developed a company to operate in the data processing industry called DCL, Inc., a holding company for Diebold Computer Leasing, Inc. Between 1968 and 1975, he operated Gemini Computer Systems, Inc. In that same period he served as a board member of Prentice-Hall, a leading publisher of data processing and engineering books, and Genesco, a shoe manufacturing firm headquartered in Nashville, Tennessee.

As a consultant, Diebold's services were also heavily used by the U.S. government. He was a member of the U.S. Secretary of Labor's Advisory Committee on Manpower and Automation (1962–1966), President John F. Kennedy's commission on the Department of Labor's fiftieth anniversary (1963), a member of the U.S. delegation to the United Nations Science Conference held in Geneva, Switzerland, in 1963, and a member of the advisory council for the Society for Technological Advancement of Modern Man (1963–1976). Between 1965 and 1968, he was associated with the commission on human values established by the Society for Advancing Technology of the National Council of Churches. He was a presidential appointee to the national advisory council on the Peace Corps (1965–1970) and also found time to be a member of the Commission on the Second Regional Plan for New York City (1966–1970).

In addition to these responsibilities, Diebold became interested in international affairs. He was a trustee and treasurer of the executive committee of the National Committee on U.S.-China Relations (beginning in 1969), a trustee and secretary of the Business Council for International Understanding (1970–1983), a member of the prestigious International Institute for Strategic Studies in London (1971 to the present), a member of the board of consultants of the advisory committee of the U.N. called We Believe (1972 to the present), a member of the Advisory Council on Japan-U.S. Economic Relations (1972–1974), a member of the steering committee of the Atlantic Conference (1972–1982), vice-chairman of the American Council on Germany (1970–1980), a trustee (1980–1983) of the Council on Foreign Relations since 1967, and advisor to the U.S. Secretary of State on budgeting (1966–1967).

Diebold was very active in the general field of academic studies of business. Between 1961 and 1966, he was a member and chairman of the visiting committee of the School of Business Administration at Clarkson College, and served in a similar capacity at Harvard University (1963–1969, 1970–1976), Columbia University (1968–1975), and Vanderbilt University (1969–1974). Beginning in

1957, he was also active with the Institute for Crippled and Disabled in New York. In 1965, he participated in the affairs of the advisory committee of the European Institute of Business Administration and soon after joined the business advisory committee of the Graduate School of Industrial Administration at Carnegie-Mellon University (1969). Other organizations with which he was associated included Freedom House (since 1969) and the Committee for Economic Development (since 1970); he founded the Diebold Institute for Public Policy Studies in 1967; he became trustee for the Carnegie Institution in Washington, D.C. in 1975 and the Hudson Institute in 1967; and, between 1971 and 1974, he was a member of the visiting committee for Research in Computing Technology and the Office for Information Technology, Harvard University.

Remarkably, Diebold was a member of nearly two dozen other organizations, primarily within the United States, interested in business education, economic development, technology, or international affairs, particularly during the 1960s and 1970s. During World War II he served in the U.S. Naval Reserve.

This energetic and intelligent consultant has published a number of books dealing with automation in manufacturing, the role of technology as agent of social change, and the data processing industry in general. Diebold has long advocated specific ways computers could be used in manufacturing, education, and communications. In numerous speeches and books, he has argued that organizations would have to be restructured to use the tools of data processing effectively. This theme, developed by the early 1960s, was evident in his work over the next fifteen years. By the early 1970s, he was also articulating the social impact which technology might be expected to have on education, entertainment, and society's values, changes which he saw both positively and in cautionary terms.

Diebold was recognized for his work in defining the characteristics of technology and data processing, particularly as they effected business. The Data Processing Management Association (DPMA),† for example, honored him with its Distinguished Information Science award in 1980. He was a Fellow of the International Academy of Management, and beginning in 1957 he was a director of the International Cybernetics Association.

In many ways, this business executive's career exemplified that of the American consultant. He established consulting firms, specialized in several areas (management, technology transfer, data processing), participated in numerous organizations, spoke out on industry-related topics, published, and served as a spokesman for industry in general. The difference between his role and that of other consultants within American business was probably one of degree. Diebold was the best known of a large cadre of consultants, and he was the most important of the early consultants in the new world of data processing.

For further information, see: John Diebold, *Beyond Automation: Managerial Problems of an Exploding Technology* (New York: McGraw-Hill, 1964), *Man and the Computer: Technology as an Agent of Social Change* (New York: Praeger Publishers, 1969), and *The World of the Computer* (New York: Random House, 1973).

D'OCAGNE, MAURICE (1862–1938). This French engineer turned mathematician is best remembered for writing a classic work on computation, *Le Calcul Simplifié*. His book, originally published in the early 1890s, appeared in its most important edition in 1928. The volume reminded historians that few aids to computation existed in the late 1800s and that much effort had been devoted to the development of such tools and methods, efforts which seemed elementary in the 1890s. By the time d'Ocagne published his final edition, there existed many mechanical tools to assist in mathematical calculations, the most important of which were desktop calculators.

Born in Paris in 1862, d'Ocagne also grew up in the French capital. Most of his professional life was spent in French government agencies and academic institutions. He received an appointment as engineer in the Corps de Ponts et Chaussées in 1885, a position he kept until 1889. Next d'Ocagne focused attention on hydrological problems near Cherbourg. In 1894 d'Ocagne became a professor at the École Nationale des Ponts et Chaussées. In 1901 the French government appointed him Director General of the nation's mapping service. He went back to teaching as a professor at the École Polytechnique in 1912 and finally served as Inspector General of the Ponts et Chaussées, beginning in 1920.

He joined the prestigious Academie des Sciences in 1922. Membership in the Academic indication was a clear of d'Ocagne's administrative and intellectual achievements. His scientific peers had already realized that his work in mathematics and in describing mechanical aids to calculation were more than useful. His primary area of interest was nomography, and he published a major work in the field, *Traité de Nomographie* (1899; 2nd edition, 1921).

However, d'Ocagne's most relevant contribution to the history of data processing was *Le Calcul Simplifié*. The third edition, published in 1928, was the most complete. It offered a wealth of information about computational techniques and equipment that existed as of about 1920. The improvements in scientific equipment and knowledge that took place since the earlier editions had appeared were reflected in, for instance, the use of nomograms to solve problems in ballistics during World War I. He devoted one chapter to descriptions of mechanical devices and their use, another to graphical calculations, and others to the mathematics of grapho-mechanical calculation, nomography, and other types of arithmetic problems. His second edition (1905) had devoted more pages to older mechanical aids to calculation. D'Ocagne died in 1938.

For further information, see: Maurice d'Ocagne, *Le Calcul Simplifié: Graphical and Mechanical Methods for Simplifying Calculation,* translated by J. Howlett and M. R. Williams (Cambridge, Mass.: MIT Press, 1986).

E

ECKERT, JOHN PRESPER, JR. (1919–). Eckert was the co-builder, with John W. Mauchly,** of the ENIAC,* EDVAC,* BINAC,* and UNIVAC* computers in the 1940s and early 1950s—the first electronic digital computers.* His work at the Moore School of Electrical Engineering† at the University of Pennsylvania and later at the Eckert Mauchly Computer Corporation ushered in the era of the electronic computer by taking such technology out of the laboratory and into the world of commercially viable products. He is considered one of the most important developers of computers of the twentieth century.

Eckert was born in Philadelphia in 1919 and spent his childhood there where his father was a real estate developer. Eckert attended the University of Pennsylvania where he studied engineering at the Moore School, graduating with a B.S. degree in 1941. In 1943, he completed his master's at the same institution. While a graduate student, he became interested in computational devices. As early as 1941, he met Mauchly who shared some of his ideas about the construction of an electronic computer. By this time, Eckert was regarded as a highly talented electrical engineer. Eckert soon began to study intensely the question of an electronic computer, and he quickly learned at the school all that was then known about counting circuits, for example. He seriously approached the subject of electronic computing and gained the respect of his peers and faculty. Some thought him intimidating, and others characterized him as a ''no nonsense'' individual. He and Mauchly were ideally suited to work as a pair on the ENIAC: Mauchly was outgoing and visionary, whereas Eckert proved more intensely concerned about the engineering challenges at hand.

At that time, the Moore School was under contract to the U.S. government to develop a machine that would perform war-related calculations. A great deal of attention was therefore focused on the ENIAC project at the Moore School, a project that rapidly absorbed all of Eckert and Mauchly's energies. Eckert was at first a graduate assistant and later a staff member at the school just prior to the start of the project. Mauchly was already on the staff when Eckert graduated from college.

After the Moore School signed the contract for the construction of the ENIAC on June 5, 1943, Eckert was named chief engineer on the project. Mauchly served as a consultant, and Herman H. Goldstine** served as the contact with the government, particularly with the U.S. Army Ballistics Research Laboratory. Eckert quickly developed a management style that kept him involved in all aspects of the machine's design and construction. Thus, for instance, Eckert checked all the calculations made for the ENIAC's resisters and watched every action taken by fellow engineers. His strategy was to adapt existing technology where possible, for example, vacuum tubes, in order to construct a computer quickly in response to pressure from the government to move rapidly. His concern for detail meant that the overall design of the machine would be the result of direct work done by both himself and Mauchly.

Years later, controversy erupted (and persists) over who invented the ENIAC and electrical computers in general. Between 1971 and 1973, Honeywell† and Sperry Rand† were in court, battling over patent rights to their computers which reached back to patents on the ENIAC. The strategy of taking existing technologies and binding them together to build the ENIAC was the source of most controversy since the question of ''firsts'' was ever present. Eckert and Mauchly had submitted patent applications on the ENIAC and sold rights to them to Remington Rand, which later became Sperry Rand.

Honeywell argued that prior inventions were incorporated in the ENIAC, the most important of which included the technology in the Atanasoff-Berry Computer, better known as the ABC. John Vincent Atanasoff** had invented parts for a computer he designed in the late 1930s that used such common ENIAC components as vacuum tubes. He too worried about how to make electronic counting circuits. His machine was never completely built and was only designed to handle linear equations, whereas the ENIAC was fully constructed and operational and handled a much broader range of applications. Although the court judged that Eckert and Mauchly had developed the ENIAC, Mauchly's role was controversial because in early 1941 he had corresponded with Atanasoff, and in that same year he had visited Atanasoff to see the partially made ABC. That was also the year that Mauchly articulated his thoughts on the construction of a general-purpose electronic computer at the Moore School and captured Eckert's interest in the subject.

For years afterwards, especially in the 1970s, Mauchly argued that he had not taken any ideas from Atanasoff and that he had in fact suggested how the ABC's design could be improved. A leading historian of computing, Nancy Stern, has concluded that Mauchly was an innovator of technology and Eckert a true inventor. Others close to the project were in agreement that Eckert was the key technical resource who came up with many of the developmental tasks that saw construction through to a successful conclusion.

The ENIAC was so successful that the Moore School was asked to work on an improved model proposed by the staff even before the first machine was fully operational. The new device, called the EDVAC,* was constructed in 1945 and 1946. With the war coming to an end, Eckert and Mauchly began to consider the commercial applicability of the ENIAC and in the fall of 1944 began the long process necessary to secure patent rights. Fellow engineers at the Moore School offered no serious protest: they either accepted the legitimacy of their claim or saw no potential for a commercial version of the machine. The University of Pennsylvania wanted the patent rights assigned to the school, but in the early stages, its position was that the staff could file for patent rights in their own names. Some controversy thus developed over who had the right to file. The U.S. Army supported Eckert and Mauchly, knowing they would cooperate in sharing access to this technology with the government. The University of Pennsylvania protested mildly but did not block the action. Finally, in March 1945, the University formally allowed the two to file for patents as inventors of the ENIAC. With that, the two engineers moved forward on their application. At last, they believed that their contribution would be fully protected by having patent rights. Little did they realize that they faced years of concern over patents. But at least at the time, colleagues at the Moore School had recognized that Eckert had provided the design genius and the amiable Mauchly the project its vision and management.

Partly because of the patent controversy at the Moore School, following World War II, the University of Pennsylvania changed its policies toward government-sponsored research. As a result, many engineers who had worked on the ENIAC and the EDVAC left the university, among them were Eckert and Mauchly who formed their own company to sell computers. In January 1946, Irven Travis took charge of research and development at the Moore School where he had once taught years earlier and moved quickly to establish a policy on patents. He decided that employees of the Moore School should assign all patent rights to the University for inventions made while on the job; hence, individuals who had completed projects at the school would be denied financial gain from them. This policy was consistent with that in existence at some universities and in many companies. On March 15, 1946, Travis asked all members of the Moore School to sign patent release forms; Eckert and Mauchly were reluctant to sign, fearing that they would lose the opportunity to market a commercial version of the ENIAC. This crisis occurred just two months after the the ENIAC had become operational. At first, Eckert had some concern about leaving the Moore School to start a new company, but Mauchly convinced him to do so. They therefore resigned from the University of Pennsylvania effective March 31, 1946.

Eckert had options for employment. IBM, for example, had offered to establish a laboratory so that he could continue his work. Mauchly was nervous over the offer and so convinced Eckert that they should go it alone. Later, other members of the Moore School resigned, some of whom came to work for Eckert and Mauchly. In order to start their company, the two men sought contracts and eventually gained the support of the National Bureau of Standards (NBS)† and the Census Bureau. Their first contract, to study the feasibility of a computer for the NBS, brought some cash into their company, named the Electronic Control Company which they located in Philadelphia. (This was both their home and the electronics capital of the United States.)

In a subsequent contract, signed on September 25, 1946, they agreed to develop specifications for a new machine; its name as incorporated into the agreement on May 24, 1947, was the Universal Automatic Computer, or simply UNIVAC.

Eckert and Mauchly promptly went to work on the new computer, despite reservations which some scientists working or advising for the government had expressed about its feasibility. NBS encouraged other government agencies to work with Eckert and Mauchly in order to ease their company's lack of capital. Part of the method used to alleviate the lack of funding was to sign a contract with the Northrop† Aircraft Company in October 1947 to build a small computer to control missile flights on-board an airplane called the Binary Automatic Computer (BINAC). In December, the two renamed their company the Eckert and Mauchly Computer Corporation (EMCC). They spent the next year continuing to raise funds by signing up government agencies for UNIVACs while working on the BINAC. Mauchly handled sales and Eckert focused on designing machines.

The lack of adequate funding continued to plague the two engineers and finally forced them to find a buyer for their company. Assets as of June 30, 1948, equaled $206,000, but the firm had spent over $100,000 for the development of the UNIVAC, which was not yet completed. The company estimated it needed another $500,000 just to continue work on existing contracts, money it obviously did not have. After approaching various companies, including IBM and the National Cash Register Company (NCR),† the two engineers sold their firm to Remington Rand† on February 1, 1950.

A total of $100,000 was to be paid to Eckert and Mauchly and other employees holding stock in the EMCC. Eckert and Mauchly were each to receive an annual salary of $18,000 (as opposed to their then current $15,000 each) for eight years. Remington Rand also agreed to pay to the EMCC for eight years 49 percent of all net profits gained from patents. Eckert's company now became a division of Remington Rand. Mauchly left the research side of the business for sales and resigned from the company in 1959. Eckert, more interested in research than in marketing, concentrated on the design of the UNIVAC and finally on March 31, 1951, the Census Bureau accepted the first UNIVAC.

Eckert and the other engineers incorporated many of the developments that emerged from the ENIAC, EDVAC, and BINAC. The UNIVAC became the first widely accepted general-purpose computer available in the commercial arena in the United States. This computer captured the attention of many first-time users of computers and helped launch the modern age of the computer. In the early 1950s the term *UNIVAC* became almost synonymous with the word *computer*.

Eckert and Mauchly were brilliant designers who merged their knowledge and ideas with available electronics to produce useful machines. As businessmen they were far less effective, especially Eckert. Both failed to forecast true demand for their machines and, even worse, to predict accurately and then manage research costs. In fairness to them, however, other producers of computers also vastly underestimated market potential. They were late on production and always experienced severe cost overruns. Eckert seemed to dominate decisions in the 1948–1949 period, insuring that engineers would make more decisions than Mauchly and other EMCC employees interested in strictly business issues. Thus, for example, Eckert focused too much attention on the UNIVAC in 1948 and 1949 when there was money to be made in completing the less sophisticated or interesting BINAC and getting multiple copies built and sold. Instead, this machine was completed in a less than enthusiastic fashion and was never capitalized on for other than one sale. Thus, Eckert was more interested in technology and in improving on it than in getting a computer system developed, debugged, built, sold, and delivered to customers.

Yet Eckert's career as a distinguished engineer continued. He was named director of engineering in the Eckert Mauchly Division of Remington Rand when the UNIVAC was completed, and became vice-president and director of research in 1955. In 1957, the same year the UNIVAC II was announced, he became vice-president and director of commercial engineering. In 1959, the company named him vice-president and executive assistant to the general manager, and in 1963, he was made a vice-president and technical advisor to the president of Sperry Rand, the UNIVAC Division.

During the 1970s, Eckert was caught up in the controversy surrounding patent rights on the UNIVAC and even on the ENIAC. Beginning in 1947, Eckert and Mauchly had submitted patent requests on the ENIAC and, later, UNIVAC; these requests were broad and, in some instances, vague but encompassed a great deal of work done on computers. During his career, Eckert was granted over eighty patents. When Eckert and Mauchly sold their company to Remington Rand, they assigned their patent rights to the company in exchange for promised future royalties, fees which they felt were never fully paid to them. Remington Rand entered into a series of cross-licensing agreements with (IBM) International Business Machines Corporation in the early 1950s to cover computer technology derived from research on the original UNIVAC. Meanwhile, the slow patent application process ground on with the ENIAC. This computer's patent was finally issued on February 4, 1964 (No. 3,120,606), and was assigned by Eckert and Mauchly to a subsidiary of Sperry Rand, Illinois Scientific Developments.

Then in 1967 this firm sued Honeywell for patent infringement. Honeywell countersued, claiming that Illinois Scientific Developments held the patent illegally and that the deal struck by Sperry and IBM in the early 1950s violated U.S. antitrust laws. Honeywell's suit also alleged that the ENIAC patent was applied for over a year after the computer had been built and that John Atanasoff was the true inventor of the modern computer, not Eckert and Mauchly.

Most of the great scientific pioneers who helped develop computers over the previous thirty years were soon caught up in the controversy either as litigants or as witnesses. Finally, on October 19, 1973, Judge Earl Larson in Minneapolis rendered his decision: he ruled in favor of Honeywell. The judge ruled that the ENIAC had been in existence for over a year before patent was requested and thus, under the law, was not eligible for patent protection. He also stated that its technical features were well known prior to patent request and, therefore under the law, qualified as ''prior art,'' meaning that no new invention was going into the patent request. He also questioned the role of Eckert's and Mauchly's lawyers and those of Remington Rand in holding back data from the Patent Office that could have a negative impact on the application or that could have improved the financial posture of those concerned.

The machine had been publicly demonstrated before patent requests were submitted, and John von Neumann's** paper on the ENIAC, called the ''First Draft,'' had circulated in scientific circles long before 1947. The judge came close to accusing Eckert and Mauchly of fraud in their application because of the constant delays they and their lawyers caused in the application process. In the end, however, he concluded that the problem lay more with the lawyers handling the case than with the scientists. Eckert was called to task for helping to delay the patent application in hopes of improving his financial advantage. He also wrote a patent request that was too broad.

But the judge's most upsetting conclusion for Eckert and Mauchly was that Atanasoff had invented the electronic computer. The judge cited Atanasoff's work before Eckert's and Mauchly's as the final criticism of the patent. Mauchly had visited Atanasoff in Iowa, seen the ABC, and corresponded with its inventor. The question of who was first has subsequently been the subject of much debate. Historians generally agree that, while Atanasoff had done some work on electronic computing, Eckert and Mauchly were the true inventors of the modern electronic computer. Unlike Atanasoff in the 1930s, they built a complete one, and it went into productive use in the mid-1940s. In terms of impact, they clearly produced more machines and did more than anyone in their generation to usher in the age of the computer.

Following the decision, Eckert and Mauchly attempted to clarify their positions. As recently as the 1980s, many scientists still had strong feelings on the decision. Eckert finally resigned himself to the circumstances, whereas Mauchly remained bitter over the court's rulings right up to his death in 1979.

Eckert retired from Sperry Univac to Gladwyne, Pennsylvania, and in 1982, his company announced that it would no longer name any of its computers UNIVAC. Its retirement marked the symbolic close of an important chapter in the development of the modern computer.

For further information, see: John Vincent Atanasoff, ''Advent of Electronic Digital Computing,'' *Annals of the History of Computing* 6, No. 3 (July 1984): 229–282; J. G. Brainerd, ''Genesis of the ENIAC,'' *Technology and Culture* 17, No. 3 (July 1976): 482–488; A. W. Burks and A. R. Burks, ''The ENIAC: First General-Purpose Electronic Computer,'' *Annals of the History of Computing* 3, No. 4 (October 1981): 310–399; J.

P. Eckert, Jr., "The ENIAC," in N. Metropolis et al., eds., *A History of Computing in the Twentieth Century* (New York: Academic Press, 1980), pp. 525–539; Joel Shurkin, *Engines of the Mind: A History of the Computer* (New York: W. W. Norton and Co., 1984); Nancy Stern, *From ENIAC to UNIVAC: An Appraisal of the Eckert-Mauchly Computers* (Bedford, Mass.: Digital Press, 1981).

ECKERT, WALLACE JOHN (1902–1971). Eckert's role in the development of data processing technology has usually been underestimated. He first encouraged International Business Machines Corporation (IBM)† to support significant research in scientific computing; he established an important computational laboratory at Columbia University; and he developed important techniques for linking tabulating and adding equipment together into systems for the purpose of computing complex mathematical problems. He was also one of the most vocal scientists of the 1930s to advocate the use of automata for computational purposes, particularly for scientific applications. And even if he had had no part in bringing scientific applications and data processing together, he would still be remembered today as a great astronomer.

Wallace J. Eckert was born in Pittsburgh, Pennsylvania, on June 19, 1902. He received an A.B. from Oberlin College in 1925, an M.A. from Amherst College in 1926, and another M.A. from the University of Chicago a year earlier. He studied astronomy at Columbia University and completed his Ph.D. at Yale University in 1931. He did his most advanced graduate work as a student of Ernest W. Brown (1866–1938), the developer of the Lunar Theory. Brown sparked in the young scientist a life-long interest in numerical astronomy. This concern led to another growing interest: calculations and their role in celestial mechanics. As a result, Eckert would work with calculators and other related equipment for the rest of his life.

Eckert first appeared at Columbia University in 1926 as a graduate assistant and returned as a young member of the faculty after completing his doctorate under Brown. One of his earliest actions was to obtain an electric Monroe calculator for class use; he was the first professor at that school to introduce students to mechanical calculations instead of continuing to rely on logorithms. He also established a computational laboratory. In 1929, Thomas J. Watson, Sr.,** founder of IBM, had been convinced by Benjamin D. Wood, head of the Bureau of Collegiate Educational Research at Columbia, to fund the Columbia University Statistical Bureau. Watson helped with a variety of IBM tabulating equipment, and Eckert was involved from the beginning. In 1930, Watson supported the development of a tabulator called the Difference Tabulator to perform the work originally contemplated by Charles Babbage** in the nineteenth century and other inventors such as Pehr George Scheutz** and Martin Wiberg.** The device was completed in 1931, the year Eckert was promoted to assistant professor.

Mathematical jobs run on tabulating equipment varied and came from universities all over the United States. In 1931, Eckert also began to refine Brown's Lunar Theory using the university's statistical calculator on weekends.

In 1933, this handsome scientist approached Watson with the suggestion that IBM enlarge Columbia's facility with more people and machines to take on more complicated tasks. He wanted to link a variety of equipment together to do more than simple mathematics. The project interested Watson, who supplied additional equipment for use in the attic of Pupin Hall in the newly named Astronomical Computing Bureau. It was also called the Thomas J. Watson Astronomical Computing Bureau or, simply, the Watson Lab. Eckert ran this facility in cooperation with the American Astronomical Society, the Department of Astronomy at Columbia, and IBM. He served as its first director while his old mentor, E. W. Brown, sat on its Board of Managers along with T. H. Brown of Harvard, Henry R. Russell of Princeton, and, representing IBM, Charles H. Tomkinson (as eager a proponent of the lab as Eckert).

Eckert worked closely with IBM engineers to modify the equipment to meet scientific needs. This working relationship gave IBM managers in labs and manufacturing facilities considerable insight into how to deal with scientists in general. This experience proved to be an important one, especially in the late 1940s when the company embarked on a path toward new technologies, including computers. In working with IBM engineers, Eckert made the recently announced IBM 601 multiplying punch the centerpiece of his new configuration. When linked to a credit balance accounting machine and a summary punch, the punch enabled the user to read data mechanically, record them, calculate, and then generate output. Therein lay the genesis of a scientific computer. For purposes of coordinating all these activities, Eckert developed a machine called the mechanical programmer. In effect, it controlled a pluggable relay box which he stripped from Columbia's statistical calculator and used to exercise on weekends. Some twenty settings of switches allowed him to control the functions of his various tabulating equipment, taking such technology beyond the realm of card input/output into that of computers. Eckert's device told specific machines what to do and when, and used nicks in the disks installed on each that could be activated by his invention.

To put his work in perspective, little work was being done to develop computers in the early 1930s. Vannevar Bush** at the Massachusetts Institute of Technology (MIT) was working on a difference analyzer* (an analog computer*), and J. Presper Eckert** was becoming interested in calculators. The Moore School of Electrical Engineering,† where the ENIAC* would be built in the 1940s, had yet to become involved in computational projects of any significance. John V. Atanasoff** at Iowa State College was beginning his work and by the end of the decade had built only portions of a machine. In Germany, Konrad Zuse** was constructing a machine system in his parents' living room. The most advanced work on computational devices was taking place at the Bell Laboratories† where pressure was being exerted to construct sophisticated telephone message switching equipment. Even at Harvard, Howard H. Aiken** was barely getting started.

In 1940, Eckert was on the frontier of many new projects and was promoted to full professor. That same year, however, he left Columbia to become director of the Nautical Almanac Office. There he employed many of the same methods used at Columbia in lashing equipment together to produce a pilot's Air Alamanac quickly. The clouds of war were visible, and his mission was urgent: he had to prepare the necessary publication for combat pilots as quickly as possible. Eckert's agency, part of the U.S. Naval Observatory which supported many computational projects during World War II, employed his techniques on numerous projects. Meanwhile, at Columbia, his old laboratory was expanded to handle war-related calculations, including work on the famous B–29 bomber fire control system and the Manhattan Project's calculations on nuclear fission.

At the conclusion of World War II, Eckert returned to Columbia as director of the Watson Laboratory (1945–1967) and he also worked for IBM (1945–1952). IBM had asked Eckert to participate in the development of a calculator called the Selective Sequence Electronic Calculator (SSEC).* It was an electromechanical device that could handle larger volumes of data than conventional tabulating equipment, and, although not as sophisticated as the ENIAC's successors, it was IBM's first computer-like product marketed in the late 1940s. The SSEC was the first commercially available device that could modify its own commands, thereby allowing a machine to control a sequence of calculations. (This is a basis feature of all computers today.) The majority of the development work on this machine took place at the Watson Laboratory. Since a copy of the machine was now conveniently at hand, Eckert continued his earlier work in astronomy, successfully using the SSEC to plot the moon's orbit more accurately than ever before. Twenty years later, the results of his calculations were used to determine the path of the Apollo space ships.

The SSEC was an important evolution from Eckert's old "mechanical programmer." He kept some of the older machine's functions but on a more sophisticated basis. Thus, for instance, an operator sitting at the console of the IBM machine could allow the progress of calculations, making changes along the way, and all calculations could be printed either while they were taking place or as the final result of some job stream. With the announcement of the SSEC on January 28, 1948, Eckert had inextricably, if slowly, pushed IBM into the age of the computer.

Eckert's vision of the future use of sophisticated computational devices was formed in his youth. He encouraged the use of computational technologies in the 1920s and especially in the 1930s, and he realized that such machines could be used in many fields of science, not solely in astronomy. In 1940, he published *Punched Card Methods in Scientific Computation,* one of the most important books on applications for computers, particularly scientific applications. This was one of the first studies on computer-like applications, and it was read by a generation of engineers and scientists who were only then beginning to enter what would be the field of computer science. Of all the books and articles he published, this one on data processing was the most influential, read by such people as J. Presper Eckert, Jr., and John W. Mauchly,** creators of the ENIAC, EDVAC,* BINAC,* and UNIVAC.*

Eckert gained considerable recognition in his lifetime. In 1966, he received the James Craig Watson Medal of the National Academy of Science and in 1969 the IBM Award. He had already been made an IBM Fellow in 1967, which meant that he could remain on full salary for five years, working on any project he wanted. Eckert received this distinguished award the same year he retired from Columbia University. He died on August 24, 1971.

For further information, see: J. F. Brennan, *The IBM Watson Laboratory at Columbia University: A History* (Armonk, N.Y.: IBM Corporation, 1971); Charles Eames and Ray Eames, *A Computer Perspective* (Cambridge, Mass.: Harvard University Press, 1973); W. J. Eckert, *Punched Card Methods in Scientific Computation* (New York: Columbia University Press, 1940; reprinted by MIT Press, 1984) and "The Role of the Punched Card in Scientific Computation," *Proceedings of the Industrial Computation Seminar, September 1950* (New York: IBM, 1951), pp. 13–17; H. H. Goldstine, *The Computer from Pascal to von Neumann* (Princeton, N.J.: Princeton University Press, 1972).

ENGEL, FRANK AUGUST, JR. (1917–). Engel managed data processing technology and promoted portions of it. His areas of key concern were FORTRAN* and ALGOL,* two important programming languages.* He was also a president of SHARE,† an important user group.

Engel was born in 1917 in Steubenville, Ohio, and was educated in physics and engineering in the 1930s and in mathematics during the early 1950s. Between 1939 and 1962, he worked for the Speer Carbon Company, Pennsylvania Electric Company, B. F. Goodrich, Mine Safety Applicances, and the Westinghouse Electric Corporation. This manager of computing next worked as director of the Computer Center at Harvard University (1962–1964) and then moved again, this time to the Applied Scientific Department of the Electronic Data Processing Division of Honeywell, Inc.,† only to switch to MITRE Corporation in 1966, where he has remained.

During the 1950s, Engel promoted the use of International Business Machines Corporation's (IBM's)† FORTRAN, a widely used programming language for scientific computing. At the end of the decade, he participated in the U.S.-European movement within the industry to develop the specifications for a universal language which came to be known as ALGOL. Engel was a leader within SHARE, a user group made up of individuals who used IBM computers. In 1956–1957, he was its president and in 1960–1961, its vice-president. During the 1970s, Engel served as chairman of the American National Standards Committee on Fortran Programming Language Standards. He had held a similar position in the early 1960s as chairman of SHARE's IAL Committee which worked with ALGOL.

For further information, see: Jean E. Sammet, *Programming Languages: History and Fundamentals* (Englewood Cliffs, N.J.: Prentice-Hall, 1969).

ERSHOV, ANDREI PETROVICH (1931–). This Soviet computer scientist is the Soviet Union's most visible representative of data processing in the West. He has published and spoken extensively on data processing both in general and in the Soviet Union specifically.

Ershov studied at the Institute of Mathematics and Mechanics at Moscow State University from 1949 to 1954, and conducted additional studies to 1957. From 1957 to 1960, he served as a research associate for the USSR Academy of Sciences' Computing Center in Moscow. From 1960 to the present, he has been in charge of the Academy's Siberian branch Computer Laboratory located in Novosibirsk. Ershov also became a professor at Novosibirsk State University in 1967 and completed his Ph.D. there in 1968.

Ershov participated in the development of many early computers in his country. Among these projects were the programming software and procedures for the BESM* computer between 1954 and 1955, the programming system for the STRELA–3 computer from 1956 to 1958, the development of another system for the ALPHA from 1961 to 1965, and the time-sharing system known as AIST–0 from 1967 to 1971. During the 1970s, he directed the development of a universal programming tool called BETA. Ershov is a prolific writer; he has published over sixty papers, nearly half of which have appeared in the West. For this reason, he is generally perceived to be a representative of Soviet data processing. He has made presentations in the West and has served as consultant on data processing matters to his government.

Since 1964, Ershov has been a member of the International Federation for Information Processing (IFIP)† Technical Committee 2 on Programming Languages and, since 1965, part of the Working Group 2.1 on ALGOL.* Both of these projects reflect his primary interest in programming tools and the international flavor of concerted activities across the industry to develop better programming systems. Ershov has long been a member of both Association for Computing Machinery (ACM)† and of the USSR Academy of Sciences. He translated many of the ALGOL papers developed in the West into Russian where a strong interest in that language has developed along the lines evident in Western Europe.

For further information, see: Andrei Petrovich Ershov, "ALPHA, An Automatic Programming System of High Efficiency," *Journal of the Association of Computing Machinery* 13, No. 1 (January 1966): 17–24, "SYGMA, A Symbolic Generator and Macro-assembler," *Symbol Manipulating Languages and Techniques,* No. 107 (undated): 1.4.2–30, 226–246, "Theory of Program Schemata," in C. V. Freiman, ed., *Information Processing 71,* Vol. 1, Proceedings of the IFIP Congress 1971, Ljubljana, Yugoslavia, August 23–28, 1971 (Amsterdam, The Netherlands: North-Holland Publishing Co., 1972), pp. 28–45, and with Mikhail R. Shura-Bura, "The Early Development of Programming in the USSR," in N. Metropolis et al., eds., *A History of Computing in the Twentieth Century* (New York: Academic Press, 1980), pp. 137–196; R. Moreau, *The Computer Comes of Age: The People, the Hardware, and the Software* (Cambridge, Mass.: MIT Press, 1984).

ESTRIDGE, PHILIP DON (1938–1985). Estridge was the executive at International Business Machines Corporation (IBM)† most responsible for the development and manufacture of that company's microcomputer product known as the Personal Computer (PC). Introduced in the early 1980s, it ushered in a new era in computing in which such technology became available in offices, homes, and schools. Within four years after it was introduced, the PC was generating nearly 4 percent of all of IBM's revenues. The PC was the single most successful product introduction in the industry since IBM announced the S/360* in 1964.

Estridge was born in 1938 and was raised in Jacksonville, Florida, where his father was a professional photographer. He completed a B.S. in electrical engineering at the University of Florida in 1959 and promptly joined IBM at its facilities in Kingston, New York. There he participated in the development of SAGE,* an early-warning defense system that relied heavily on radar and computers. In 1963, Estridge was moved to Washington, D.C., where he did programming for the Goddard Space Flight Center as part of an IBM project. In 1969, he moved to Boca Raton, Florida, which at that time had an IBM plant site less than a year old. It would be at that location that the PC would be developed and manufactured.

In August 1980, Estridge was manager of product development for the Series/1, a minicomputer, when IBM decided it was time to develop its own microcomputer. Estridge was put in charge of the project with the title Manager Entry Level Systems—Small Systems. He was given one year to do the job, a schedule he met by combining internal development within IBM and acquiring technology from outside the firm—a first for his company which historically developed all components of its products. In 1981, IBM introduced the PC. That announcement gave respectability to microcomputers which had been around since the mid-1970s. This stamp of approval made it possible for manufacturers to sell this equipment to commercial customers, which led to an enormous growth in the use of such devices in the early to mid-1980s. By the end of 1983, for example, IBM alone had shipped over 750,000 PCs and in the following year nearly a million. By 1985, such volume constituted only about 20 percent of the total microcomputer market.

Estridge's organization also grew. In August 1980, he began the project with twelve people, a small budget, and IBM's promise of relative autonomy. A few months later, in January 1981, he had 135 people and the following January, 330. In January 1983, his payroll had swelled to 3,300 with all manufacturing still done in Boca Raton. By August 1984—the fourth anniversary of the start of his project—he managed a force of 9,500 IBMers and his department had grown from an independent unit into the Entry Systems Division (ESD). On a personal basis, his titles changed from a simple product manager to president of ESD. In January 1984, IBM also promoted him to the rank of corporate vice-president.

Estridge received as much publicity as the chairman of the board of IBM.

The American press noticed his work. *Business Week* published a photograph of him, and *Time* featured him in an article. He was also the subject of many articles in the industry's press, such as *ComputerWorld* and *InfoWorld* which characterized the soft-spoken executive as ''the Cary Grant of micros''.

Estridge and his wife Mary Ann were killed on August 2, 1985, in an airplane crash at the Dallas-Fort Worth Airport. The father of the IBM PC had become a major figure not only in the IBM of the 1980s, but also in the data processing industry. As of 1987, IBM had sold several million PCs, bringing, computing to millions of people in the industrialized world. The introduction of the microcomputer ushered in a new era characterized by heavy reliance on computers by people who knew little about its technology. Along with Steven P. Jobs,** who introduced Apple's† personal computer, Estridge had played a large role in computing.

For further information, see: Paul Freiberger and Michael Swaine, *Fire in the Valley* (Berkeley, Calif.: Osborne/McGraw-Hill, 1984); Robert Levering et al., *The Computer Entrepreneuers* (New York: New American Library, 1984).

EVANS, ROBERT (BOB) OVERTON (1927–). This International Business Machines Corporation (IBM)† engineer was the father of the S/360* family of computers, the most successful product in the history of American industry. As the chief executive on the project, Evans established many of the design criteria for these products. He made many of the decisions that determined their final characteristics, features that allowed IBM to double in size during the second half of the 1960s. The S/360 family of products (nearly 250 when fully developed) set technical and design standards for the entire data processing industry, thereby defining characteristics still dominant in the world of computers nearly a quarter of a century later.

Evans was born on August 19, 1927, at Grand Island, Nebraska, and attended Iowa State College (later Iowa State University) where he completed his undergraduate degree in electrical engineering. Earlier, between 1945 and 1946, he had served with the U.S. Naval Reserve. Between 1949 and 1951, he was an electrical operating engineer for the Northern Indiana Public Service Company. In 1951, he joined IBM as an engineer to work on the Defense Calculator, an early computational project within the company. He became familiar with the IBM 701* (as it was known) and, periodically, when a customer had difficulties with the machine, flew out to fix it. Unlike most of IBM's engineers of the early 1950s, he therefore gained experience with customers and the environments they worked in, which gave him a unique appreciation of the impact of IBM's products outside of the laboratory. He soon moved into various management positions within IBM's engineering community. At Endicott, New York, he served as the systems manager on the 7070 and the 1410 machines. He moved the effort on the 7070 to Poughkeepsie, New York, the site of a growing laboratory and manufacturing plant for IBM's largest computers.

In the late 1950s, Evans was charged with determining whether the 8000 series of processors should be developed. That project involved him heavily in the politics of product development and gained him considerable positive exposure to upper management. That assignment pitted one group of IBM scientists and engineers (led by Frederick P. Brooks**) against another community (subsequently dominated by Evans) fighting over whether the 8000 should be introduced or be leapfrogged to more advanced technologies. The internal battles led to a formal study of a proposed new family of computers (initiated by the SPREAD* report), an effort which Evans led from its inception. He forced through recommendations which called for a family of computers (a new concept in the early 1960s) that relied on advanced technologies, all of which were driven by new operating systems.* The battle ended when top management elected to side with Evans in May 1961. As was usually the case in the data processing industry, changes and major decisions were often the work of the young. Brooks was twenty-nine years old during the battle, and Evans was thirty four when he won.

Evans, seeking to appease those who had supported the 8000 series, asked Brooks to head up the effort to design the new systems, computers that would ultimately be called the S/360. In 1962, Evans was named vice-president of development for the Data Systems Division (DSD). During this period he marshalled all the resources that would be used to create the S/360. Every IBM executive was committed to it, and almost all available dollars (both internal and borrowed) were poured into the project. Over 2,000 programmers eventually worked on the software* to support it, while every other engineer at one time or another helped design new computers, disk, tape, card punch equipment, printers, communications gear, control units, and new programming language* compilers. Evans and his staff coordinated all of IBM's facilities from around the world to develop new products for announcement in the mid-1960s. As most executives of the time were wont to say, the project was a "bet your company" effort. If IBM failed, it may well have shrivelled in the face of its competitors and may have met its demise.

The S/360 was the most comprehensive project undertaken up to that time in the data processing industry. Evans had to deal with the integration of many new and emerging technologies (some of which kept shifting, such as chips*), make them cost-effective, satisfy the question of how to preserve existing installed applications customers had invested in on older computers, develop new methods of cost-effective manufacturing, write hundreds of thousands of lines of new software, pay for all these activities, and keep spirits up among various staffs when problems arose or fight off detractors. On April 7, 1964, the first of the S/360 products was announced to an industry that immediately recognized that IBM had just taken a major step forward in computing, ushering in a new era in the history of data processing for all. Only with the S/360 did the change take place from the age of nuclear power to that of information and computers.

IBM's gamble was more than worth it. Evans said it best in 1986: "In 1982 the descendants of System/360 accounted for more than half of IBM's gross income." He showed an industry how to standardize technology—a common element of today's computing. He proved that such massive projects were possible when backed by a top management courageous enough to change the form of the company and to delegate responsibility to staffs working together toward a well-defined goal.

In 1965, Evans was rewarded for his efforts by being promoted to president of the Federal Systems Division (FSD). In 1969, he became president of the Systems Development Division (SDD) (home of all of IBM's large computers and successors to the S/360, called the S/370*). In 1975, he was named president of the Systems Communications Division (SCD), and, in 1972, the board of directors of IBM elected him an IBM vice-president. He dominated IBM throughout the 1970s, guiding the company from one round of products to another as it enhanced all the S/370 architecture first developed in the 1960s. He led the firm to broaden its product line into such diverse fields as minicomputers, later microcomputers, telecommunications, and distributed processing, all of which amounted to some 3,000 different products by 1980. On July 1, 1984, he retired from IBM, relinquishing his post as IBM vice-president for engineering, programming, and technology. He subsequently became a general partner of Hambrecht and Quist Venture Partners.

Evans was active as a trustee for various educational and public service organizations. The National Aeoronautics and Space Agency (NASA) honored him with its Distinguished Public Service award for service rendered while president of FSD. IBM had provided computing services for the American space program of the 1960s. In addition, the Institute of Electrical and Electronics Engineers (IEEE)† named him a Fellow. Not sixty years old yet, he was already recognized throughout the data processing industry as one of its giants and historians had already begun to analyze how the S/360 came about. To a large extent, that was the story of Bob Evans.

For further information, see: Charles J. Bashe et al., *IBM's Early Computers* (Cambridge, Mass.: MIT Press, 1986); B. O. Evans, "System/360: A Retrospective View," *Annals of the History of Computing* 8, No. 2 (April 1986): 155–179; Franklin M. Fisher et al., *IBM and the U.S. Data Processing Industry: An Economic History* (New York: Praeger Publishers, 1983); Emerson W. Pugh, *Memories That Shaped an Industry: Decisions Leading to IBM System/360* (Cambridge, Mass.: MIT Press, 1984).

EVERETT, ROBERT RIVERS (1921–). Everett was closely associated with the construction of the WHIRLWIND* computer at the Massachusetts Institute of Technology (MIT) in the late 1940s and early 1950s. Later, he became an executive at MITRE Corporation, the home of many technological innovations in the 1960s and 1970s. Like many members of the data processing industry, he began his professional life as a scientist working in the field of computing and then went on to establish a company that manufactured products based on such technologies.

Everett was born on June 26, 1921, in Yonkers, New York. He completed his B.S. at Duke University in June 1942 and his M.S. at MIT the following year. He worked at MIT's Servomechanisms Laboratory between 1942 and 1951, at which time he became its associate director. Everett had been at MIT working in the general field of electrical engineering since he had completed his initial studies there. During the late 1940s, he became heavily involved with data processing, particularly after 1944. Between 1951 and 1956, he served as associate division head of the Lincoln Laboratory† at MIT and as its division head between 1956 and 1958. His first position at MITRE was as technical director (1958–1959), followed by his promotion to vice-president of technical operations, a position he held from 1959 to 1969. In that last year he first became executive vice-president and, soon after, president of the company.

After majoring in electrical engineering at Duke, Everett wanted to continue his studies at MIT where he also began working in various capacities in laboratories. One of his first managers was Jay W. Forrester,** the scientist who would later manage the construction of WHIRLWIND. During the early years of World War II, in addition to his studies, Everett became involved in the Servomechanisms Laboratory's war-related projects, many of which touched on computational technologies. These projects gave Everett his first taste of data processing and experience in managing technical projects. They also broadened MIT's experiences with computers from simple analog devices to digital computation. Everett began to acquire the experience that later enabled him to develop detailed technical plans, which in turn contributed directly to the construction of WHIRLWIND. That computer was the largest and fastest built between the late 1940s and early 1950s.

Everett was involved with WHIRLWIND from its inception through most of the 1950s. One of his earliest assignments related to this program involved leading others in designing all the components, their interconnections, and the sequence in which they would be used in the computer. He spent all of 1946 establishing conceptional designs and testing possibilities. By the end of 1947, he had worked out a design along with detailed statements about how a digital computer* could be used for applications relevant to the U.S. Navy. His design was much broader than the original plan had called for two years earlier. By late 1947, he, together with Forrester, had devised a plan to wed abstract numerical data to radio and radar equipment in order to develop a computer. He had therefore moved from theoretical use of electronics for complex processing to a practical vision of application, a key step in the evolution of complex computers at that time. He also went further than many of his contemporaries in articulating specific uses of such equipment, primarily for war-related applications. By the end of 1947, he was also heavily involved in the construction of components that would make up the computer. He worried more about logical circuits than about administration. Forrester focused on the management of the overall effort from budgets to people, from relations within MIT to those with the Office of Naval Research (ONR),* the sponsor of the project.

Everett's management style, which would remain essentially the same when he headed research and development at MITRE, was relaxed and friendly. He had extensive knowledge of all the various projects and experiments and appreciated the problems faced by his staff. In contrast, Forrester was rough in his dealings with engineers. Together, however, the two made up an effective team managing one of the most sophisticated technical development projects in the early history of computers. For the first time, a computer could not be invented by one or two individuals but had to emerge out of the work of dozens of people. Managing that kind of project was new in the late 1940s, and doing it successfully became one of Everett's most important contributions to computing.

The computer ran tests in real-time computing for the U.S. Air Force in 1951, proving that digital computers could be used for a large defensive network. While he worked on refining this system and even on developing a follow-on (WHIRLWIND II) in the mid-1950s, his contribution to the project had been made. He had shown that large, complex general-purpose computers had to be built out of very reliable electronic components and that engineering problems had to be fixed concurrently if all of them were to be resolved and to lead to the construction of a large machine.

In 1959, Everett left MIT to help establish MITRE, taking with him many engineers who had worked on WHIRLWIND and on other projects under his management. At MITRE he worked on the development and manufacture of various systems during the 1960s and 1970s. But his greatest years from an historical point of view had been spent at MIT.

For further information, see: Kent C. Redmond and Thomas M. Smith, *Project WHIRLWIND: The History of a Pioneer Computer* (Bedford, Mass.: Digital Press, 1980).

F

FAIRCHILD, GEORGE WINTHROP (1854–1924). Businessman, member of the U.S. Congress, and leading figure of many business ventures in the State of New York, Fairchild also served as vice-president and chairman of the board of the Computing-Tabulating-Recording Company (C-T-R), which became International Business Machines Corporation (IBM).† A poor leader of the company who battled with Thomas J. Watson,** founder of IBM, Fairchild was nonetheless a highly respected business manager in his day.

Fairchild was born on May 6, 1854 in Oneonta, New York, and dropped out of school at the age of thirteen. His first job was as a printer, and over the years he developed a company called International Time Recording in Endicott, New York, later the home of IBM's most important manufacturing facility of the 1920s and 1930s. He served in Congress from 1907 to 1919.

A skilled business leader named Charles R. Flint* put together a series of companies in order to make profits on stock transactions associated with the deal called the C-T-R. It consisted of four companies chartered in New York on July 5, 1911: Computing Scale Company, Tabulating Machine Company (Herman Hollerith's** old firm), International Time Recording Company (Fairchild's), and the Bundy Manufacturing Company. Bundy was located in Endicott, as the International Time Recording company would be later. When Flint organized this new company called C-T-R, he hired Fairchild to manage it as its first chairman of the board. By then Fairchild had vast experience in business and was serving on the boards of several companies (White Plains Development Company, Peoples Trust Company, and Citizens National Bank in his home town). He had also been head of International Time, making it a successful venture. Thus, he had become a highly respected member of the New York business community, giving C-T-R credibility just by his presence.

Some historians argue that Fairchild had too many interests and therefore could not develop the new firm effectively. During the early years of its existence, for example, he was serving in Congress. In 1910, President Theodore Roosevelt asked him to be diplomatic envoy to Mexico at a time when relations between the United States and that country verged on war. The best Fairchild could do was to lend respectability to the new company and serve a figurehead role. He was not interested in either consolidating the product sets of three companies making up the bulk of C-T-R or in controlling costs. Hardly any research and development took place, particularly with that portion of the business which was selling tabulating equipment in a highly competitive marketplace. He favored lowering the company's debt and in 1913 allowed stockholders to be paid a dividend so great that basic projects had to be cut back to cover costs.

Flint hired Watson as chief executive officer of the company in 1913. Watson, eager to consolidate product lines, expand business, and make the firm a thriving concern, soon clashed with Fairchild who was more interested in short-term profits than in long-term growth. Watson's position in the company gradually became impregnable. In 1919, when Fairchild left Congress with the intent of focusing more attention on business matters, Watson dominated the company. Watson managed to change the name of the firm to IBM in 1922 without much resistance from Fairchild. On December 31, 1924, Fairchild died, leaving Watson master of the company.

For further information, see: Geoffrey D. Austrian, *Herman Hollerith: Forgotten Giant of Information Processing* (New York: Columbia University Press, 1982); Robert Sobel, *IBM: Colossus in Transition* (New York: Times Books, 1981).

FANO, ROBERT MARIO (1917–). Fano, a professor at the Massachusetts Institute of Technology (MIT), was responsible for the development of significant real-time computing facilities during the 1960s. As a result of his efforts, MIT was able to contribute to the science of online computing while training many in the field of data processing in general. By the early 1970s, it was not unusual for more than fourteen students to complete their Ph.D. in the subject. Fano is more specifically remembered for the effort known as Project MAC.*

Fano was born on November 11, 1917, in Torino, Italy, moved to the United States in 1939, and became a naturalized citizen in 1947. He completed his B.S. in electrical engineering at MIT in 1941 and his Sc.D. in 1947. He spent the rest of his professional career at MIT. Fano was a teaching assistant at MIT (1941–1943), an instructor in electrical engineering (1943–1944), and a staff member (1944–1946), a research associate also in the electrical engineering department (1946–1947), assistant professor (1947–1951), group leader (1950–1953), associate professor (1951–1956), full professor (1956–1962) and then Ford Professor. His greatest contributions to real-time computing were made while he was director of Project MAC (1963–1968) and first associate head of the Computer Science and Electrical Engineering Department (1971–1974). During these years, in addition to teaching and administering projects, he conducted research on electromagnetic fields. His distinguished career earned him membership in the National Academy of Science, the National Academy of Engineering, and the American Academy of Arts and Sciences.

Fano was introduced to computer-like projects during World War II when MIT conducted research on radio-frequency modulators and microwave filters. He also taught measurements and communications during these years. As a graduate student, he had been interested in radio transmissions, but after completing his doctoral thesis in June 1947 he took up some of the ideas of Norbert Wiener* (father of *cybernetics*) who was also at MIT. Fano was especially concerned with the quantity of information within communications analogous to "negative entropy in statistical mechanics." In other words, he wanted to study how information was transmitted. This research came at a time when many scientists were trying to develop a theory of information at MIT and elsewhere. Thus, he was in the mainstream of that effort. Fano studied how speech and coded messages were transmitted. Throughout the late 1940s and the next decade, he taught courses on how information flowed and he with time published a number of works on the subject.

Fano became increasingly immersed in computer science. When the Lincoln Laboratory† was established at MIT in August 1951, he became part of its staff, working on continuous-wave components and tubes. These efforts were part of the larger project to create WHIRLWIND,* a massive computer for the U.S. military community in the 1950s. It also fed technology into the SAGE* project. As a member of the study group formed in 1960 to examine the overall computing requirements of MIT, Professor Fano participated in the development of recommendations in 1960 and 1961 calling for a campus-wide network for real-time computing. The result of those suggestions, and the work of many engineers and computer scientists, was the use of the Compatible Time-Sharing System (CTSS).*

In 1962, MIT decided to expand its time-sharing capabilities and named Fano director of the project, effective with the start of the next school year. The effort would become known as Project MAC. Fano's staff included Fernando J. Corbató** (one of the developers of CTSS) and Marvin L. Minsky** (a well-known specialist in artificial intelligence*). Project MAC was an important success. It involved the establishment of a new data processing facility, initially with CTSS, and a network which, by the spring of 1964, had over 200 users. Continuous enhancement of the system allowed Fano to train students in the science of computerized technologies while making such services available to an increasing number of users. Fano also led the effort to create a follow-on system to Project MAC called Multics in the mid-1960s. That project went live in 1969 and within two years was used by over 500 people. The system ran twenty-four hours a day, seven days a week, year-round. By then Professor Fano had concluded that there was no longer any purpose in doing research on real-time systems. Software subsystems were already available commercially to provide such services. Therefore, he used his influence to have the system moved to the Information Processing Center as early as October 1969. He next turned his attention to narrower computational research projects during the 1970s and 1980s.

For further information, see: Robert M. Fano, *Electromagnetic Fields, Energy, and Forces* (New York: John Wiley and Sons, 1960), "Project MAC," in *Encyclopedia of Computer Science and Technology* (New York: Marcel Dekker, 1979), 12: 339–360, and Robert M. Fano et al., *Electromagnetic Energy Transmission and Radiation* (New York: John Wiley and Sons, 1960); Karl L. Wildes and Nilo A. Lindgren, *A Century of Electrical Engineering and Computer Science at MIT, 1882–1982* (Cambridge, Mass.: MIT Press, 1985).

FELT, DORR EUGENE (1862–1930). Felt invented one of the first key operator calculating machines and one of the first practical adding and listing machines, both of which became very popular in American offices in the last quarter of the nineteenth century.

Felt was born in Beloit, Wisconsin, on March 18, 1862. In addition to his inventing, he also established a company to sell his devices called the Felt and Tarrant Company in 1886. The following year he expanded its mission to include manufacturing and renamed the firm the Felt and Tarrant Manufacturing Company, headquartered in Chicago. He remained president of the company until his death on August 7, 1930.

With America's full entrance into the Industrial Revolution after the Civil War (1861–1865), the need for more accurate and faster recordkeeping, particularly of numbers for population studies, insurance, larger businesses, and so on, increased the demand for mechanical tabulation methods. The result was the development of a variety of adding and tabulating equipment marketed by Felt, as well as by other firms such as The National Cash Register Company (NCR)† and Burroughs†—both of which came into existence in the same era as Felt's company. Reflecting the view of his day, Felt argued that adding long columns of numbers was "turning men into veritable machines." His devices substituted successfully for some of that tedium. His most important machine, introduced in 1890, was called the Comptometer which, along with Burroughs' Adding and Listing Machine, became the two most popular accounting devices in the world.

Felt's Comptometer was a black box with keys representing numbers, much like adding machines had in the twentieth century. Felt developed the earliest version of this machine starting on Thanksgiving Day, 1884. He used a macaroni box, metal staples, rubber bands, and meat skewers to create the key-driven machine. In 1887, he had his first patent. Revisions of this same design resulted in more patents throughout the 1890s and the early twentieth century. Almost from the beginning, Felt called his machine the Comptometer. Many years after his invention he would recall that the only way such devices could be successful was if they worked faster than accountants, and that meant building machines that could work faster and more accurately than an individual who could add four columns of numbers at the same time.

Felt's machine was so successful that Burroughs had a difficult time penetrating the market. From about 1887 until about 1902, the Comptometer virtually dominated all sales of desktop calculators. By the early 1900s, however, Burroughs had finally developed good products and was marketing them effectively. Felt's machines were bought by businesses, especially railroad companies, banks, and insurance firms, as well as by government agencies (such as the U.S. Treasury, U.S. Navy, and the New York Weather Bureau). Thus, the applications used on these machines were both commercial and scientific. From the beginning, customers used mechanical aids to calculation for more applications than the designers originally imagined. This pattern of use continues even today. Thus, even in the era of the modern computer, manufacturers still underestimate the variety of uses to which their equipment will be put. That fact often explains why demand for such devices usually exceeds supply.

Felt was a conservative Republican, a highly successful manufacturer, and an important member of the Chicago community. His Comptometer was not the perfect machine. While it could calculate, it was not designed to record the results of such efforts. Burroughs was able to invent a machine that did print all the numbers entered and the results in a grand total. With that improvement in technology, he was able to enter the office machine market with a product that could successfully compete against Felt's. All the same Felt's machine contributed to the automation of the office and reaffirmed the notion that machines could assist humankind. His part in that process was to help develop a variety of calculators in the United States and in Europe that would increasingly lead scientists toward the electronic computer.

For further information, see: "An Improved Calculating Machine," *Scientific American* 59 (1888): 265; C. V. Boys, "The Comptometer," *Nature* 64 (1901): 265–268; Charles Eames and Ray Eames, *A Computer Perspective* (Cambridge, Mass.: Harvard University Press, 1973); D. E. Felt, "Mechanical Arithmetic," *Scientific American* 69 (1893): 310–311; Joel Shurkin, *Engines of the Mind: A History of the Computer* (New York: W. W. Norton, 1984).

FISCHER, ERNST GEORG (1852–1935). Fischer invented a tide predicting machine, which was a calculator with computer-like capabilities, just prior to World War I. It was electromechanical and it worked, making it one of the earliest of these machines in the twentieth century. The machine allowed the U.S. government to make tidal wave forecasts, information which, during World War I, allowed captains to plan when to sail in shallow waters. A German version of his machine produced data useful to U-boats. The results of Fischer's work were not lost on scientists who after World War I worked on other electromechanical devices in universities (such as the Massachusetts Institute of Technology [MIT] and Harvard), government agencies (particularly the military and Bureau of the Census), and businesses (such as the newly established Bell Laboratories†).

Fischer was born in Baltimore, Maryland, on August 6, 1852. At the age of thirteen his family sent him to school in Dresden, Germany, where he attended the Zschogg Real Schule (1865–1867) and the Engineering Works of Moritz Kleber (1867–1870), and finally tutored privately. His education, particularly in science, mathematics, and engineering, was outstanding. He returned to the United States and went to work for the U.S. Coast and Geodetic Survey on June 1, 1887, becoming its chief of the Instruments Division on March 1, 1898. He retired from government service on August 22, 1922, and died in Washington, D.C., in September 1935.

While working for the government, Fischer invented a wide variety of devices used for measuring phenomena and he improved earlier machines. A partial list of the engines and measuring equipment which he either invented or improved includes: plane table alidade (used to determine direction) with a rule and telescopic function, base bars, spring balance for base tape measurement, an apparatus for stretching tape, a tide gauge and several tide indicators, a compass declinometer, a camera, an interferometer used to measure the fixture of gravity pendulum supports, an electric signal lamp for triangulation, a magnetometer, a transit micrometer, plane tables, direction theodolite for primary triangulation, and an astronomical transit. His interest in measuring physical phenomena also led him to design and build such devices as a geodetic invar level rod, a pressure sounding tube, a modern artificial horizon for sextants, and, of course, his most important machine, the tide predictor.

Because of his responsibility for developing equipment to measure events, Fischer focused on his agency's most important tasks, which at the time centered on the oceans of the world. Soon after he became chief of the Instruments Division, his staff began to work on the development of a new machine to predict ocean waves. The project took over fourteen years to complete successfully. In essence, it computed data using thirty-seven different tidal components, generating the results of its calculations on dials. Nicknamed the Great Brass Brain, it took simple waves and added them together to create a tide prediction curve which it could plot on sheets of paper with the days of the year marked out. Thus one could see what the tides would be for any day. Built with the help of a colleague, R. A. Harris, it was a massive device with gears and pulleys that accumulated totals and numbers as they moved. The machine was effective. Today, the same application is performed on electronic digital computers* in record time. For a full year's tidal forecast, computers today need operate only less than two hours.

During World War I, the accuracy of the information made possible by the tide predictor and its rapid availability made it possible for British sea captains to elude German U-boats by sailing in shallow waters as early as 1915. In response to this tactic, the Germans developed their own version of Fischer's machine; it was completed in 1916 and managed by the Imperial Observatory. Thus, the tidal device may represent the first modern use of computers in warfare.

For further information, see: C. H. Claudy, "A Great Brass Brain," *Scientific American* 110 (March 7, 1914): 197–198; R. A. Harris, "The Coast and Geodetic Survey Tide Predicting Machine," *Scientific American* 110 (June 13, 1914): 485; and on the German version, see H. Rauschelbach, "Die Deutsche Gezeitenrechenmaschine," *Zeitschrift für Instrumentenkunde* 44 (July 1924): 285–303.

FLANDERS, DONALD ALEXANDER (1900–1958). This American mathematician participated in the management and development of computing services at the national laboratory at Los Alamos, New Mexico, during and just after World War II and at the Argonne National Laboratory in the late 1940s and 1950s. During his tenure at Argonne, the AVIDAC computer was built for Argonne and the ORACLE* for the Oak Ridge National Laboratory.

Flanders was born in Pawtucket, Rhode Island, on August 14, 1900. He completed an A.B. in mathematics at Haverford College in 1922, with minors in Greek and philosophy. His Ph.D. in mathematics was earned at the University of Pennsylvania in 1927. Between 1927 and 1929, he was a National Research Council Fellow at Princeton University. He then became an instructor of mathematics at New York University (NYU), remaining until 1949. While at NYU he expanded the school's faculty in mathematics with experienced and well-respected professors. In 1937, he was on sabbatical at the University of Copenhagen, Denmark. Between July 1943 and September 1946, he established a computing department at the Los Alamos National Laboratory. He hired and trained people in the general field of computing to staff his department. In 1946, Flanders returned to NYU's Institute for Mathematics and Mechanics where he conducted research on nonlinear vibrations.

In 1948, Flanders went to work for the Argonne National Laboratory, and the following year he convinced the laboratory to construct a computer like the digital device then under construction at the Institute for Advanced Study at Princeton (known later as the IAS Computer*). His efforts led to AVIDAC, sponsored by the U.S. Atomic Energy Commission. That project in turn led to the ORACLE for the laboratory at Oak Ridge, Tennessee, and later to the replacement of AVIDAC with the GEORGE. Flanders encouraged the construction of each of these machines and the acceptance of the IAS version. In November 1956, he was promoted to director of the Applied Mathematics Division. Only a year and a half later, on June 27, 1958, he died through a self-administered overdose of prescribed medicine. Richard Courant, a mathematician at the University of New York, described him as "a thoroughly saintly personality, but in reality he was a full-blooded human being," arguing that Flanders "suffered profoundly from the conflict of these two sides of his personality."

For further information, see: J. C. Chu, "Computer Development at Argonne National Laboratory," in N. Metropolis et al., eds., *A History of Computing in the Twentieth Century* (New York: Academic Press, 1980), pp. 345–346; Constance Reid, *Courant in Gottingen and New York. The Story of an Improbable Mathematician* (New York: Springer-Verlag, 1976).

FLINT, CHARLES RANLETT (1850–1934). This brilliant businessman built trusts and conglomerates in the late nineteenth and early twentieth centuries and was the creator of the Computing-Tabulating-Recording Company (C-T-R) which later became the International Business Machines Corporation (IBM)†. He was also responsible for hiring Thomas Watson,** who was IBM's chief executive officer from the 1920s through the early 1950s. In the process of creating C-T-R, Flint made the inventor of card punch equipment and tabulators, Herman Hollerith,** a wealthy man.

Flint was born on January 24, 1850, at Thomaston, Maine, and graduated from the Polytechnic Institute in Brooklyn, New York, in 1868. This enthusiastic outdoorsman formed his first company in 1871, Partner Gilchrest, Flint and Company and joined the W. R. Grace and Company in 1872. From 1877 to 1879, he served as the Chilean consul in New York City and later as the consul general of both Nicaragua and Costa Rica. In 1885, he joined the company formed by his father in 1837, Flint and Company, which owned and managed ships. He continued to play a role in Latin American affairs, as a smuggler of guns to two opposing sides in a war and as a delegate to the International Conference of the American Republics, 1889–1890. He served on other international commissions which studied banking relations between Latin America and the United States. In 1893, he created a battle fleet for Brazil and sold a ship to Japan during its war with China in 1895. The following year in San Francisco he formed the Pacific Coast Clipper Line which shuttled between that city and New York. During the Spanish-American War he acquired vessels for the U.S. Navy. During the Russo-Japanese War (1904–1905), Flint sold Russia twenty submarine and torpedo boats and represented Moscow in some diplomatic negotiations in Turkey.

Known as the "father of trusts," Flint, a man of boundless energy, also found time to create the American Chicle Company (which made chiclets), the U.S. Rubber Company, and, in 1911, C-T-R. He was quite well known in business circles on both sides of the United States. Although he began his career in international trade, his great mark on American industry was obviously made through his consolidations. His idea was simple: he took companies with compatible products that were small and weak, and from them he created stronger firms; he did this successfully even though trusts were then unpopular. Then in 1900 he swept together several time-clock manufacturers into the International Time Recording Company, headquartered in Binghamton in upstate New York. By 1910, this company was the dominant force in time clock markets. In 1901, Flint created the Computing Scale Company in Dayton, Ohio. This second company did not do as well as Time Recording, and so in 1911 he sought to merge it with others to strengthen its position. He decided to take another firm, called the Tabulating Machine Company (Hollerith's firm which was then experiencing capital shortfalls and stiff competition from other tabulating firms, especially Powers), and merged it into his other two concerns. He added a fourth venture, the Bundy Manufacturing Company (Endicott, New York), to form C-

T-R. To complete the transaction which would result in considerable profit through stock manipulations, Flint hired a well-respected businessman to run the company, George W. Fairchild.** And, of course, he brought in Watson as general manager, allowing him to operate the company as he saw fit.

For further information, see: Geoffrey D. Austrian, *Herman Hollerith: Forgotten Giant of Information Processing* (New York: Columbia University Press, 1982); C. R. Flint, *Memories of an Active Life* (New York: Putnam, 1923); Joel Shurkin, *Engines of the Mind: A History of the Computer* (New York: W. W. Norton, 1984); Robert Sobel, *IBM: Colossus in Transition* (New York: Times Books, 1981).

FORRESTER, JAY WRIGHT (1918–). Wright was the father of the WHIRLWIND* computer, the largest processor built up to the mid-1950s. It was constructed on behalf of the U.S. government at the Massachusetts Institute of Technology (MIT) and represented one of the major early digital computer* projects. WHIRLWIND was one of the first computers built by a large development team—a normal course of events for the creation of new computers after WHIRLWIND. The management necessary to complete this project, along with the technologies that grew out of it, represented a milestone in the history of data processing. To a large degree, credit for the results could be given to Forrester.

Forrester was born in Anselmo, Nebraska, and completed his B.S. degree in engineering at the University of Nebraska in 1939. He moved to MIT to continue his studies in electrical engineering and almost immediately began working on government-related research projects. These involved feedback mechanisms, analog equipment, and mechanical and electrical devices, almost all of which were entirely the preserve of the Servomechanisms Laboratory. World War II provided the opportunity for considerable work at MIT on behalf of various military branches. Forrester quickly gained experience working on projects, one of the earliest of which was a fighter-director radar control system for the U.S. Navy. This project gave him the technical and administrative background needed for his later appointment by the U.S. Navy as manager of project WHIRLWIND.

The first step toward WHIRLWIND was taken when Forrester began working on the Airplane Stability and Control Analyzer program, the nucleus of WHIRLWIND, during World War II. The objective of this program was to build a flight simulator complete with calculating controls that could be used for testing new aircraft and training exercises. These involved feedback mechanisms, electrical engineering, and the use of nascent qualities of computation. The project grew in both size and complexity during 1944. In 1944 and 1945, Forrester converted the naval project into WHIRLWIND, a general-purpose digital computer.* He was head of the project for nearly twelve years.

Forrester retained overall responsibility for the management of the WHIRLWIND from 1944 to its conclusion. Specifically, he recruited engineers, managed an ever growing budget, negotiated contracts for funding with the U.S. Navy (later with the U.S. Air Force), dealt with MIT's administration, and ran his laboratory. Early in the project, he and his staff decided to broaden the effort to develop a general-purpose computer rather than simply a flight simulator. The necessary negotiations and canvassing for funds and support for this new direction fell largely on his shoulders.

Although Forrester had to leave many of the details concerning the design of specific components to others, he had personal experience with many of the technical issues involved in the development of this computer. From the time he entered graduate school at MIT, he began work with various analog devices (MIT's strength in the late 1930s) and thus appreciated the potential of computational equipment, including digital devices. That explains why, in 1945, he was able to conclude that a switch from an analog-based system to digital was possible and why. This switch was his most important technical decision because it opened the door to many new technologies and ultimately a very large and sophisticated digital computer.

In the two years immediately following World War II, Forrester's project grew larger and more complex. His most difficult years were no doubt 1947 through 1949 when the Office of Naval Research (ONR)†—the sponsor of the project—increasingly complained about the slowness with which the work was progressing, about the sacrifice of mathematical content to engineering concerns, and about its increasing cost. Ultimately, WHIRLWIND would be the most expensive computer of that era. This project was ongoing during the years when research budgets at the Navy Department were being cut back. Government pressure on MIT forced Forrester's management into scrutinizing his project. Yet he persisted and development progressed.

In November 1950, the U.S. Air Force supplemented ONR's funding of the project with additional funds and thus opened a new era for MIT. From then on other computer-related projects came to MIT involving large national defense work. MIT responded by reorganizing various computer-based development projects; one consequence was the establishment of the Lincoln Laboratory† and the Digital Computer Laboratory for Project WHIRLWIND. The U.S. Air Force provided a new lease on life for engineers. The Servomechanisms Laboratory became part of the Digital Computer Laboratory established in September 1951, and Forrester became the laboratory's first director. Construction of the computer continued, and early tests were conducted. The computer finally became operational in 1951; that followed nearly eight years of difficult work and at an expense of nearly $5 million. All previous computer projects supported by the U.S. government had cost less than $1 million. The closest in price had been the Harvard Mark III (about $695,000) which was completed soon before WHIRLWIND.

In 1951, after initial tests, the U.S. Air Force was convinced that WHIRLWIND could play a major and specific role in a national defense network. Therefore, it enthusiastically embraced the expensive project. During the first half of the 1950s, the computer became operational for the U.S. Air Force. Forrester saw it to completion and actually began work on WHIRLWIND II.

Then in 1956 he gave up the project, joined the teaching faculty at MIT, and turned his attention to the study of industrial and engineering organizations. He no longer administered computer development projects; his role in the history of data processing had come to an end.

Forrester personally participated as the inventor of coincident-current magnetic core memory,* which became one of the basic technologies imbedded in computers throughout the 1950s and early 1960s. In effect, he and his staff invented random-access storage—a basic feature of all computers down to the present. His staff also developed marginal checking, which allowed a computer to identify deteriorating components before they failed entirely to perform. He encouraged and supported the extensive use of cathode ray tube (CRT) displays, along with advanced programming techniques.

Forrester also trained a generation of engineers at MIT. Many of those who worked on WHIRLWIND went on to invent other computers and to establish their own companies. Robert R. Everett,** for example, helped to establish the MITRE Corporation, while Kenneth Olsen set up the Digital Equipment Corporation (DEC)†. MIT's engineers worked for these two companies, as well as for International Business Machines Corporation (IBM),† Burroughs, and other computer manufacturing firms.

Until Forrester undertook the project, MIT's role in data processing had been limited to analog devices, some research on feedback mechanisms, and radar. The size, complexity, and volume of dollars involved with WHIRLWIND forced MIT into the mainstream of computing research at American universities. As a consequence, it built up the administrative and technical support structure which made possible many developments at MIT in the 1950s, ranging from defense systems growing directly out of WHIRLWIND to real-time computing networks on campus to the near-creation of artificial intelligence* as an independent field of study—all within a decade of WHIRLWIND becoming a usable machine.

The project energized a sufficiently large number of people and organizations (MIT, the U.S. Air Force, and the U.S. Navy initially and such companies as DEC and IBM, for example) to help advance and speed up the development of computers in the early 1950s in the United States. Forrester had also shared with his staff the complexities of managing such a project. That in turn gave his people the courage and experience necessary to develop their own projects and to accept their own management challenges in subsequent decades. Despite continuous problems with the ONR (and a temper that would cause him to turn on an engineer if something did not develop as he wanted), Forrester had the vision and staying power to turn a small project into a major contribution (i.e., the development of the modern digital computer). Consequently, he is considered one of the giants of early digital computing.

For further information, see: Kent C. Redmond and Thomas M. Smith, *Project WHIRLWIND: The History of a Pioneer Computer* (Bedford, Mass.: Digital Press, 1980); Karl L. Wildes and Nilo A. Lindgren, *A Century of Electrical Engineering and Computer Science at MIT, 1882–1982* (Cambridge, Mass.: MIT Press, 1985).

FORSTER, JAMES FRANKLIN (1908–1972). Forster was a high-level executive with Remington Rand† and Sperry Rand Corporation† in the 1950s and 1960s. He became chairman of the board of Sperry, a major supplier of mainframes to the data processing industry in the 1960s. During the 1950s, Remington Rand had introduced the UNIVAC.*

Forster was born on May 20, 1908, in Higginsville, Missouri. He graduated from the U.S. Naval Academy in 1930 and, after a tour of duty with the U.S. Navy, went to graduate school at Harvard University. He completed an M.B.A. there in 1936. From 1936 to 1939, he was a staff accountant for Arthur Anderson and Company. He then joined Remington Rand, remaining there until his death on July 1, 1972. Forster rose through management ranks to become a director in 1964, a position he held until he died. He was also president of Sperry Rand (1965–1972) and chairman and chief executive officer (1967–1972). One of his earlier management positions within the firm included working in the Sperry Gyroscope Company between 1939 and 1941. He next worked as treasurer and executive vice-president of Vickers, Inc., located in Detroit, Michigan, from 1941 to 1964. He was president of the Univac Division between 1964 and 1966.

Under his presidency, Sperry's gross revenues grew from $1.279 billion to over twice that amount in the 1970s. The Univac Division's revenues increased from $635 million in 1970 to over $2 billion by 1980. While president of the Univac Division, Forster had to respond to International Business Machines Corporation's (IBM's)† S/360* announcement with new products, including a series of processors known as the 1100, the company's large computer offering of the late 1960s and 1970s. In the fall of 1971, Radio Corporation of America (RCA)† decided to stop marketing computers and sold its data processing operations to Sperry. Forster attempted to merge the two sets of products as a means of enhancing his company's share of the computer market. The process was not completed before he died.

For further information, see: Franklin M. Fisher et al., *IBM and the U.S. Data Processing Industry: An Economic History* (New York: Praeger Publishers, 1983); Katharine D. Fishman, *The Computer Establishment* (New York: Harper and Row, 1981); Nancy Stern, *From ENIAC to UNIVAC: An Appraisal of the Eckert-Mauchly Computers* (Bedford, Mass.: Digital Press, 1981).

FOURIER, JEAN-BAPTISTE-JOSEPH, BARON (1768–1830). This brilliant French mathematician developed a theory of heat that relied on mathematics to describe physical phenomena. His work contributed to the development of mathematical procedures for defining many complex physical occurrences. That knowledge encouraged scientists in the nineteenth and twentieth centuries to find faster and better ways to perform large numbers of calculations for such common uses of the computer as wave and tidal analyses and weather prediction. All of these applications motivated early builders of computers to refine the calculators of their day. The requirement to perform large and complex calculations thus played an important role in the development of the modern computer.

Joseph Fourier was born on March 21, 1768, in Auxerre into a poor family in which his father was a tailor. By the time Fourier was fourteen years old and attending a local school run by Benedictine monks, he had already displayed considerable talent and interest in mathematics. He was also a gifted young writer of religious sermons. He continued his studies, became a teacher of mathematics in his home town, and, in 1794, when the École Normale was established to train instructors, he enrolled. The following year he taught at the school and soon also at the École Polytechnique where he met other important French mathematicians. In 1798, he accompanied Napoleon and some French scientists to Egypt where he conducted research on Egyptian history until 1801. He also advised the French government on antiquities and diplomatic issues. While in Egypt, he served as secretary of the Institut d'Egypte (1798–1801). Upon his return to Paris, he edited the publication of a twenty-one volume study of Egypt, *Description de l'Egypte* (Paris, 1808–1825), which summarized the French studies on Napoleon's expedition. That publication established Egyptology as an independent field of study. While living in Grenoble, Fourier served as the local prefect (1802–1814) and continued to study mathematics and Egyptian culture.

In 1815, the Baron Fourier (Napoleon granted him the title in 1809) became the head of the Statistical Bureau of the Seine, giving him the opportunity to continue work in mathematics. He became a member of the Academie des Sciences in 1817 and its secretary in 1822. In 1826, he was elected to the Académie Française and to the Académie de Médecine, both for his Egyptian studies.

Fourier's major work, the study that would most influence mathematics and consequently the development of computers, was published in 1822, *Théorie analytique de la chaleur*. (The first English edition appeared in 1878 as *The Analytical Theory of Heat*.) He described heat conduction in two dimensional objects, relying on mathematics. Thus, Fourier illustrated what happened to thin sheets of material by way of a differential equation:

$$\frac{\partial u}{\partial t} = k\left[\frac{\partial^2 u}{\partial x^2} + \frac{\partial^2 u}{\partial y^2}\right]$$

u represented temperature, t a time at a point on a plane (represented by x and y), while k was the constant of proportionality which he titled diffusivity of the particular material in question. For the mathematician, the issue at hand was to identify the temperature in a conducting plate of material if time was 0 and the temperature that at the boundary and at all points on the plane. He developed a series of sines and cosines to offer a solution to the problem in a single dimension, which is now known as the Fourier series. This is best reflected by a sample of his work:

$$y = \tfrac{1}{2} a_0 + (a_1 \cos x + b_1 \sin x) + (a_2 \cos 2x + b_2 \sin 2x) + \ldots.$$

This work made an important contribution to mathematics and was well received by the scientific community at large.

Many of the early designers of computers had either been trained as mathematicians or needed Fourier's work in conducting research in physics. Thus, they understood his ideas when they began building their machines. Even in his own time, Fourier's concepts were understood by Charles Babbage,** an important inventor of analytical engines who was known to other builders of computational machines of the period. In an attempt to challenge, correct, and confirm Fourier's work, mathematicians in the late 1800s and early 1900s advanced the work on differential equations and related algebraic calculations while also focusing on the concept of real functions, a relatively new area of modern mathematics. Fourier died on May 16, 1830, in Paris.

For further information, see: J. B. J. Fourier, *Oeuvres de Fourier,* 2 vols. (Paris: n.p., 1888–1890); H. H. Goldstine, *The Computer from Pascal to von Neumann* (Princeton, N.J.: Princeton University Press, 1972); I. Grattan-Guinness, *Joseph Fourier, 1768–1830* (Cambridge, Mass.: MIT Press, 1972).

FRIEDMAN, WILLIAM FREDERICK (1891–1969). Friedman invented a variety of cryptographic devices that relied on computer-like technology. In the 1930s, he described how mathematics could be used to break codes, and he encouraged the use of computational equipment. During World War II his work proved vital to the success of the Allied efforts, particularly in the Pacific.

Friedman was born on September 24, 1891, at Kishinev, Russia, and came to the United States in 1893. He completed his B.S. at Cornell University in 1914 and continued his study of biology the following year. From 1915 to 1918 he was director of the Department of Genetics at the Riverbank Labs in Geneva, Illinois. He also became director of the Department of Ciphers (1917–1921), where he first was exposed to the field of cryptology. Between 1921 and 1947, he served as the chief cryptoanalyst at the U.S. War Department in Washington, D.C. It was during this tenure that he made his most important contributions. He held a variety of positions while working for the government: chief of Signal Intelligence Service (1930–1940); director of communications research, Army Security Agency (1942–1949); chief of the technical division, Armed Forces Security Agency (1949–1950); chief of technical consultants (1950–1952); special assistant to the director of the National Security Agency (1953–1955); cryptologist at the U.S. Department of Defense (1947–1955); and consultant at the Department of Defense (1955–1969). He also lectured and consulted for the Armed Forces Service Schools. He was a commissioned officer in the U.S. Army: first lieutenant, military intelligence (1918–1919); captain, Signal Corps (1924–1926); major (1926–1936); and lieutenant colonel (1936–1951).

Although code making and breaking had been around for thousands of years, during World War I governments applied mechanical and electronic means to conduct cryptography. The value of this approach encouraged various governments to pursue additional research on the subject following World War

I. Considerable work was done in this field in the 1920s and 1930s in Poland, England, Germany, Japan, and the United States. Friedman was the most important American worker in this field. Throughout the 1920s and 1930s, he invented a number of devices, and following World War II, he directed the efforts of others. In 1922, he published an important paper arguing the case for using mathematics in cryptography. This paper opened the door to the use of computer-like technology in this field. The paper had such an impact that by the early 1930s standard card-tabulating equipment was being rigged into systems to perform code making and breaking. In the 1930s, the government installed such equipment at Pearl Harbor, Corregidor, and Washington, D.C., to monitor Japanese activities.

Friedman established his reputation long before the war. In addition to his important paper of 1922, in 1924 he broke a coded message during the Teapot Dome case. That activity provided him with additional support to continue his work at the War Department; this work led to the use of card input/output equipment in the 1930s and, ultimately, to the heavy reliance on computers during World War II. During World War II, Friedman encouraged various government agencies to support research whose outcome was the general-purpose electronic digital computer.*

Friedman made his greatest contribution during World War II when the American government was desperate to break Japanese codes, and the British were working on Germany's communications systems. The Allies were able to break codes on both fronts largely through the brilliant efforts of a small group of scientists working in England and the United States almost independently of each other. This success was ultimately based on mathematics and computers. As early as 1920, U.S. cryptoanalysts broke a Japanese diplomatic code. In 1934, Japan began using a German machine called the Enigma* and continued refining its procedures, even using one of Friedman's inventions, the *Sigaba* (M–134–C), a device that relied on a system of rotors and printed out messages after decoding them. Friedman developed various models which came into common use in Europe during the late 1930s (the B–21 was bought by Sweden in 1926, the C–36 by France in 1936, and the M–209 was used by the U.S. Army). The Japanese modified and enhanced their own equipment while Friedman tried to keep up. Then in 1937, the Japanese began using a machine called the 97-*shiki-0-bun In-ji-ki* whereby a user typed the message to be transmitted in code, and the machine coded it and printed out a coded copy on another typewriter. This device, nicknamed *Purple* by the Americans, was the cryptographic workhorse of the Japanese on the eve of World War II.

During 1940, Friedman devoted long hours, month after month, to cracking *Purple*. Finally, on September 25, 1940, he began deciphering part of the Japanese messages. By early 1941, he had built several copies of the *Purple* which were shared with the British. By trial and error and, finally, reconstruction of the Japanese rotor system, he proved successful. The intelligence gained from breaking the Japanese code was massive; the U.S. government, for example, learned of Japan's plans to attack Pearl Harbor. Other military and secret communications were studied throughout the war with the same open-book success achieved with Enigma* and Ultra* in reading German communications.

In 1956, the U.S. Congress gave Friedman an award for his work; ten years earlier he had received the Medal of Merit; and in 1955, he was awarded the National Security Medal. His first award for work done on *Purple* was the U.S. War Department's Exceptional Service Award (1944). He wrote a number of articles, papers, and monographs on crytography and won the literary prize of the Folger Shakespeare Library (1955) for his book that explained Shakespearean ciphers. He died on November 12, 1969.

For further information, see: W. F. Friedman, *The Index of Coincidence and Its Application in Cryptography* (Geneva, Ill.: Riverbank Laboratories, 1922); Józef Garliński, *The Enigma War* (New York: Charles Scribner's Sons, 1979).

G

GALLER, BERNARD AARON (1928–   ). Galler was both a scientist and leader in the data processing industry, contributing to the development of programming languages* during the 1950s and 1960s and serving in key positions within the industry's own organizations.

Galler was born and raised in Chicago where he attended the University of Chicago, completing his B.S. in 1947. In 1949, he earned an A.M. from the University of California at Los Angeles (UCLA) and a Ph.D. in mathematical logic from the University of Chicago in 1955. He then joined the faculty at the University of Michigan where he has remained. By the early 1980s, he had risen to the rank of full professor of Computer and Communication Sciences and served as associate director of the Computing Center. Between 1968 and 1970, he was the president of the Association for Computing Machinery (ACM),† one of the key organizations of the entire data processing industry. He has also served on the Board of Governors of the American Federation of Information Processing Societies (AFIPS).† His key industry-wide role was within the ACM where he held a variety of positions during the 1960s: vice-president, regional representative to the National Council, chairman, ACM National Lectureship Program Committee, and member of the editorial staff that helped produce the large number of publications from the ACM.

In Galler's own research and writing, he has concentrated on programming languages. Among his contributions is the Michigan Algorithmic Decoder (MAD),* a programming language for numerical scientific problem-solving developed in 1959. MAD became a popular language at many American universities during the 1960s but was hardly used by American companies or government agencies. For its day, it was a popular and powerful language.

For further information, see: Bernard Aaron Galler, *The Language of Computers* (New York: McGraw-Hill Book Co., 1962) and *A View of Programming Languages* (Reading, Mass.: Addison-Wesley Publishing Co., 1970); Jean E. Sammet, *Programming Languages: History and Fundamentals* (Englewood Cliffs, N.J.: Prentice-Hall, 1969).

GATES, WILLIAM H. (1955–). This software* engineer became a well-known celebrity in the U.S. data processing industry when, before reaching the age of thirty, he developed the most widely used operating systems* for microcomputers. Gates was the founder of Microsoft Corporation,† which by the mid-1980s was the largest vendor of programs for such computers in the industry. Along with Steven P. Jobs,** of Apple Computers,† to mention only one other contemporary, Gates represented dozens of bright, technically skilled individuals who developed data processing products and established firms to sell them. Many of these software and hardware engineers became millionaires while still very young men.

Gates, born in 1955, was raised in Seattle, Washington, the son of a prominent local lawyer. He began his college education at Harvard University but dropped out in 1975 before completing his degree in order to spend his time writing programs. By the mid-1980s, over two million micros used his software as their operating systems, along with programming languages* and other miscellaneous products he had developed. Gates was first exposed to computers while in the seventh grade, and along with a friend, Paul Allen, he worked on various projects throughout high school. These self-admitted hackers broke into existing computing networks, including those run by Burroughs Corporation† and Control Data Corporation (CDC).† While in high school they formed a company called Traf-O-Data and used the Intel† 8008 microprocessor to construct a device to control traffic patterns in Seattle. They sold the system to the city for $20,000. Gates was only fifteen years old.

Gates dropped out of high school for one year to work with his friend for TRW,† made $30,000 that year, and bought a speedboat. At this point Gates was seventeen.

In 1975, Gates and Allen became excited over an article in *Popular Electronics* which described a computer kit from Altair that could be acquired for $350. The device, called MITS, was being built by a company in Albuquerque, New Mexico. That article excited a number of young people interested in computing, including Jobs. Allen, then working for Honeywell Corporation,† urged Gates to join with him in developing a compiler for the programming language BASIC* to operate on this machine. They wrote a draft of the compiler during February and March 1975 in Gates' dorm room at Harvard, contacted MITS to see if the company was interested in their work, and then demonstrated it to the firm's engineers. MITS saw the value of its machine now enhanced with a programming language that was relatively easy to use. That combination made it possible for millions of people with little or no knowledge of computers to write programs in the next decade.

Encouraged, the two men moved to New Mexico and began Microsoft in a hotel room. In 1977 MITS, one of the first companies to introduce a microcomputer product, went out of business, and the two software writers moved back to Seattle. Gates was now twenty-one years old. During the next ten years, microcomputing exploded into its own growth industry. Millions of machines were built by over 150 companies, while dozens of software firms emerged to supply them with software. Microsoft emerged as the largest and most important of these Companies, and Gates ran the enterprise. The greatest stimulus to the firm came when the International Business Machines Corporation (IBM)† decided to enter the microcomputer market and selected Microsoft's operating system, known as the Microsoft Disk Operating System (MS-DOS), or DOS, as the operating system for the Personal Computer (PC). In August 1981, IBM announced its PC with Microsoft's operating system. By 1983, after the IBM PC had become the industry standard, Gates licensed his software to over 100 vendors, making his the dominant one in the world. In the early 1980s, Gates expanded the company's offerings to include M-BASIC, Microsoft Word, Multiplan, Adventure, Flight Simulator, and Olympic Decathalon. Adventure and the Flight Simulator became two of the most popular games for use on a PC. By 1984, his company had the broadest set of successful products for micros in the industry.

The size of Gates' company grew from two people to thirty-two by the time IBM approached him in 1980. In early 1984, the firm employed 620 people, enjoyed annual sales of $100 million, and had an estimated profit of some $15 million. This privately held company was largely the property of Bill Gates.

Gates has been characterized as a brilliant programmer. He has also exhibited entrepreneurial instincts that have been critical to the establishment of companies to sell products with no prior market history. Furthermore, his programs have set standards for the software industry as a whole. By 1982, he was already commenting publicly that future technological breakthroughs in the data processing industry would be in the field of software and that he intended to lead the charge. He began moving in the direction of developing software that could make more decisions about its own characteristics, what he termed "softer software." He envisioned programs in the near future that would have some of the qualities attributed to artificial intelligence.* These included the capability of taking rules, employing a reasoning engine, and then deriving new rules and data. Hence, Gates had not only become a leading manufacturer of software but had also defined its future direction.

For further information, see: Paul Freiberger and Michael Swaine, *Fire in the Valley* (Berkeley, Calif.: Osborne/McGraw-Hill, 1984); Suan Lammers, *Programmers at Work, 1st Series* (Redmond, Wash.: Microsoft Press, 1986); Robert Levering et al., *The Computer Entrepreneurs* (New York: New American Library, 1984).

GILL, STANLEY (1926–1975). Gill worked with Maurice V. Wilkes** and others at Cambridge University to build EDSAC,* and he also worked on the Pilot ACE.* Both computers were early British research efforts resulting in the construction of stored-program machines in Great Britain. With Wilkes he also co-authored one of the first textbooks on programming.

Gill was born in 1926 and studied mathematics at St. John's College at Cambridge in the mid-1940s. He joined the Mathematics Division of the National

Physical Laboratory in 1946 and worked on the Pilot ACE. In the fall of 1949, he was back at Cambridge as a research student working for Maurice V. Wilkes at the University Mathematical Laboratory. He completed his Ph.D. in 1952, writing his thesis on the EDSAC. Between 1952 and 1955, he served as a Research Fellow at St. John's College, and between 1953 and 1954, he was assistant professor at the University of Illinois, teaching computer science. From 1955 to 1964, he was head of computing research in the Computer Department of Ferranti, Limited,† a company that built and sold commercial versions of the EDSAC, one of the earliest stored-program computers built in Great Britain. During the 1960s, Gill periodically taught computer science at the University of Manchester and at the University of London.

Gill helped mold the direction of British computing. In addition to his early work in the development of digital computers* and, later, in the manufacture of commercial machines, he advised the Ministry of Technology (1966–1970) and was president of the British Computer Society (1967–1968).

Gill's most important publication, cited below, became the first widely read book on computing during the early 1950s. He died on April 5, 1975.

For further information, see: the introduction by Martin Campbell-Kelly to the reprint of Gill's book, co-authored with David John Wheeler and M. V. Wilkes, *The Preparation of Programs for an Electronic Digital Computer, with Special Reference to the EDSAC and the Use of a Library of Subroutines* (Los Angeles: Tomash Publishers, 1982, originally Cambridge, Mass.: Addison-Wesley Press, 1951).

GLUSHKOV, VICTOR MIKHAYLOVICH (1923–1982). Glushkov, a leading Soviet computer scientist, was instrumental in the development of computer languages and techniques for the design of computers.

Glushkov was born on August 24, 1923, in Rostov where he studied mathematics at Rostov University. He completed his doctoral dissertation in 1955 on the theory of graphics. In 1956 he joined the Ukrainian Academy of Sciences in Kiev and became head of its computer center. In 1962, his department was renamed the Institute of Cybernetics of the Ukrainian Academy of Sciences, and he was named director, a position he held until his death. Glushkov's institute proved to be a major force in introducing computer technology to the Soviet Union. His personal research spanned a broad range of data processing-related areas from abstract algebra to the theories of automata, computer architecture, and programming languages.* He helped to design Soviet computers and teleprocessing networks. Glushkov published two books: *Synthesis of Computing Automata* (1962) and *Introduction to Cybernetics* (1964). His most important contribution to software* was the development of ANALYTIC, which was more sophisticated than a similar language called FORMAC and was used to write analytical algebra in machine-(computer-) readable form. His primary concern in the area of applications involved those that would help the Soviet economy and in process control. During the late 1960s and 1970s, he participated in many international congresses and encouraged visits to his institute by European and American data processing scientists. He died on January 30, 1982.

For further information, see: Heinz Zemanek, "Eloge: Victor Mikhaylovich Glushkov," *Annals of the History of Computing* 4, No. 2 (April 1982): 100–101.

GOLDSTINE, HERMAN HEINE (1913–). This mathematician and computer scientist is one of the best known of the early pioneers in the field of computing. Goldstine directed U.S. government aid to support the development of the ENIAC,* the first electronic digital computer,* and he wrote one of the most widely known and useful technical histories on computers. He also influenced the development of other computers, including those designed at International Business Machines Corporation (IBM).†

Goldstine was born on September 13, 1913, in Chicago. He studied at the University of Chicago, completing his B.S. in 1933, his M.S. in 1934, and his Ph.D. in 1936, all in mathematics. He stayed at Chicago first as a research assistant in 1936–1937 and then as an instructor. After moving to the University of Michigan in 1939 as an assistant professor, he taught mathematics, serving until the start of World War II when he became involved in war-related activities. Soon after he joined the Army, he was transferred to the Ballistic Research Laboratory (August 2, 1942) as a first lieutenant to work on ballistic computations. The primary mission of this organization was to produce firing and bombing tables along with other gun control information. During the war, attempts were made in the United States to automate these processes, which directly motivated the development of the earliest electronic digital computers.* Like others at the laboratory, Goldstine sought to produce these tables quickly; the method then in vogue—using female graduates of American colleges with desktop calculators—had proved to be too tedious

In September 1942, Goldstine heard about computer-like research being done at the University of Pennsylvania's Moore School of Electrical Engineering,† which at that time was considered to be second only to the Massachusetts Institute of Technology (MIT). Goldstine used his connections with the Ballistics Laboratory to persuade the government to sponsor what amounted to the development of the ENIAC, and later a follow-on computer, in order to further the creation of firing tables quickly. As a result of his encouragement and influence, a series of computers were built that culminated in the UNIVAC* series during the 1950s; this series riveted the public's attention and showed the potential of computers. He participated in the planning sessions held at the Moore School in the last two years of the war while the ENIAC was being built.

After the war, Goldstine joined the Institute for Advanced Study at Princeton, New Jersey, working for John von Neumann** whom he had met and worked with during the war on the ENIAC project. Goldstine remained there from 1946 to 1955, during which time an electronic digital computer was built with the capability of stored programs (called the IAS Computer*). Goldstine was acting project director from 1954 to 1957 and was appointed a permanent member of the Institute in 1952. In 1957, he joined IBM as the director of its Mathematics Sciences Department, and during the 1960s, he consulted on various computer-related design projects, eventually becoming an IBM Fellow. In 1972, he published *The Computer from Pascal to von Neumann,* an important technical history of computers in which he provided a strong defense of von Neumann's role in the development of the modern computer and discussed his own part in the evolution of electronic computers during the 1940s and 1950s.

For further information, see: H. H. Goldstine, *The Computer from Pascal to von Neumann* (Princeton, N.J.: Princeton University Press, 1972); Joel Shurkin, *Engines of the Mind: A History of the Computer* (New York: W. W. Norton and Co., 1984).

GOODMAN, RICHARD (1911–1966). This British mathematician publicized the work of computer scientists in the 1950s and 1960s, particularly in Great Britain. Trained as a mathematician and a professor at Brighton College of Technology in East Sussex, Great Britain, Goodman took his first major step in the field of computing in 1958 when he organized the Conference on Automatic Programming at his school. The following year he organized the Automatic Programming Information Center (APIC) which became a forum for other computer scientists in Great Britain. He edited the *APIC Bulletin,* which carried articles on computing and programming in general, and he actively fostered support for ALGOL* in the early 1960s as a universal language. Goodman is perhaps best known for publishing the *Annual Review in Automatic Programming,* a collection that was continued after his death under the editorship of Mark I. Halpern and Christopher J. Shaw. Goodman launched another series of publications called Studies in Data Processing, also under the auspices of APIC. This series published monographs on key European languages, including ALGOL. Both series are still being published today.

Goodman took a particular interest in ALGOL which, in the early 1960s, promised to be a universal language supported by most European data processing organizations. His support, while it came earlier than that of many others, was a very European reaction to the language, reflecting the views of his colleagues. Goodman died in 1966 while serving as chairman of the Technology Department of Computing, Cybernetics, and Management at Brighton College.

For further information, see: Richard Goodman, *Annual Review in Automatic Programming* (New York: Pergamon Press, 1960–).

GORE, JOHN K. (1845–1910). This life insurance actuary and executive at Prudential Insurance Company invented a series of card-tabulating devices in the late 1800s and early 1900s. After working long hours manipulating data manually, he began to search for mechanical methods to handle such tasks.

Gore was born in 1845 and became a teacher. At the age of twenty-eight (in

1892), he resigned his teaching position at the Woodbridge School in New York to take a position with Prudential as a life insurance actuary. At the time, Prudential was rapidly becoming a major insurance company and, like other insurance firms, railroad companies, and government agencies, depended on fast and accurate manipulation of growing amounts of information.

Gore quickly recognized that Prudential's many statistical departments needed more efficient ways to manipulate data. When he became a manager, he taught his employees to count cards by listening to the noise they made as they were riffled under their thumbs. Next, he began designing hardware to handle data cards. During the early 1890s, he and his brother-in-law (a mechanical engineer) experimented with a series of machines which, beginning in 1895, were installed at Prudential in Newark, New Jersey. Prudential used variations of these machines as recently as the late 1930s.

One of Gore's machines was a multiple-key punch that automatically ejected punched cards when new ones were brought into position for punching. It looked like a primitive typewriter. He also invented a sorter that looked like a model of a metal tower on a table. Each of its four tiers contained ten compartments to house cards. These cards revolved over pins into selected categories or classifications preset by a user. With such equipment, clerks could tabulate insurance premiums relatively quickly. Each card had the name of the insured, and round holes were punched to trap numerical information. In many respects, these cards looked like those which International Business Machines Corporation (IBM) could introduce with its System/32 minicomputer over half a century later.

Gore achieved great success at Prudential, rising to the vice-presidency of the company. He publicized his machines, which he believed had a competitive edge over operations at other insurance companies because they increased the accuracy of data while holding down costs. Decades later, these same reasons would be used to acquire data processing equipment. His early successes also established Prudential's tradition of relying on technology, a tradition that persists today. This company, for instance, became an extensive and early user of digital computers.*

For further information, see: Charles Eames and Ray Eames, *A Computer Perspective* (Cambridge, Mass.: Harvard University Press, 1973).

GRANT, GEORGE BARNARD (1849–1917). Grant was an American inventor of difference engines* and calculators in the tradition of Charles Babbage.**

Grant was born in 1849 and grew up in Maine. He attended Harvard College, where he completed his B.S. degree in 1873 and became interested in calculating devices and began designing them. While at Harvard, he also learned about Babbage's efforts in Great Britain to build a difference engine and expanded these efforts, even applying for patents while still in school. Grant first designed and built a calculator and printer as an extension of Babbage's work. Next, he attempted to construct a calculator that was better than the most widely used machine available in the United States: the arithmometer* made by Charles Xavier Thomas. Since Grant's work built on Babbage's (whose efforts were not well known in the United States), Grant's ideas appeared to be new.

Grant exhibited his first machine at the 1876 Centennial Exhibition. The device was 8 feet long and 5 feet high, weighed 2,000 pounds, and was made up of 15,000 parts. Although obviously impractical, he may have established the record for having built the largest calculator in history. He subsequently designed arithmetic calculators and adding machines. The most widely known of his devices were the Barrel, also known as the Centennial, and the Rack and Pinion. The Pinion was a mild commercial success; he actually sold 125 copies.

Grant's rack and pinion device was reliable, proving that calculating machines could be usable in the American economy. It was also relatively easy to use. The user entered figures into the machine on its wheels which were mounted on a drum much like a child today would "enter" numbers on a combination lock by rotating wheels with numbers etched on them until they were in a precise position. The user then turned the handle on the right of the shaft interconnected to another with numbered wheels connected by gears. The results appeared on wheels mounted on the second shaft. It was not a perfect machine, even though it was obviously much smaller than the monster he exhibited in 1876. Multiplication still required various steps, and the wheels still had to be jiggled to get them to show the correct numbers for calculation. Yet *his* gears worked; Babbage's did not.

By the early 1880s, the technical manufacturing problems that had frustrated Babbage had been solved. New methods of calculation created by other inventors also contributed to the general availability of useful calculators, particularly adding and subtracting machines.

Grant made money indirectly out of his interest in calculators. As part of the effort to design and build them, he learned a great deal about engineering techniques, especially about gears. In time, he manufactured these gears. During his fruitful life, he established the Grant Gear Works, the Philadelphia Gear Works, and the Boston Gear Works, and he rightfully gained the title of father of the American gear-cutting industry.

For further information, see: Christopher Evans, *The Making of the Micro: A History of the Computer* (New York: Van Nostrand Reinhold, 1981).

GREEN, JULIEN (1924–). This specialist in programming languages* and operating systems* helped develop systems that ran on the key International Business Machines Corporation (IBM)† computers of the 1950s and early 1960s, including some for the System/360* set of computers.

Green was educated at Columbia University, receiving his B.S. in 1949 and his M.S. in 1950. In June 1953 he worked as a development engineer for the Statistical Methods Department of General Electric (GE)† at its General Engineering Laboratory at Schenectady, New York, and remained there until

June, 1957. While working for GE, he programmed an IBM 650 to do work for other engineers.

In June 1957 Green joined IBM, remaining with that firm until May 1965. His years with IBM were also devoted to programming projects. His first assignment, for example, was as a programming specialist with the Data Systems Division, the manufacturing home of IBM's large computers. Between June 1957 and May 1960, he worked with the IBM 610 general-purpose board which was used to simulate stored-program logic on a 610. He participated in the development of the IBM 305 RAMAC Assembly program, the IAL and ALGOL* compilers that ran on the IBM 709/7090 computers, and worked with SHARE† on other ALGOL-related projects. In May 1960, he was named manager of Advanced Programming Development within the General Products Division (GPD), remaining in that capacity until March 1964. While with GPD, he developed XTRAN, a programming language that could implement systems programs. The language was used on IBM's 709/7090/7094, 1410, and 1620—all important computers of the period. As a byproduct of this project, GPD established a large advanced programming development staff, which in subsequent years produced compilers and advanced programming languages for IBM computer systems. The same organization also did research on the mathematics of sorting, machine-to-machine communications, algebra of flow, and optimization of object programs.

In June 1957 and continuing to May 1965, Green served as manager of Industry System Analysis within the Systems Development Division, moving from White Plains to Endicott, New York. In this new job he participated in the development of software* for the System/360. For instance, he produced the emulator which allowed programs written for use on the IBM 1620 to run on an IBM 360/30. This was the first emulator that used microprogramming within a computer and 360 machine language together for input/output tasks.

Green left IBM in May 1965 to become director of research for National Computer Analysts, Inc., in Princeton, New Jersey. In March 1966, he started his own firm: Julien Green Associates, Inc., which, in October 1968, became part of Scientific Resources Corporation. He remained with this organization until June 1970, as president of the portion he had established in 1966. His company consulted for Radio Corporation of America (RCA),† which at the time was active in the data processing industry, developing computers to compete with IBM's. Green participated in the specifications for the design of the Spectra 70/46 time-sharing system. Spectra was RCA's direct competitor to the S/360. He also worked on the IBM 360 Remote Job Entry System; designed software for Chase Manhattan Bank to do financial analysis of balance sheets; signed a contract with Burroughs Corporation† to design an airline passenger service system; and did work for Bendix Corporation and General Electric.

In June 1970, Green first joined RCA as manager of the Systems Programming Staff, then as manager of Control Systems and finally manager of Data Management Systems. He left the firm in October 1971 as a result of RCA's decision to get out of the computer business and its consequent dismantling of the Computer Division in which Green worked. Like thousands of other RCA employees, Green had to find employment elsewhere within the data processing industry after that surprise announcement. First, he joined First National City Bank of New York as assistant vice-president of Technical Audit and Review, and then, in March 1972, he went over to Equitable Life Insurance where he has continued to serve as Director of Development. All such job changes in jobs, though frequent, were widespread in the data processing community as engineers moved from company to company and from technical to business and managerial responsibilities. Product development people frequently moved from one organization to another, and so Green was not an exception. Green played an active role in industrywide organizations. He was a member of the original ALGOL Group which in the early 1960s studied the possibilities of a universal language, working on ALGOL 60. He was also active in American Federation of Information Processing Societies (AFIPS)† task forces, particularly in the 1960s. He was a member of the Association for Computing Machinery (ACM)† Standards Committee (January 1963–March 1964) and that committee's chairman in 1966. In the early 1960s, he was also an assistant editor for the Standards section of the *Communications of the ACM*, an important data processing industry technical journal. Green also published on ALGOL and XTRAN among other subjects.

For further information, see: C. J. Bashe et al., *IBM's Early Computers: A Technical History* (Cambridge, Mass.: MIT Press, 1985); Julien Green, "Symbol Manipulation in XTRAN," *Communications of the ACM* 3, No. 4 (April 1960): 213–214; Jean E. Sammet, *Programming Languages: History and Fundamentals* (Englewood Cliffs, N.J.: Prentice-Hall, 1969).

GRILLET, RENÉ (1600s). This French clockmaker built a calculating machine in the same century as Samuel Morland,** Gottfried Wilhelm von Leibniz,** Blaise Pascal,** and Wilhelm Schickard,** all of whom were part of a great burst of creativity in the seventeenth century in the fields of mathematics and science.

Little is known of Grillet's background, except that he was King Louis XIV's clockmaker and that he constructed a calculator which he took to local fairs, charging admission to see it. Some historians believe that Leibniz adopted Grillet's ideas when he began building his own mechanical calculator.

Grillet wrote an article about his device in 1678 in which he suggested that the machine combined the technology of Pascal's calculator with John Napier's** bones, a precursor to logarithms. Grillet's only illustration of the gadget suggests that wheels possibly stored data and performed some mathematical functions but historians are not sure. The papers of another French mathematician, Michel Chasles (1793–1880), contain a lengthy document suggesting how the machine was used. If that file can be believed, then Grillet's machine only did simple addition and subtraction. Napier's tables may also have been usable, but precisely how is not clear. To do an addition, a user simply lined up the numbers on the upper row of wheels and a second number to be added on the middle row of wheels, added the numbers in his head, and set the answer numbers up on a third row of wheels. It could handle eighteen digits, more than Morland's machine, but the Englishman's device performed calculations better with Napier's tables than with Grillet's. Its most interesting aspect, however, was that, if made small enough, it could be carried in a pocket without requiring the use of paper, quill, and ink. Perhaps it was an unrelated precursor to the pocket calculator of the twentieth century.

For further information, see: René Grillet, "Nouvelle machine d'Arithmetique," *Journal de Savans* (1678): 164–166; Michael R. Williams, *A History of Computing Technology* (Englewood Cliffs, N.J.: Prentice-Hall, 1985).

GRISWOLD, RALPH E. (1934–). This computer scientist was a principal developer of SNOBOL,* a string and list processing language that has been widely used for solving problems in artificial intelligence (AI),* manipulation of formal algebraic expressions, linguistic data processing, and picture processing. SNOBOL is one of the most widely used programming languages* in the field. In addition to SNOBOL, Griswold created Icon, which modernized and extended the functions of earlier SNOBOL languages.

Griswold was born on May 19, 1934, in Modesto, California, and completed his B.S. degree in physics at Stanford University in 1956. He served in the U.S. Navy before returning to Stanford where he completed his M.S. in 1960 and his Ph.D. in 1962. These last two degrees were in electrical engineering.

For the next decade, Griswold worked for Bell Telephone Laboratories† at Holmdel, New Jersey. Between 1962 and 1967, he was a member of the technical staff within the Programming Research Department where his earliest projects involved programming manipulation of symbolic expressions. This exposed him to string listing applications, which in turn led to the development of SNOBOL in collaboration with Ivan Polonsky and Dave Farber. Their first version of the language became available in 1963 and was followed by enhanced versions called SNOBOL2, SNOBOL3, and SNOBOL4 during the 1960s. Between 1967 and 1969, Griswold served as supervisor of the Computer Languages Research Group at Bell Labs and from 1969 to 1971, he managed the company's Programming Research and Development Department. His staff conducted research on programming languages* in general and ran a time-sharing system used for the preparation of documents.

In 1971, Griswold joined the faculty of the University of Arizona within the Department of Computer Science where he continued to do research on programming languages into the 1980s. During this period he published dozens of papers and continued writing books on programming languages, especially on SNOBOL and on Icon. Icon was part of his broader work in the general area of high-level facilities for manipulating nonnumeric data and document preparation through the use of computers. While teaching and doing research on programming languages, he also served as chairman of the University Committee on Computer Science (1971–1974), acting head of his department (1972–1974), chairman of the department (1974–1981), and chairman of the University Computing Committee (1978–1979). Between 1973 and 1981, he also produced six Ph.D. graduates in the general field of computer science, all of whom had research interests in programming languages.

For further information, see: Ralph E. Griswold, ''A History of the SNOBOL Programming Languages,'' in Richard L. Wexelblat, ed., *History of Programming Languages* (New York: Academic Press, 1981), pp. 601–660 and his books in the order of their publication, *The SNOBOL4 Programming Language* (Englewood Cliffs, N.J.: Prentice-Hall, 1971) with James F. Poage and Ivan P. Polonsky, *The Macro Implementation of SNOBOL4; A Case Study in Machine-Independent Software Development* (New York: W. H. Freeman, 1972), *A SNOBOL4 Primer* (Englewood Cliffs, N.J.: Prentice-Hall, 1973) with Madge T. Griswold, *String and List Processing in SNOBOL4; Techniques and Applications* (Englewood Cliffs, N.J.: Prentice-Hall, 1975), *The Icon Programming Language* (Englewood Cliffs, N.J.: Prentice-Hall, 1983) with Madge T. Griswold, and *The Implementation of the Icon Programming Language* (Princeton, N.J.: Princeton University Press, 1987) with Madge T. Griswold; Jean E. Sammet, *Programming Languages: History and Fundamentals* (Englewood Cliffs, N.J.: Prentice-Hall, 1969).

GROVES, LESLIE RICHARD (1896–1970). Groves was head of the Manhattan Atomic Development Project. He also had a distinguished military career and was a vice-president of Sperry Rand Corporation† when it became a major force in the general field of data processing.

Groves was born on August 17, 1896, in Albany, New York, attended the University of Washington (1913–1914) and the Massachusetts Institute of Technology (1914–1916), but graduated from the U.S. Military Academy at West Point (1918). During his career in the U.S. Army, he attended various schools, including Army Engineering School (1921), Command and General Staff School (1936), and the Army War College (1939). He rose to the rank of lieutenant general and retired from active service in 1948. Between 1942 and 1947, he managed the Manhattan Project, which spearheaded American military use of atomic energy, particularly of the first atomic bomb. In 1948, he joined the Remington Rand Company† which later became a division of the Sperry Rand Corporation. He joined the firm as a director and worked in that portion of the company that brought out the UNIVAC,* one of the first successful commercially available computers of the early 1950s.

Groves participated in his company's decision on February 1, 1950, to buy the Eckert-Mauchly Computer Corporation, which was in the process of building the UNIVAC. Groves, who was in charge of research for the company, was given responsibility for the new enterprise. Groves joined Remington as a director of research and became vice-president at the time that the acquisition was made. His primary task was to see that the UNIVAC was brought out as a product.

On March 31, 1951, his division learned that the first UNIVAC had been formally accepted by the U.S. Census Bureau, an agency with a long history of using information processing technology (dating back to just after the U.S. Civil War). Groves saw his division build forty-six UNIVAC's for the U.S. government alone, not to mention others for commercial customers, for a while making the term *UNIVAC* almost synonymous with the word *computer*.

Remington did not take advantage of that success, however. Subsequent models of the UNIVAC came out too slowly, and so, by 1957, and possibly as early as 1955 or 1956, International Business Machines Corporation (IBM)† had wrestled technological leadership, and hence marketing initiative, away from Groves' company. Remington never regained it. Thus, while at the start of the 1950s it appeared that the UNIVAC would make Remington, and later Sperry, a dominant force in the computer market, it lost, trailing in second or third place by the end of the decade. It kept that position on and off for the next two decades. Finally, in June 1986, Burroughs Corporation† bought controlling interest in Sperry's stock and merged that organization with its own, now called Unisys Corporation.

General Groves died on July 13, 1970, and was buried in Arlington National Cemetery. He is most remembered for the Manhattan Project rather than for his work with UNIVAC.

For further information, see: Nancy Stern, *From ENIAC to UNIVAC: An Appraisal of the Eckert-Mauchly Computers* (Bedford, Mass.: Digital Press, 1981).

H

HARTREE, DOUGLAS RAYNER (1897–1958). This British professor of physics encouraged the development of computer science in England during the 1930s and 1940s. He built a differential analyzer* similar to Vannevar Bush's** machine at the Massachusetts Institute of Technology (MIT), and he trained other British scientists in computers. Hartree was also the first scientist to solve problems in atomic energy using calculating devices.

Hartree was born on March 27, 1897, at Cambridge, England, where he attended Cambridge University, earning his B.A., M.A., and Ph.D. degrees in physics, and an M.Sc. from the University of Manchester. He served as a research fellow at St. John's College (1924–1927) and at Christ's College (1928–1929), both at Cambridge University. He then moved to the University of Manchester as professor of applied mathematics (1929–1937) and, later, as professor of theoretical physics (1937–1945). Hartree subsequently moved to Cambridge University in 1946 as professor of mathematical physics, where he stayed until the end of his life. He visited the United States several times, establishing a dialogue between scientists interested in computers in both countries. His most important U.S. visit was to MIT in the early 1930s where he saw Bush's computer. In 1948, he became acting chief at the Institute of Numerical Analysis of the U.S. Bureau of Standards at the University of California. He was also a visiting professor at Princeton University in 1955.

Hartree first became involved in scientific issues professionally during World War I. In 1916, he joined an antiaircraft experimental group at the Munitions Inventions Department in the Ministry of Munitions, serving until 1919. During World War II he was associated with the science research section of the Ministry of Supply.

In 1933, during his summer holiday from the University of Manchester, Hartree visited MIT with the intent of using Bush's differential analyzer. He ran a series of problems on the machine concerning the functions of mercury and was so pleased with the computer's efficiency that upon his return to England he built a copy of it with the help of his young protégé, Arthur Porter. Hartree first thought that Bush's machine looked like a large toy made out of children's building supplies. He built his own machine with Meccano parts costing approximately £20. With his pro-type, Hartree illustrated the principles of the Bush differential analyzer, making his own function as accurately as MIT's machine. He therefore built a fully completed version of the device and employed it in his work on wave mechanics. He expanded the machine at the University of Manchester. As a result of his own work and his encouragement of others to build similar devices, by 1939 there were four in Great Britain. His was at the University of Manchester, and the others were constructed at Cambridge University, Queen's University at Belfast, and the Royal Aircraft Establishment at Farnsborough. Between his own work on atomic theory (dating back to the early 1930s) and replication of differential analyzers, he singlehandedly increased British expertise in computer science on the eve of World War II. He was the father of modern British computing, responsible for training a whole generation of scientists and encouraging others. The most important of this new generation was probably Maurice V. Wilkes,** builder of the first stored-program computer in Britain incorporating design characteristics of American machines such as the ENIAC* and the EDVAC.*

Hartree's work with analog equipment and, later, his move to Cambridge University, put him in a position to continue his enthusiastic, decade-long campaign to foster computer development in Britain within scientific circles, government agencies, and even companies. Yet he was also an accomplished mathematician/physicist. Although recognized as the first British scientist to use a computer to solve differential equations, his concern for the subject of mathematics dates back farther. Early in his professional career, he attacked the problem of how to solve partial differential equations—the concern that eventually led him to use Bush's machine. The need for work in this area was already clear to many mathematicians by the 1920s. Ballistics research depended on new breakthroughs in mathematics and computation, whereas the study of variables such as pressure, air density, and temperature on airplane wings was another obvious need. One of the problems to be solved was how to use partial instead of total differential equations in calculating and measuring physical phenomena. Hartree developed some calculations using differential analyzers while at Manchester.

Hartree's concern for mathematics dominated his own research throughout the 1930s and 1940s. His most important investigation for the history of data processing was his study of very intricate physical and engineering situations using computational means rather than experimentation and simple observation. That single change in tactics illustrated how scientists could benefit from the use of computational equipment while encouraging others to develop such instruments in the late 1930s and throughout the 1940s.

Following World War II, Hartree used his prestige and influence to foster the expansion of computer capability, particularly within universities and government agencies. Although he did not believe that the demand for such devices was significant, particularly in a commercial arena, he was convinced that much work could be done to improve their effectiveness in science. Between April 20 and July 20, 1946, he visited the United States and participated in the final work done on the ENIAC. In England he lectured on computers and wrote articles in the popular press on the subject; he supported work on computers at Cambridge University; and he helped establish the Mathematics Division at the National Physical Laboratory (NPL) in 1945, which rapidly became an important development center for computer technology. Through his extensive contacts with the American scientific community, Hartree established active lines of communication between American and British designers of data processing equipment and software,* especially in the late 1940s and early 1950s. Professor Hartree died on February 12, 1958.

For further information, see: Herman H. Goldstine, *The Computer from Pascal to von Neumann* (Princeton, N.J.: Princeton University Press, 1972); D. R. Hartree, "Approximate Wave Functions and Atomic Field for Mercury," *Physical Review* 46 (October 15, 1934): 738–743, *Calculating Instruments and Machines* (Urbana: University of Illinois Press, 1949), *Calculating Machines—Recent and Prospective Developments* (Cambridge: Cambridge University Press, 1947), and "The Mechanical Integration of Differential Equations," *The Mathematical Gazette,* 22 (October 1938): 349; Simon Lavington, *Early British Computers* (Bedford, Mass.: Digital Press, 1980).

HAZEN, HAROLD LOCKE (1901–1980). This professor of electrical engineering at the Massachusetts Institute of Technology (MIT) helped develop analog computational devices during the 1930s and 1940s and, as an administrator at MIT during the 1950s and 1960s, obtained additional facilities for the continued work on electronics and computing, thereby insuring his university's importance in the evolution of computer technology over four decades. He is best remembered, however, for his work on analog computers,* both as a student and as a colleague of Vannevar Bush.**

Hazen was born on August 1, 1901, in Philo, Illinois, and attended public schools in Three Rivers, Michigan. He graduated from MIT with a degree in electrical engineering in 1924. His thesis focused on how to build a machine to model currents in an electric network such as that operated by large power companies. Immediately upon graduation, he worked for General Electric (GE)† (1924–1925), trying to implement some of his ideas at the urging of his professors. He returned to MIT to study under Bush who was then just becoming interested in analog devices. Hazen next became a research assistant in the Department of Electrical Engineering while doing graduate study. Between 1926 and 1931, he served as an instructor. His S.M. degree was awarded in 1929 and his Ph.D. in 1931, the same year he became an assistant professor. While a student, he continued to work on devices to model electrical currents and on portions of what eventually would come to make up portions of Bush's differential analyzer.* In his early years, Hazen built a Product Integraph as part of his continuing interest in analyzing traveling electrical waves. He helped Bush by designing torque amplifiers and built what became known as the MIT Network Analyzer.

This device, whose purpose was to model electrical flows in an electrical network, filled a room by 1932. It was the first such device to do the required mathematics and measurement accurately, and it was used for nearly twenty-five years. Forty copies of the machine were built. It had many of the features of an analog computer. Hazen also developed more efficient electrical servomotors essential for the electrical industry, and for this project, he was given the Levy Medal in 1935 by the Franklin Institute. In 1939, he became chairman of the Department of Electrical Engineering, the largest department at MIT.

In June 1940, the government established the National Defense Research Committee (NDRC) with the mission of applying science to ready the United States for war. In 1942, Hazen became head of Division 7 within the agency whose mission it was to develop better rapid-fire guns (mainly cannon at first) for use by the U.S. armed forces. At that time, the military had few modern weapons, and fire control research was limited. Through a combination of grants to companies and universities, Hazen's group was able to generate considerable new technology, much of it based on computational equipment developed by such universities as MIT and Pennsylvania. After the war, he returned to MIT and eventually became dean of the graduate school, retiring in 1967. He died on February 21, 1980.

For further information, see: Gordon S. Brown, "Eloge: Harold Locke Hazen, 1901–1980," *Annals of the History of Computing* 3, No. 1 (January 1981): 4–12; H. L. Hazen, "Design and Test of a High Performance Servomechanism," *Journal of the Franklin Institute* 218 (1934): 543–580.

HOLLERITH, HERMAN (1860–1929). This American engineer invented a large number of punched card data processing equipment, inventions that are generally considered to mark the birth of the data processing industry. He also established a company to sell these items which eventually became the International Business Machines Corporation (IBM).† In addition, from 1890 to about 1914, his devices dominated the electronic information handling market. Although other vendors would sell similar products, punched card equipment was frequently called Hollerith-type devices until World War II. In the late 1800s and early 1900s, Hollerith held the basic patents on punched card technology. He pioneered many of the applications for which his equipment was also used, including tabulation of the U.S. Census, population counts in other countries, the capture and analysis of medical and other vital statistics, accounting in public utilities and railroad companies, inventory control, and cost accounting in manufacturing firms. Many of his marketing practices later became IBM's fundamental policies.

The form that card equipment would take (and continues to take today) was largely the product of Hollerith's design decisions of the 1880s and 1890s. Thus, for example, today's computer punch card is the size of nineteenth-century U.S. dollar bills because Hollerith could use existing money cabinets to store them. The positioning of code on cards today stems directly from Hollerith's designs from the early twentieth century when he developed a column-by-column keypunch for use by the U.S. Department of Agriculture in its Census of 1901.

Hollerith was born in Buffalo, New York, in 1860 and graduated from Columbia University in 1879. Immediately thereafter, he took a job in Washington, D.C., in the U.S. Census Bureau where he worked for Colonel John Shaw Billings, director of the Division of Vital Statistics. In 1881, Billings remarked to the young engineer that somehow a way ought to be devised to mechanize the bureau's tabulating process. Indeed, the suggestion emerged that data on individuals could also be kept and analyzed in the same manner. Billings was concerned about the large number of people it took every decade to calculate the census, all of whom had to do laborious counting and tabulating. As the population grew, the time needed to complete the census would also grow and perhaps even exceed the ten years between censuses. After a study of the problem, Hollerith reported that such a machine could be built.

In 1882, General Francis Walker, head of the Census Bureau, moved to the Massachusetts Institute of Technology (MIT) and invited Hollerith to join the faculty. Hollerith moved to Boston and taught mechanical engineering while working on a "census machine." After only one year at MIT, he decided to return to Washington and spend more time inventing. He obtained a job at the U.S. Patent Office where he could learn about patents and their protection. A year later he opened up his own patent consulting firm in Washington while continuing to work on his machine. On September 23, 1884, he applied for his first patent.

During the 1880s, Hollerith also worked on a number of other devices, the most interesting of which involved railroad cars. The railroad industry encouraged the introduction of new technologies and could afford to invest in them. So it was logical that he would turn his attention to this industry. He developed an air brake but was never able to convince the industry to adopt it. As a consequence, he made little positive impact on railroad companies until years later when he improved their accounting procedures with his card punches.

Despite his work with air brakes, Hollerith's real interest lay in inventing a census machine that would tabulate totals automatically. By the late 1880s, he had built a device that could tabulate and handle aggregates through the use of cards, read by electrical sensing. By aggregates he meant that a hole in one of his cards could represent more than one piece of information. He demonstrated his machine in Baltimore and later in New York. When both cities tested its capabilities for handling vital statistics, he was able to start selling his machine. During the 1880s, his device was also changing from a simple card reader into a full system that could punch, read, and tabulate. They were made out of oak, and looked like desks and boxes with rows of clocklike counters, each of which could count up to 10,000 occurrences. He even developed an electrical sorter for grouping cards by predesignated types.

Hollerith's first major government contract came in 1889 with the Army Surgeon General's Office to handle military statistics. Following a pattern that would become common in the 1940s, government agencies were frequently the first to install new data processing technology and to encourage the development of computing devices. Hollerith wrote a description of this first government system, which he submitted to Columbia University as a dissertation and which in return granted him a Ph.D. "for achievement" in 1890. While obtaining his government contract, he also tested his machines against those of other inventors and won the bid to supply the U.S. Census Bureau with equipment for the Census of 1890. This was the first national census that would use data processing equipment, and it was this particular census that made it possible for Hollerith to convince other governments to use his devices in Austria, Canada, Italy, Norway, and Russia. These early systems tallied totals and were later equipped with the capability of accumulating totals (adding).

The U.S. Census of 1890 represented the single most important event in Hollerith's early career and a major milestone in the history of data processing. No other event so clearly signalled the start of the age of data processing. This census measured a large country and a greater variety of issues than ever before. It also marked an age when the public had fallen in love with technology. Hollerith's machines were seen as symbols of progress. They made it possible for one person to count thousands of people in a day, keypunching data captured by tens of thousands of census takers throughout the United States. Thousands of families were tabulated with his machines daily, and each day between 10,000 and 15,000 people were counted. That year his machines tabulated 62,622,250 people. The use of Hollerith's inventions saved the Census Bureau $5 million over manual methods of the past and did the job before the end of the year. Additional analysis of other variables, again using his machines, meant that the Census of 1890 could be completed within two years. The speed and convenience of his methods received considerable publicity both in the United States and in Europe.

Hollerith's success in the United States led him to spend the rest of the 1890s trying to sell his machines to other countries for the same purpose while refining his products. Between 1900 and 1914, he made major enhancements to his technology. He modified his cards shortly after 1900, so that numeric data appeared in columns which in turn permitted the design of a simpler keypunch, an automatic-feed card sorter, and a new tabulator that worked faster and more reliably than his earlier machines. As would become characteristic in the data processing industry after World War II, competition provided much impetus for enhancements of the machines. By the start of World War I, Hollerith's machines could accumulate numbers of any size, enabling him to sell the use of his machines to many industries. The U.S. Army, for example, used his devices in

France—a first for war; in another first, the United States used typewriters at the Versailles Peace Conference.

In 1905, the U.S. Census Bureau concluded that Hollerith's profits for the machines were excessive and so began developing its own based on his technology while also buying from other vendors. The bureau and other companies took advantage of the expiration of some of the basic patents covering Hollerith's work. The demand for such devices, when coupled with the less restricted impact of patents on the industry, encouraged the spread of card punch equipment across a number of government agencies and led to the dramatic increase of their use by industries in the years just prior to World War I and after. Thus, by 1940, Hollerith-type devices were common.

To handle his growing business affairs, Hollerith established a company, chartered on December 3, 1896, called the Tabulating Machine Company. This company eventually became part of IBM. He continued to modify and enhance his products while marketing them and so business grew. For the U.S. Census of 1900, he supplied the Bureau with 311 tabulating machines, 20 automatic sorters, and 1,021 keypunches. He gave more attention to inventing and less to the management of his company. While he had good products, many customers increasingly complained that they could not get timely delivery and so service became of some concern. Friends urged him to sell out. These suggestions came at a time when his company also needed to expand if it were to preserve its market position, and that required capital. He agreed and sold his firm for $1,210,500 in 1911.

The sale of his company, combined with the sale of several others, resulted in a newly chartered organization on July 5, 1911. It consisted of Hollerith's; the Computing Scale Company of Dayton, Ohio; International Time Recording Company of Binghamton, New York; and the Bundy Manufacturing Company of Endicott, New York. The new firm was now called the Computing-Tabulating-Recording Company, or simply C-T-R. The entire arrangement had been executed by an expert in mergers and financial dealings, Charles R. Flint,** who also arranged for additional funding to finance the new venture. Hollerith remained on the board of directors and continued improving his machines. On May 1, 1914, Thomas J. Watson, Sr.,** recently fired from the National Cash Register Company (NCR),† came to C-T-R as general manager. It was Watson who would convert the small company, with its wide variety of products, into the mighty IBM.

When Hollerith sold his company, he was fifty-one years old and a millionaire. In retirement, he continued to tinker with his machines and to build up his farm in Virginia. His biographer notes that from 1911 to 1914 he obtained a variety of patents improving on the design of his machines. He died on November 15, 1929, a victim of heart failure.

For further information, see: G. D. Austrian, *Herman Hollerith: The Forgotten Giant of Information Processing* (New York: Columbia University Press, 1982); F. H. Garrison, *John Shaw Billings: A Memoir* (New York: Putnam, 1915); L. E. Truesdell, *The De-*

velopment of Punched Card Tabulation in the Bureau of the Census, 1890–1940 (Washington, D.C.: U.S. Government Printing Office, 1965); C. Wright, *The History and Growth of the United States Census* (Washington, D.C.: U.S. Government Printing Office, 1900).

HOPPER, GRACE BREWSTER MURRAY (1906–). This U.S. naval officer is considered to be one of the most important historical figures in the development of programming languages.* Beginning in the 1940s, she was both a mathematician/programmer and administrator in the development and use of many major computer languages. She did more to advocate the development and use of higher level languages than perhaps any other individual in the data processing industry. The acceptance of such languages as COBOL* can largely be attributed to her.

Grace Hopper was born Grace Murray in New York City on December 9, 1906. She graduated from Vassar College in 1928 in Poughkeepsie, New York, with a B.A. In 1930 she completed her M.A. at Yale University in mathematics and married Vincent Foster Hopper. She completed her Ph.D. in mathematics at Yale in 1934 and taught at Vassar between 1931 and 1943, rising from instructor to associate professor. In December 1943, she joined the U.S. Naval Reserve as a lieutenant j.g. Upon completion of her initial military training, leading to her commission at the U.S. Naval Reserve Midshipman School, she was assigned to duty at the Bureau of Ordnance Computation Project at Harvard University. The assignment was unique because fewer than six computer projects were then underway in the United States and one of these was at Harvard. While there, Hopper wrote her first programs, supplying work to the Mark I* computer. It was the earliest automatic sequence digital computer* in the world. In 1946, she left the Navy and went to Harvard as a research fellow in engineering and applied physics within the Computation Laboratory, center of Harvard's computer development work. While there, she participated in the development of programs for the machines that made up the Mark II and Mark III projects funded by the U.S. Navy.

In 1949, Hopper joined the Eckert-Mauchly Computer Corporation with the title of senior mathematician, and in 1950 she became a senior programmer—probably one of the first persons to hold what today is a common title within the data processing industry. She served in this capacity until 1959, long after the company had been absorbed first into the Remington Rand Corporation† and next into the Sperry-Rand Corporation.† While at the Eckert-Mauchly Corporation (one of fewer than twelve private computer-related firms in existence in the late 1940s/early 1950s), she participated in the design of the UNIVAC I.* This computer became the first large commercially available general-purpose machine. Historians agree that it was the single most influential computer in launching the modern age of data processing. During her work on UNIVAC, Hopper managed a department that produced the first language compiler, called the A-0 and then the A–2.* Until this time no adequate software* had been available to translate a programmer's problem-oriented language to machine-readable instructions. The use of compilers made programming dramatically easier and faster. The creation of A-0 thus signalled the dawn of a new era in programming languages.

Hopper also developed a programming language called MATH-MATIC.* An experience that led her to develop a commercially oriented high-level language by the standards of the day called FLOW-MATIC,* completed in the mid-1950s. This language came out at a time when programming specialists were calling for standardization in languages. While there were many candidates for standardization, particularly FORTRAN,* ALGOL,* and COBOL, COBOL was the most popular candidate in the late 1950s. It was easy to use and was applicable to commercial applications. As a result of that development, her FLOW-MATIC was very important in forming opinions about how COBOL should look. Thus, it influenced the most widely used language of the 1960s and 1970s. In 1959, Hopper was named director of automatic programming development at the UNIVAC Division of Sperry Rand Corporation. She remained associated with the company until her official retirement in 1971.

In 1967, Hopper returned to active duty with the U.S. Navy. In 1973, she reached the rank of captain, serving as special assistant to the Commanding Officer, Naval Data Automation Command. She served as a spokesperson for the industry as well as one of its most respected pioneers. By the time she retired from active service in August 1986, she had also attained the rank of rear admiral. During the entire span of her professional career, she published a series of articles on the subject of computer languages and used her positions in management within industry and the U.S. Navy to promote the development and use of higher level languages.

Hopper's promotion of COBOL was particularly significant. She used FLOW-MATIC to write the first COBOL compiler; in fact, many of the verbs from the earlier language survived many releases of COBOL in the 1960s and 1970s. During the 1960s, Hopper's staff attempted to standardize all versions of COBOL used by the Navy and, as a consequence, was active in the data processing industry's attempt to standardize the language for commercial applications. That effort proved highly successful. By 1980, nearly 80 percent of all commercial applications in the United States had been written in COBOL.

Hopper gained early recognition. In 1946, she received the Naval Ordnance Development award, in 1968 the Connelly Memorial award, and in 1972 the Wilbur L. Cross medal from Yale. Yet more recognition and honorary degrees came in the 1970s. In 1969, she had already been named Man of the Year by the Data Processing Management Association, and in the following year she received the Harry Goode Memorial award from the American Federation of Information Processing Societies (AFIPS)†—the largest data processing organization.

For further information, see: G. M. Hopper, "The Education of a Computer," *Proceedings, ACM* (1952): 243–249, and "Keynote Address, ACM SIGPLAN History of Programming Languages Conference, June 1–3, 1978," in Richard L. Wexelblat, ed., *History of Programming Languages* (New York: Academic Press, 1981), pp. 7–24; Jean E. Sammet, "Introduction of Captain Grace Murray Hopper," ibid., pp. 5–7, and *Programming Languages: History and Fundamentals* (Englewood Cliffs, N.J.: Prentice-Hall, 1969); Henry S. Tropp, "Grace Hopper: The Youthful Teacher of Us All," *Abacus* 2, No. 1 (1984): 7–18.

HOUSEHOLDER, ALSTON SCOTT (1904–). This mathematician published a number of works on data processing technology, particularly on software,* and was active in conventions and other meetings held to discuss computing. Like many of his peers, he was first a mathematician and then moved to computing science in general. In an industry with many organizations and series of publications, he typified the careers of many professors and technocrats in the general field of computing.

Householder was born and raised in Rockford, Illinois, and completed his B.S. in mathematics at Northwestern University (1925), his M.A. at Cornell University (1927), and his Ph.D. at the University of Chicago (1937). He was an instructor in mathematics at Northwestern University, (1926–1927); a tutor for the Miss Harris Schools in Chicago (1929–1930); an instructor in mathematics at Washburn College in Topeka, Kansas (1930–1931); and an assistant professor in 1931. In 1937, he became a Rockefeller Foundation Fellow at the University of Chicago where, in 1944, he was promoted to assistant professor in mathematics and biophysics. During 1944 and 1945, he served as a senior research psychophysiologist for the National Defense Research Committee at Brown University. In 1946, he moved to Washington, D.C., where he became a consultant in mathematics to the Naval Research Laboratory, home of computer-related projects in the 1940s. That same year he moved to Oak Ridge National Laboratory in Oak Ridge, Tennessee, where much work was being done on atomic energy and weaponry. He was a mathematician there until 1969. Between 1955 and 1958, he also was a member of the mathematics division of the National Research Council. He began teaching at the University of Tennessee as a professor of mathematics in 1964.

With regard to his data processing career, between 1954 and 1956 Householder served as president of the Association for Computing Machinery (ACM),† one of the larger organizations within data processing. He has also been president of other societies, such as SIAM, and has served on the editorial boards of various publications, including *Psychometrika, Computers and Automation,* and *Computers in Biomedical Research*. He served on the editorial boards of various mathematical publications, such as the *ACM Journal*. His primary areas of research have involved mathematical biology, numerical algebra, and mathematics in data processing. In 1959, he worked with Isaac Auerbach** and Samuel Alexander, two well-known computer scientists, to establish the

International Computation and Information Processing (ICIP) meeting in Paris. This session, held in June 1959, was the first of a series of international technical conferences later sponsored by International Federation for Information Processing (IFIP)† Congresses. He also helped establish the Gatlinburg Symposia for the same purpose of allowing computer scientists to share the results of their work.

Householder was awarded the Harry Goode Memorial award in 1969 and became a Fellow of the American Association for the Advancement of Science. He is a member of the American Academy of Arts and Sciences.

For further information, see: I. L. Auerbach, ''International Federation for Information Processing (IFIP),'' in Anthony Ralston and Chester L. Meek, eds., *Encyclopedia of Computer Science* (New York: Petrocelli/Chester, 1976), pp. 729–732.

HULL, CLARK (1884–1952). This American statistician sought to reduce the drudgery required to prepare statistics for calculations while lowering the error rate. He did this by building his ''automatic correlation calculating machine'' which was operated by feeding it paper tape. Such an approach was very progressive for the 1920s and 1930s. He built the machine in 1925 while a professor at the University of Wisconsin. The device itself was a tangle of shafts and gears that sat on top of a table. At the time, he claimed that it could perform hundreds of times faster than someone operating (calculating) manually. His was one of many such devices developed in the early twentieth century by scientists as aids to calculations.

For further information, see: Clark L. Hull, ''An Automatic Correlation Calculating Machine,'' *Journal of the American Statistical Association* 20 (December 1925): 522–531.

HURD, CUTHBERT C. (1911–). This engineering executive was involved with the design and development of the IBM 704* and 705,* the Defense Calculator (IBM 701*), LARC,* the Magnetic Drum Calculator (IBM 650*), and even FORTRAN*. At the end of his career he managed International Business Machines Corporation (IBM),† important early work in the area of process control. During the 1960s and 1970s, he managed consulting operations outside of IBM.

Hurd was born in Estherville, Iowa, on April 5, 1911. He completed his A.B. at Drake University in 1932, his M.S. at Iowa State College in 1934, and his Ph.D. at the University of Illinois in 1936, all in mathematics. Between 1936 and 1942, he was an assistant professor in mathematics at Michigan State College. From 1942 to 1945, he was in the U.S. Coast Guard Reserve, rising to the rank of lieutenant commander. At the end of World War II, he became a dean at Allegheny College, but in 1947 he served as head of technical research for Union Carbide and Carbon Corporation at Oak Ridge where he was asked to set up a computing service for the Oak Ridge National Laboratory. He had already become familiar with card punch equipment while a graduate student and then as a consultant after World War II in establishing card punch facilities at the U.S. Naval Academy. While at Oak Ridge, he became familiar with the IBM 604* and the SSEC.* With these experiences behind him, he applied for a job with IBM.

On March 1, 1949, Hurd became an employee of what was still primarily a card punch, office systems manufacturing company just beginning to inch its way into the world of computers. Hurd soon became involved in the use of the IBM CPC,* the company's first major electronic computational product. Through his highly successful efforts to convince customers to use this equipment, he was named director of the Applied Science Department. In order to promote the use of IBM's machines for technical applications, he hired people to help with the process, held seminars, and published technical newsletters. During his tenure in this position (until 1953), he made suggestions on how to improve the technical features of the 604, the Magnetic Drum Calculator, and other equipment. He supported the 604, assigned staff members to define enhancements, and fought to see the Magnetic Drum Calculator brought out as a product, which was announced in 1953 as the 650 Magnetic Drum Calculator.

In 1953, Hurd was named director of the Applied Science Division, a position he held until 1955. This change represented a growth in IBM's commitment to developing computers. Hurd had been actively involved with the Defense Calculator, which was needed for the Korean War, and now looked to other projects as well. In 1955, Hurd was named director of electronic data processing machines, remaining in that post until 1956. In his new capacity, he participated in the company's STRETCH,* a significant project for developing new technologies that could be incorporated in computers. It was funded partially by defense contracts and partially by IBM. In his new role, Hurd pulled together technical recommendations, accounted for the realities of marketing, and made recommendations to senior management on what IBM should produce in the way of computers.

In his new job, one of Hurd's primary concerns was to obtain a major contract with the U.S. government to build a large computer based on solid-state technology. That technical effort would provide IBM with the technical leadership necessary to maintain the firm's position within the data processing market which at that moment was being threatened by Remington Rand.* While negotiating with potential customers for a supercomputer, Hurd's engineers worked to devise technical plans and proposals. After Hurd left this position, STRETCH continued and generated much technology that appeared in products during the late 1950s and early 1960s. It did not, however, lead to any major contract with the U.S. government while Hurd was in power.

Following a major reorganization late in 1956, Hurd was named director of automation research and remained in that position until 1960. From 1961 to 1962, he was director of control systems, and from 1956 until he left IBM in 1962, he managed research on the use of computer simulation and control. The industrial control system, as later used in petroleum refining, and in paper and steel manufacturing, was long recognized as an obvious need, but it was only partially addressed during the 1940s and early 1950s. Aware that this area was not taking advantage of the technologies then becoming available, Hurd elected to use the IBM 1620* computer as the basis for automation systems. His team of engineers developed analog-to-digital conversion equipment that could be used with the 1620, in effect making it the first digital computer* of any consequence used in control systems. In March 1961, it was announced as the IBM 1710 Control System. It was recognized that computers had come far enough along to be trusted to remain "up" and thus reliable enough to manage a manufacturing or processing operation. Hundreds of these systems were sold in the early 1960s.

Hurd left IBM in 1962 to become chairman of the board of Computer Usage Company, Inc., in New York City, a position he left in 1974 to become chairman of Solar Energy Research. He was associated with that firm for two years after which he established Cuthbert Hurd Associates, a consulting company. He subsequently added to his group by establishing Picodyne Corporation in 1978 and Quintus Computer Systems in 1984.

For further information, see: Charles B. Bashe et al., *IBM's Early Computers* (Cambridge, Mass.: MIT Press, 1986); Cuthbert C. Hurd, "Computer Development at IBM," in N. Metropolis et al. eds., *A History of Computing in the Twentieth Century* (New York: Academic Press, 1980), pp. 389–418, and his "Early IBM Computers: Edited Testimony," *Annals of the History of Computing* 3, No. 2 (April 1981): 163–182.

HUSKEY, HARRY DOUGLAS (1916–). This computer scientist participated in the creation of the modern electronic digital computer.* As data processing became an industry, Huskey became a leader in its organizations. As a professor, he trained students who in turn went on to become computer scientists. Because of his work on both the ENIAC* and the British ACE* computer he is closely linked to the dawn of the modern computer.

Huskey was born in Whittier, North Carolina, completed his B.S. at the University of Idaho (1937), and his M.A. (1941) and Ph.D. (1943) at Ohio State University. While on the faculty of the University of Pennsylvania, (1943–1946), he first became interested in computers, having become acquainted with work in progress on the first electronic digital computer at the Moore School of Electrical Engineering.† As an engineer, he learned a great deal about building computers there, knowledge he would later apply to other projects. In 1947, he lived in Great Britain, working at the National Physical Laboratory (NPL) when it was sponsoring projects leading to the construction of computers. During that year, he worked with Alan M. Turing** on the ACE computer. When he returned to the United States, he went to work for the National Bureau of Standards and then for the Institute for Numerical Analysis (INA, located at the University of California at Los Angeles—UCLA). He worked for the INA from 1948 to 1954. While with both organizations, he worked on SWAC,* which was yet another early digital computer. Huskey was also technical director of Wayne University's Computation Laboratory (1952–1953). From 1954 to 1966, he was back at UCLA as a professor in electrical engineering and served as vice-chairman of the Electrical Engineering Department in 1966. In the mid-1960s, he was offered a position at the University of California at Santa Cruz which he accepted in 1968, becoming professor of information and computer science, a position he still holds today.

In addition to his normal academic responsibilities, Huskey was an advisor to other organizations. He worked with scientists at the Bendix Computer Division of Bendix Corporation, designing the Bendix G–15 computer, and he participated in the logic design of the G–20 (computers of the 1950s). He advised the United Nations on the computer's potential role in developing nations, and on several occasions in India (1963–1964, 1971), Chile, and Nigeria. During the 1980s, he served as a member of the Advisory Panel of the Institutional Program concerning computers of the National Science Foundation, along with helping the Naval Research Advisory Committee. Huskey also served as chairman of a committee created by the U.S. National Academy of Sciences to advise Brazil on how to teach data processing. Between 1960 and 1962, he was president of the Association for Computing Machinery (ACM),† one of the industry's most important organizations and one that actively sought to encourage the exchange of technical information concerning computing science. Between 1965 and 1970, Huskey was editor for publications of the Institute of Electrical and Electronics Engineers (IEEE)† Computer Society. Huskey was a Fellow of the IEEE and of the British Computer Society. He has published over fifty papers and has produced many students in the field.

For further information, see: Harry D. Huskey, "Characteristics of the Institute for Numerical Analysis Computer," *Mathematical Tables and Other Aids to Computation* 4, No. 30 (1950): 103–108, *The Development of Automatic Computing. Proceedings, First U.S.A.-Japan Computer Conference, 3–5 October 1972, Tokyo* (Montvale, N.J.: AFIPS, 1972), "Electronic Digital Computers in England," *Mathematical Tables and Other Aids to Computation* 3 (1948): 213–216, "Electronic Digital Computing in the United States," in *Report of a Conference on High Speed Automatic Calculating Machines, 22–25 June 1949* (Cambridge: Cambridge University Mathematics Laboratory, 1950), pp. 109–111, "The National Bureau of Standards Western Automatic Computer (SWAC)," *Annals of the History of Computing* 2, No. 2 (April 1980): 111–121, and with et al., "The SWAC—Design Features and Operating Experience," *Proceedings of the IRE* 41, No. 10 (1953): 1294–1299; Maurice V. Wilkes, *Memoirs of a Computer Pioneer* (Cambridge, Mass.: MIT Press, 1985).

J

JACOBS, WALTER W. (1914–1982). This scientist was a computer/mathematics specialist who worked for the U.S. government during the 1940s and 1950s and employed computers in military projects. His career illustrates how military applications became an important method by which computers became government tools.

Jacobs was born in Newark, New Jersey, on September 26, 1914, and was raised in New York City. He graduated from the City College of New York in 1934 with a B.S. in mathematics. As early as elementary school, like many contemporaries who went into computer science as a profession, he showed brilliant natural abilities in mathematics and played an outstanding game of chess. By 1937, he found work as an actuarial mathematician with the Railroad Board in Washington where he also worked on his M.A. in mathematics, completing the degree at George Washington University in 1940. In 1941, he joined the Army Security Agency (ASA) and soon after was drafted into the Army. Late in 1944, he was sent to Bletchley Park† in England to work on cryptographic projects which employed computer-like equipment developed there. After World War II he returned to ASA, and in 1947 he went to work for the U.S. Department of Commerce. He completed his Ph.D. at George Washington University, in 1951. He next went to work for the U.S. Air Force, staying until 1957.

Jacobs' first job with the Air force was as deputy chief of the computational division where he used the second UNIVAC I* built. He became one of the first managers to employ computers to solve managerial planning problems, such as the cost and requirement for aircraft engines. He developed TRIM, a computer-based war plans model that defined what it would take to implement a particular plan. These kinds of applications were unique for their time. He did this kind of work during a period when computers were either used to calculate census data or for sophisticated scientific applications (e.g., weather forecasting).

In 1957 Jacobs became deputy at the Office of Mathematical Research at the

National Security Agency (NSA) at Fort Meade, Maryland. This job allowed him to return to his wartime work in cryptoanalysis. NSA was then working with International Business Machines Corporation (IBM)† to build a large computer called Harvest. Jacobs convinced the agency that a similar effort would have to be launched for software* if the system were to be effective, and he was made head of the project. Although he worked on the software development project for only one year before being transferred to the Office of Research and Development at NSA, he brought it alive with a staff of fifteen to twenty people and another thirty from IBM. In 1961, he became chief of the Office of Machine Processing for NSA where he was able to fully support Harvest when it came on-stream. In 1963, Jacobs joined the Institute for Defense Analysis as a visitor where he became exposed to artificial intelligence.* He returned to NSA in 1964. In 1966, he became head of NSA's National Cryptologic School, where he stayed until his retirement in 1969.

Jacobs next joined the faculty at American University and soon after became chairman of the Department of Mathematics and Statistics. While there he established B.A. and M.A. programs in computer science. He relinquished the chairmanship in 1972 and retired from teaching in 1981. This government scientist and pioneer in the use of computing has been described as soft-spoken and brilliant. He died of cancer on February 11, 1982.

For further information, see: Joseph Blum et al., "Eloge: Walter W. Jacobs, 1914–1982," *Annals of the History of Computing* 6, No. 2 (April 1984): 100–105; W. W. Jacobs, "Military Applications of Linear Programming," in H. A. Autosiewicz, ed., *Proceedings of the Second Symposium in Linear Programming* (Washington, D.C.: U.S. Air Force, 1955), II: 1–27.

JACQUARD, JOSEPH-MARIE (1752–1834). Jacquard invented an attachment for a weaving loom that automated the weaving process by using cards with holes to direct the process. His was one of the first devices to contain rudaimentary sequence control mechanisms with primitive program control. His use of punched cards to direct the actions of a machine later influenced the work of Charles Babbage** and Herman Hollerith.** Both relied on similar approaches for the design of their calculators and data handling equipment.

Jacquard was born and raised in Lyon, home of France's silk-producing industry in the eighteenth and nineteenth centuries. His family had been involved in the silk industry for some time and continued to be during his lifetime. The young Jacquard therefore felt the impact of the Industrial Revolution at first hand and learned about the business of weaving very early in his life. In an attempt to improve the efficiency of existing looms and the procedures of weaving various patterns, in 1801 he finished the construction of a device that could be attached to a conventional weaving loom that made possible automated pattern weaving. The machine was fed with cards punched out with rectangular holes directing the feeding of various threads into the loom, resulting in a specific predefined pattern, a pattern determined by the arrangement of the holes on the cards. Wire hooks passed through the holes to grab the specific thread to be weaved into the cloth. The holes determined which threads were used because if a rod came up against the stiff card at a point where there was no hole, no thread would be pulled up.

The procedure for directing the series of cards into automating the patterning of the weaving process mimics that of a programmer to a certain extent. Both have to decide in advance what the ultimate result has to be. Each has to determine the series of steps that must be executed in order to achieve the goal. The weaver has to decide what pattern to use, produce the cards that would instruct the machine as to which thread to use and when, and finally when to repeat the process again.

Jacquard's device made it possible to develop complex patterns while speeding up the weaving process itself. It still took considerable time to establish the order of the cards, a process which itself was finally computerized during the 1960s. Historians have linked Jacquard's device to the early history of programmable machinery since his was fed instructions for automatic function. Moreover, because of its wide acceptance, it was understood by those individuals, such as Babbage who developed other devices later. At the time, workers feared the looms would eliminate their jobs. Despite this concern, the invention radically and quickly altered weaving. By 1812, some 11,000 Jacquard looms were in operation in France. This technology spread to northern Spain where thousands more were used in the early nineteenth century in the manufacture of cotton cloth.

Jacquard's achievement did not emerge out of a vacuum; some work on looms had been done before him. The idea of sequence-controlled machinery had been the subject of research for hundreds of years. Rotating pegged music boxes represented an early version of such a machine. The oldest of these seems to date back to Heron of Alexandria (circa 100 A.D.). By the Middle Ages, large clocks with figures were controlled by similar technologies. An important step that led to Jacquard's work was taken by Basile Bouchon who, in 1725, used perforated tape to control the weaving of ornamental patterns in silk. He devised a means whereby the cords to be pulled up through the material being weaved could be selected through a combination of perforations on a tape. Working later with Bouchon in Lyon, Falcon helped to expand the capability of the original design to allow some 400 cords to be controlled by several rows of needles driven by the perforated tape. The tape was eventually replaced with perforated cards strung together in a desired sequence. By the time of Falcon's death in 1765, some forty such devices were in use.

Other important preliminary work was done by Jacques de Vaucanson, a great French engineer who is credited with originating the idea that cards be strung in sequence. Vaucanson developed various types of automata. His most famous machine played the flute in 1736 by recreating the gestures of lips and fingers to accomplish the task. He turned his attention to the draw loom, perhaps because of his friendship with Falcon, in about 1750, developing one that did not require any human intervention and thus was fully automatic. Until the mid-1700s, all industrial looms required the services of two people—one to handle the shuttle and another to control the warp threads used at each pass of the shuttle on the loom. Vaucanson used perforated cylinders instead of the more effective perforated cards which Bouchon and Falcon already had developed and which he was undoubtedly familiar with by the late 1740s.

Jacquard to a large extent simply had to improve on the work of earlier people and then install the new equipment. He accomplished both tasks. He sold his device, proving that its acceptance was widespread. It was accepted first because it worked well. A second reason for its acceptance was connected with the Industrial Revolution, a time when businessmen were looking for labor-saving devices to increase their productivity and thereby drive down the unit cost for products. Third, Jacquard lived in an age of reverence for scientists; in contrast, in the previous century Blaise Pascal** and Gottfried Wilhelm von Leibniz's** inventions received little more than amused curiosity. Indeed, Napoleon's government awarded Jacquard a pension as a reward for his work on the loom. This action represented an early example in Europe of government support for applied research in a nonmilitary project. The next such example occurred in the 1820s when the British funded part of the work which Charles Babbage did on the difference engine.*

Jacquard's work had great impact on Babbage, who was impressed by the technology that went into the Jacquard loom. The Countess of Lovelace,** Babbage's friend who articulated the Englishmen's ideas for an analytical engine, wrote that ''We may say most aptly that the Analytical Engine weaves *algebraic pattern* [her italics] just as the Jacquard-looms weaves flowers and leaves.'' In the late nineteenth century, the idea of using punched cards in equipment that automated some of the work of the U.S. Census Bureau also drew on Jacquard's efforts. In fact, Herman Hollerith's equipment was entirely centered around the concept of input/output conducted with cards.

For further information, see: A. Doyen and L. Liaigre, *Jacques Vaucanson, Mecanicien de Genie* (Paris: Presses Universitaires de France, 1966); E. A. Posselt, *The Jacquard Machine Analyzed and Explained: With An Appendix on the Preparation of Jacquard Cards, and Practical Hints to Learners of Jacquard Designing* (Philadelphia: Pennsylvania Museum and School of Industrial Art, 1887).

JACQUET-DROZ, PIERRE (1700s). This Swiss clockmaker and his sons made automata with great capability, including one called The Scribe which could write with a mechanical hand and represented an interesting milestone in the history of automata or robots.

The clockmaker and his family lived at La Chaux-de-Fonds in the mountains of Switzerland next to France. He and his two sons made a living constructing clocks, but they also enjoyed inventing devices driven by clock-like mechanisms. In 1768, they completed one that had taken four years to build and that they called The Scribe and Charles. This automaton looked like a human male and could write any kind of message, providing it was forty characters or less in length. It appeared as an early programmable device. The machine motioned much like a human being, even dipping its pen into an inkwell periodically and, if it had too much ink, shaking the excess off. It could move to a new line on the paper when appropriate, and the figure's head and eyes followed what was being written. Charles, shaped like a human, had both a right and left hand but wrote with its right. The left hand pulled the paper being written on sidewise. One student of the figure believed that this action eliminated ''the curvature of the line that would result if the paper remained in one place.'' It took the builders six hours to change the message it wrote. The device was apparently very sensitive to changes in temperature which caused it to make spelling errors. It had brass cams which controlled motions. Messages were coded by using removable metal disks which were placed in the figure's back. The Scribe has survived and is exhibited in the art museum of the city of Neuchâtel, Switzerland.

The Scribe was one of three major known devices invented by this craftsman and his two sons. A second, called ''Henri, the Draftsman,'' could be programmed to draw four different pencil sketches, including a boy riding in a cart pulled by a butterfly; profile sketches of King Louis XV of France, or England's King George III and his wife; and ''My Doggie'' *(Mon Toutou)*. The figure in ''My Doggie'' was programned and drew its picture, after which it would bow his head. If any surplus graphite remained on the page, it would blow it away. Henri was clearly the earliest example of a working CAD (Computer Aided Design) application at work.

A third and better device was called ''Marianne, the Musician.'' This android represented a young woman playing a small organ with ten fingers. She had personality because, while playing, she reflected social customs of the day, such as coyness and flirtation in the style of the eighteenth century. She could lean forward, imitating someone wanting to look at the music sheets more carefully while playing, and she would periodically glance at both the audience and at her hands. One witness of her performance was careful to note that ''her breast rises and falls in an uncanny simulation of restrained breathing,'' and that she ended her performance with a curtsy.

The Swiss clockmaker's reputation spread even to Spain, where the royal family invited him to visit Madrid. Jacquet-Droz was unable to make the trip because apparently at this time, peasants in his community, perhaps thinking his automata were the Devil's work, accused him of sorcery. But even this threat to his physical safety did not stop him from building a bleating lamb. One of his last known devices was a playful dog which could guard a basket of fruit. If someone picked up either an apple or banana, this dog-like android would bark until the fruit was put back in place.

These projects were incredible works of engineering given the limitations of machining in the eighteenth century and perhaps the limits in the knowledge of human behavior. They were also very clever. Others constructed automata in the 1700s, just for show. The most famous of these were made by Jacques de

Vaucanson*; the best known of his works, the Vaucanson Duck, was built in 1738 and entertained audiences all over Europe. It ate, drank, and even defecated. The other famous builder of automata of the period was the Hungarian Baron Wolfgang von Kempelen** who, in 1769, went on tour with a chess-playing mechanism. The long-term influence—if any—of these automata on the construction of what would be known as robots is uncertain. They may have simply been affectations that grew out of the skills available in clockmaking, a fashionable art form and skill during that century, or perhaps they represented advances in the use of machine-like devices. Clearly, the automata described in this entry were meant for entertainment, and perhaps therein lay the real motivation for the work of the Jacquet-Droz family.

For further information, see: Tom Logsdon, *The Robot Revolution* (New York: Simon and Schuster, 1984).

JEVONS, WILLIAM STANLEY (1835–1882). Jevons developed a machine that could solve logical problems. Both by training and by profession he was an economist and a logician. In 1864, he published *Pure Logic: or, the Logic of Quality apart from Quantity* in which he followed George Boole's** system of logic and yet found much to criticize in Boole's mathematical approach. Building on ideas presented in the book, Jevons proceeded to build a logical machine to obtain conclusions mechanically derived from premises. It looked like a portable piano and in fact is often called the logical piano, each key marked with letters and results displayed above by other ivory keys. Jevons first demonstrated his device before the British Royal Society in 1870. Others expanded on his original conception, including Allan Marquand** in the United States in 1881 and Charles L. Dodgson (a fellow mathematician and logician better known by his pen name, Lewis Carroll, author of *Alice in Wonderland*) with a ruled board and counters. Dodgson's machine was designed in 1886 and resembled an abacus* more than a piano.

Jevons' machine can be called the first working logic solver that operated faster than a human. Boole's introduction of his own ideas in 1854 and Jevons' completion of his machine in 1869 indicates the intensity of interest in such mechanisms in Great Britain.

For further information, see: B. V. Bowden, ed., *Faster Than Thought* (London: Sir Isaac Pitman and Sons, Ltd., 1953); Margaret Harmon, *Stretching Man's Mind: A History of Data Processing* (New York: Mason/Charter, 1975).

JOBS, STEVEN PAUL (1955–). As founder of Apple Computer, Inc.,† Jobs typified a generation of young entrepreneurs within the data processing industry in California who established technology-based firms in an area known as Silicon Valley. Jobs virtually launched the era of the personal computer by the successful operation of Apple Computer.

Jobs was raised in Los Altos, California, and while in high school became interested in electronics. The story is told that at the age of thirteen he called William Hewlett, then president of Hewlett-Packard Company,† to ask for spare parts he could use for a project. The president gave him what he needed, and the following summer, impressed with the boy's boldness, found him a job in one of H-P's factories. Jobs and a friend who would later help him establish Apple Computer, Stephen G. Wozniak,** built microcomputers in the garage of Jobs's home. By the mid-1970s, the Apple I had been born. At first, microcomputers were built for members of the Homebrew Computer Club† (made up primarily of students at Stanford University) followed by an additional fifty machines they made for the Byte Shop in Mountain View—one of the first computer stores in the United States. This last order represented the first major business enterprise by Jobs and Wozniak which they financed by Jobs selling his Volkswagen and Wozniak two H-P calculators.

Jobs was not interested in going to college and dropped out of Reed College after one semester (1972) in order to focus all his attention on the business of making and selling micros. He did hold one job before establishing Apple. At the age of nineteen he went to work for Atari, Inc.† (1974) to design video games. During the early 1970s, he pressured his friend Wozniak to leave his job at H-P to form their own microcomputer manufacturing company. Wozniak was the more technical of the two and concerned himself primarily with the design of machines while leaving Jobs to figure out how to create a business. By 1976, when Jobs was twenty-one, Apple was on the verge of becoming a company. The following year, he was chairman of the board of what would become one of the most talked about firms in the industry.

The story of the establishment of a company that quickly became worth hundreds of millions of dollars, which started in a garage and was led by a young entrepreneur who dropped out of college, has now become an important part of the folklore of the data processing industry.

Apple's short yet rapid rise was typical of many companies formed in the Silicon Valley. In 1976, Wozniak designed and built the Apple II, and with this product in hand in the following few months Jobs found enough venture capital to support the creation of the Apple Computer Company. Almost from the day it opened its doors in January 1977, the company did extremely well and grew. The microcomputer market expanded so rapidly that in late 1981 even International Business Machines Corporation (IBM)†, long the large computer manufacturer and now diversifying into many submarkets within the data processing industry, entered the microcomputer market with its PC. From then on, Apple and IBM dominated the multibillion dollar small computer market. Competition between the two was extensive. Along the way Jobs introduced the Apple IIe and c, Lisa in 1983, and Macintosh in 1984. The company's growth was phenomenal. In 1977—its first full year of operation—sales reached $775,000 and, in 1984, nearly $1.5 billion. When the company went public in

December 1980, Jobs' share of stock was valued at $165 million and, by the mid-1980s, had exceeded $200 million.

His success could be attributed to several factors. First, the demand for microcomputers was moving out from the hobby market to a broader one at a time when the cost of such technology was rapidly dropping in the early 1970s. Second, startup costs for such companies was quite low. In the case of Apple, it was originally capitalized at approximately $1 million. Third, Jobs, an intelligent, hard-working achiever, did not hesitate to seek aid from financial advisors. From the earliest days of Apple, he hired presidents and managers. Michael Scott, from National Semiconductor, became Apple's first president, followed by Mike Markkula who was a venture capitalist within the industry and who came out of ''retirement'' at the age of thirty-eight to help establish Apple. In 1983, John Sculley, who had been president of Pepsi-Cola and was in his early forties, became the next president of Apple. Fourth, Jobs was aggressive and believed in the future of his company. By early 1984, he told the press that by the end of the 1980s or early 1990s, Apple would be a $10 billion enterprise. He had a reputation for abrasiveness and lack of humility, characteristics often evident in successful entrepreneurs. Colleagues typically characterized him as intense and very ambitious. In September 1985, he resigned from Apple after losing a power struggle with the president, John Sculley.

Although others had introduced microcomputers in the early 1970s, it was Steve Jobs who had the greatest success in creating and defining the market for them. The success of Apple, and later of IBM, in this area meant that computing was becoming more available at the office, on the plant floor, in classrooms, and at home. The penetration of computing power into the lives of many people in the 1970s and 1980s was largely due to the success he had with the Apple. The impact of the microcomputer on industrial society and the growth of the data processing industry was at least as great as the introduction of digital computers* in the 1950s or of the S/360* in the 1960s. No other development in the industry during the late 1970s or early 1980s had as dramatic an impact as the introduction of the microcomputer. A main character in that story was Steve Jobs.

For further information, see: Paul Freisberger and Michael Swaine, *Fire in the Valley: The Making of the Personal Computer* (Berkeley, Calif.: Osborne/McGraw-Hill, 1984); Robert Levering et al., *The Computer Entrepreneurs: Who's Making It Big and How in America's Upstart Industry* (New York: New American Library, 1984); Michael Moritz, *The Little Kingdom: The Private Story of Apple Computer* (New York: William Morrow, 1984).

JOHNSON, REYNOLD B. (1906–). Johnson invented a machine that could detect hand-written pencil marks in the 1930s. That development made possible widespread testing and soon came to the attention of International Business Machines Corporation (IBM)† which converted his work into a product. By the

1950s, this kind of technology had become a common feature of the American educational community and led directly to the establishment of national testing programs such as achievement tests, the College Boards, and the Graduate Record Examination (GRE). Less known to historians was Johnson's second contribution, namely, the development under his management of IBM's first disk drives* during the 1950s. Within IBM he was also known as the founder of the San Jose Laboratory.

Johnson studied education at the University of Minnesota and, in 1932, while teaching in a Michigan high school, devised an electrical gadget that scored multiple-choice tests. It counted pencil marks on an answer sheet much as they would be done during the next sixty years. Ben D. Wood, who taught at Columbia University, heard about it and suggested to his friend, Thomas Watson, Sr.,** then head of IBM, that he pursue its commercial potential. In 1934, IBM approached Johnson, who at the moment was unemployed. The company had tried to build such a gadget for Wood earlier but with little success. The company therefore decided to develop a machine based on Johnson's work and offered him a job which he accepted at the firm's laboratory at Endicott, New York. In 1935, Johnson had a new version of the machine ready which he tested in 1936. In 1937, IBM announced it as the 805 Test Scoring Machine.

Johnson continued his work with mark-sensing technologies at IBM and during World War II managed various projects for the military on behalf of IBM. In the late 1940s, he studied wire-matrix printing, notched-card mechanisms, and other card-retrieval systems. On January 15, 1952, IBM made him manager of a new laboratory to be established in San Jose, California. It was to be IBM's first applied-research laboratory and, under Johnson's direction, one of the most successful within the American data processing community. He introduced ideas that were new at IBM's laboratories. For example, he insisted that all engineers in the laboratory be familiar with everyone else's projects. Moreover, when engineers were asked to consult for each other or for other parts of the company, those requests were to be treated as the highest priority, not one's own pet project. Within a staff that initially consisted of fewer than a dozen people, these principles worked well. In the 1960s, when the staff had grown to thousands, successive plant and laboratory managers introduced other management principles. His laboratory's primary mission was to conduct research on digital computing.

The most important of Johnson's projects involved launching IBM into the world of disk storage. The primary area of focus at this laboratory remained disk storage for the next thirty years. In June 1953, his staff had a working model of a disk drive and demonstrated it publicly during 1955 and 1956. It eventually became IBM's initial disk product, the 350 Disk Storage (RAMAC* Project), the first movable-head disk unit in the data processing industry.

Johnson continued to manage research and development of disk storage technology throughout the 1950s. By 1962, the laboratory at San Jose employed nearly 3,000 people and had manufacturing facilities as well. Yet for the purpose of data processing's history, his major contributions had been made: pencil-sensing equipment and the RAMAC.

For further information, see: Charles J. Bashe et al., *IBM's Early Computers* (Cambridge, Mass.: MIT Press, 1986).

K

KATZ, CHARLES (1927–). This computer scientist helped develop important computers in the 1950s and 1960s; he participated in the creation of the UNIVAC* series of the 1950s and later products from General Electric.† This mathematician turned computer scientist did a great deal of his work in the field of software* but also made contributions to the logic design of computers, typical tasks of many mathematicians who became involved in the development of computers over the previous forty years.

Katz was born in Philadelphia in 1927 and completed his B.S. in mathematics at the local Temple University in 1950. Philadelphia had been the radio and electrical center of the United States during the 1930s and 1940s, the place where numerous projects relevant to data processing were conducted, including the ENIAC* which is considered the first electronic digital computer.* Against this background Katz continued his graduate work, completing an M.S. in mathematics from the University of Pennsylvania in 1953. His first foray into data processing came in June 1953 when he went to work for Sperry Rand* to work for Grace M. Hopper,** who at the time directed a group working on software for the UNIVAC computer. Her organization sought to create a programming compiler and high-level languages for use in the new computer.

This mathematician first worked on the A–2 Compiler* which ran on the UNIVAC I. Next, work on new releases of the compiler was required. Yet the A–2 was a fixed form algebraic compiler which became the first commercially available device. Katz subsequently led the development effort for MATH-MATIC,* an early high-level programming language* for the UNIVAC that employed free-form algebraic and English expressions. The system became available in late 1956 and subsequently ran on the UNIVAC II as well. Katz worked on the other early UNIVAC software—the FLOW-MATIC* compiler—which was a programming language intended for business applications. In contrast, MATH-MATIC was designed for engineering, scientific, and mathematical applications. FLOW-MATIC was one of the first English-like programming languages, and it served as a useful experience for those who developed COBOL* in the early 1960s. Katz and three other U.S. citizens representing the American data processing community, met in 1958 in Zurich, Switzerland, where the initial idea of developing a universal programming language emerged. It was called the International Algebraic Language (IAL) and later became ALGOL 58.* Throughout the 1960s, Katz participated in various task forces manned by the data processing industry involved with that language and reported on some of the key developments in policy papers.

In 1959, Katz left the UNIVAC group and joined General Electric as Manager of Systems Software within its Computer Department. There he had full responsibility for all the software GE developed for the GE–200 family of computers and later for the GE–400 and GE–600. The packages produced for these systems over which Katz had direct control included a compiler for FORTRAN,* as well as one for COBOL, WIZ (a one-pass fast ALGOL-dialect compiler), and GECOM* which represented a marriage of features from both ALGOL and COBOL. In 1966, GE asked Katz to stop producing compilers; he now turned to the creation of real-time software and application packages.

In the late 1960s, Katz was working for Burroughs Corporation† developing software. Replicating some of the successes at GE, he developed and implemented software for the B8300 system, making specific products for those who wanted real-time multiprocessing. He also played a role (which was not fully understood at this time) in the design of an airline reservation system for Trans World Airlines (TWA). During the 1970s, almost all of his work involved the development of real-time systems, including a Postal Source Data System, Attendance System, and the Work-Load Recording System all for the U.S. Post Office. During the 1980s he joined Xerox Corporation† where he was working on BISCOM (a system for switching messages), and on a project to build an order entry system for the Bell Telephone System.

For further information, see: Charles Katz, "GECOM: The General Compiler," in *Symbolic Languages in Data Processing* (New York: Gordon and Breach, 1962): 495–500; Jean E. Sammet, *Programming Languages: History and Fundamentals* (Englewood Cliffs, N.J.: Prentice-Hall, 1969); Nancy Stern, *From ENIAC to UNIVAC: An Appraisal of the Eckert-Mauchly Computers* (Bedford, Mass.: Digital Press, 1981).

KILBY, JACK ST. CLAIR (1923–). Along with Robert N. Noyce,** Kilby is considered one of the developers of the "chip,"* the building block of all computerized technology since the mid-1960s. The development of the chip (and consequently of the modern computer) is considered to be one of the single most important events in the evolution of technology. The chip brought "intelligence" to many pieces of equipment from rocket guidance systems to microwave ovens to computers. Kilby, along with Noyce, ushered in a multibillion dollar industry (modern electronics) and possibly a post-industrial era. Little known outside of a small circle of computer scientists, these two men have had an influence equal to that of Thomas Edison and possibly more than Albert Einstein.

Kilby was born in Jefferson City, Missouri, on November 8, 1923, and graduated from the University of Illinois with a B.S. in electrical engineering in 1947. In 1950, he completed a master's degree at the University of Wisconsin. He worked as a program manager for Globe-Union, Incorporated in Milwaukee and then joined Texas Instruments, Incorporated† as a vice-president in 1958, staying until 1970. Subsequently, he consulted, did research, and in 1978 joined the faculty of Texas A & M University as distinguished professor of electrical engineering, remaining there at least until 1986.

Kilby's most important accomplishment has been his invention of the monolithic integrated circuit. While working for the Centralab Division of Globe-Union, he became interested in electronics of the sort that eventually resulted in improvements in electronic circuitry. In 1952, he attended a seminar at Bell Laboratories† devoted to the transistor which had been developed there several years earlier. Back at Centralab, he put this newfound knowledge to work trying to develop hearing aids based on germanium transistors. During this period he concluded that silicon would be a better raw material on which to structure the development of transistors. He found the electrical conduction properties of silicon to be better, even if they were still semiconducting; however, Globe-Union had committed itself to germanium, and, probably more for that reason than for any other, Kilby decided to leave the company and join Texas Instruments in 1958. It had already started work on silicon.

Kilby wanted to provide an evolutionary path beyond the existing solid-state transistors then available. What in effect happened, however, was that the chip represented a revolutionary change from the past. He sought to miniaturize circuits while reducing costs. As a result of his research in 1959, Texas Instruments filed for a patent on what would become some of the earliest integrated circuits built in the United States. Soon after, Texas Instruments began manufacturing what it called a ''semiconductor solid circuit.'' The work was marred, however, by its lawsuit against Robert Noyce and Fairchild Industries† for infringement of patent. As it turned out, Noyce had filed for a similar patent in 1959. The debate over which firm really developed the first integrated circuit has therefore raged to the present; however, in 1969 the courts ruled in favor of Noyce. Meanwhile, the two firms had licensed their techniques to other companies entering the electronics field.

Historians credit Kilby with building the first working version of an integrated circuit but agree that Noyce made it practical enough for industrial use. Kilby was eventually inducted into the National Inventors Hall of Fame as a co-inventor, and his work has also been recognized by others. In 1969 he received the National Medal of Science and in 1967, the Ballentine Medal from the Franklin Institute. Other awards bestowed on him throughout the 1960s and 1970s are testimonials to the importance of his work. The controversy surrounding who invented what and when accents a pattern of behavior in the field of science. Often, particularly in the case of computerized technologies, several people will be working on a particular topic and at about the same time. Patent laws and rivalries repeatedly document this pattern. The case of Kilby and Noyce, while interesting and important for themselves, once again confirms the typically evolutionary and sometimes obvious next step that has to be taken with a particular technology.

Kilby, described as an engineer excited at the prospect of creative research, was gentle, unhurried, and went about his work quietly in the 1950s. This lanky, casual individual stumbled across the idea of how to put multiple devices on one sliver of silicon by figuring out how to interconnect electronic devices (in this case functions). This resulted in reduced size, less dissipation of heat, the elimination of mechanics, and much cabling or wiring, while shrinking labor and materials costs. The development of chips became possible because they replaced the more expensive transistors and were more reliable. Although transistors had represented a quantum leap forward in improvement when compared to vacuum tubes in the 1940s, a decade later chips were even more impressive—so impressive and effective that within ten years of its invention, no manufacturer of computers still used transistors. Chips were simply far more reliable, took up less space, and from the beginning were considerably cheaper. The chip was largely responsible for bringing the cost of data processing down and its availability up during the 1960s, often at measurable rates of costs versus productivity in excess of 30 percent each year. This trend, while slowing in the 1970s and 1980s, continued dramatically in double digit figures.

For further information, see: Dirk Hanson, *The New Alchemists* (Boston: Little, Brown and Co., 1982); T. R. Reid, *The Chip: How Two Americans Invented the Microchip and Launched a Revolution* (New York: Simon and Schuster, 1984); Everett M. Rogers and Judith K. Larsen, *Silicon Valley Fever: Growth of High-Technology Culture* (New York: Basic Books, 1984).

KILDALL, GARY A. (1942–   ). Like many other young American experts in computing, this entrepreneur founded a company to market software* which he developed for microcomputers. Kildall established Digital Research, Inc., to sell the first and most popular of the early operating systems* for micros, called the Control Program/Monitor (CP/M). He was one of several dozen young programmers who early in the era of personal computers (1970s) saw their potential, created some of their technology, and became wealthy before the age of forty.

Kildall was born on May 19, 1942, in Seattle, Washington, the son of a merchant marine barge captain. His family operated the Kildall Nautical School, established in 1927. Gary Kildall taught part-time there and eventually wrote software applications for the school using a micro. While in high school in Seattle, he tinkered with automobiles, built devices to do such things as practice Morse code with a tape recorder, operate flip-flop binary switches, and operate burglar alarms in automobiles. He attended the University of Washington, majoring in mathematics, he next earned an M.S. in computing, and, in 1972, he completed his Ph.D. there in computer science. Between undergraduate and graduate school Kildall also served in the U.S. Navy, teaching at the U.S. Naval Postgraduate School before and after his tour of service. At that school he instructed naval personnel in computer programming, and in turn he became interested in the use of microchips.

Kildall's first contact with such technology came in 1972 when he began programming with Intel's† 4004 microprocessor (computer on a chip*). His initial applications were primarily navigational. Although these did not interest Intel, various mathematical routines he had developed did. In exchange for these, Intel gave him a prototype of a personal computer called the Sim-04. This happened in 1972, three to four years before such technology became available even in kit form, thus making him one of the first people ever to work with micros. One of the first programs he wrote for the machine was Programming Language for Microcomputers (PL/M) which ran on the 8008 chip, the follow-on to the 4004. He next created Control Program for Microcomputers (CP/M) which was of no interest to Intel. By 1975, hardware technology for micros had progressed far enough to make CP/M a practical operating system. Kildall began marketing this software through mail orders generated from advertisements he ran in *Dr. Dobbs' Journal,* an early microcomputer magazine.

In 1976, Kildall established Intergalactic Digital Research to sell the software but soon after changed its name to the more familiar Digital Research. His firm sold what amounted to the most popular of the early operating systems. In 1981, International Business Machines Corporation (IBM)† chose to use an operating system for its Personal Computer (PC) developed by another native of Seattle—William H. Gates,** founder of Microsoft Corporation†—called MS-DOS. These two operating systems were the industry standards of the early 1980s and remained the most popular as of 1987.

Kildall resigned his teaching position at the naval school in 1977 to devote full time to his new business. By the end of 1983, his company had 665 employees and sales of $44.6 million. In addition to serving as chairman of the board of Digital Research, in 1984 he formed Activenture Corporation (now called KnowledgeSet Corporation) to produce optical-disc publishing products. He announced the firm's first major project in 1985: publication of *Grolier's Encyclopedia* in CD ROM format.

For further information, see: Paul Freiberger and Michael Swaine, *Fire in the Valley* (Berkeley, Calif.: Osborne/McGraw-Hill, 1984); Susan Lammers, *Programmers at Work, 1st Series* (Redmond, Wash.: Microsoft Press, 1986); Robert Levering et al., *The Computer Entrepreneurs* (New York: New American Library, 1984).

KNUTH, DONALD ERVIN (1938–). Knuth is one of the best known of the early computer scientists in the field of programming languages.* He has written a seven-volume study on the subject and has played an important role in some of the data processing industry's most important organizations.

Knuth was born in Milwaukee, Wisconsin, on January 10, 1938. He studied mathematics at the Case Institute of Technology (today called Case Western Reserve University), completing his course of study in 1960. Because he took so many graduate-level courses as well, he was awarded the B.S. and M.S. degrees simultaneously; he graduated first in his undergraduate class. While at Case, he first encountered computers and wrote assembler and compiler code for the school's IBM 650.* He developed a basketball scoring system used on the computer that was described by *Newsweek* Magazine and later was also shown on CBS television. Knuth went on to the California Institute of Technology to do graduate work, completing his thesis in mathematics in 1963. He remained there as an assistant professor in mathematics and was promoted to associate professor in 1966. During 1968–1969, he was a mathematician at the Institute of Defense Analysis, Communications Research Division, located in Princeton, New Jersey. In the fall of 1969, Knuth joined the faculty of Stanford University where he was a full professor as of 1986. From 1960 to 1968, he also consulted for the Burroughs Corporation† on various computer-related subjects.

Knuth wrote a compiler for ALGOL 58* in 1961 that ran on a Burroughs 205 computer. The following year he prepared compilers for FORTRAN II* used on the UNIVAC solid-state 80 and 90 systems. He helped promote the study and use of ALGOL, and in 1963–1964, he was chairman of a committee looking at ALGOL for the Association for Computing Machinery (ACM).† Between 1964 and 1967, he edited the *ACM Journal* and in 1966, *Communications of the ACM,* perhaps the most widely read technical journal in the field of data processing. In 1966–1967, he was an ACM National Lecturer and Visiting Scientist. This prolific writer has published over 100 articles on computer science, but the majority are on software. He is most remembered for his seven-volume study, *The Art of Computer Programming* (1968–). Knuth has done research on analyzing algorithms (the subject of Volume 1), programming languages in general, and their history.

In 1971, Knuth became the first recipient of the Grace Murray Hopper** award, one of the most prestigious awards given by the data processing industry. In 1979, he received the National Medal of Science. Knuth was a Guggenheim Fellow (1972–1973) and a Fellow of the American Academy of Arts and Sciences. In 1974, he received the Alan M. Turing award, and in 1975 the Mathematical Association of America honored him with its Lester R. Ford award. Even more honors were to come: the Institute of Electronics and Electrical Engineers (IEEE),† the McDowell award in 1980 and the Computer Pioneer award in 1982—all for his work with programming languages.

For further information, see: Donald E. Knuth, *The Art of Computer Programming* (Reading, Mass.: Addison-Wesley, 1968, Foreword, with the first three volumes in 1968, others through 1973), and with L. T. Pardo, "The Early Development of Programming Languages," in N. Metropolis et al., eds., *A History of Computing in the Twentieth Century* (New York: Academic Press, 1980), pp. 197–274.

L

LAKE, CLAIR D. (1888–1958). Lake, an engineer and inventor at International Business Machines Corporation (IBM),† developed IBM's first printing tabulating equipment and worked on other card punch products for over thirty years. He is most remembered for his work in helping to build the Harvard Mark I* for Professor Howard H. Aiken** in the early 1940s. Lake was also responsible for the standard computer card used in the data processing industry from 1928 to the present.

Lake's professional career developed in an ad hoc manner. After completing the eighth grade, he elected not to attend high school but rather a manual training program. His earliest work involved designing parts for automobiles. In 1915, Thomas Watson, Sr.,** head of C-T-R (which would become IBM in the 1920s), hired him to help design and build a printing tabulator machine. Between slow progress, the eruption of World War I, and other duties, he was not able to complete the task until the end of the war. He improved existing tabulating products which enabled him to develop a printing version (first introduced in 1919). Lake continued to refine this and other tabulating equipment for IBM during the 1920s, thereby helping to expand the acceptance of the company's products. As a reward for making the printing tabulator a successful product, he was made the superintendent of all tabulating machine activities at IBM's plant at Endicott, New York. He was now in charge of the company's most important product line. He next developed what became known as the eighty-column "IBM" card, introduced in 1928. For generations this was the standard "computer card" that one could not "fold, spindle, or mutilate."

As in the 1920s, Lake continued to enhance and develop IBM's card punch product line during the 1930s. This included the very popular IBM Type 405 alphabetical accounting machine which remained a profitable product until after World War II. During this second decade of his career at IBM, tabulating equipment was developed which allowed the newly formed Social Security

Administration to keep records and print checks. That contract with IBM allowed the company to enter the 1940s as one of the most respected and fastest growing corporations in the United States.

Lake's next important contribution came with the development of Aiken's calculator. In the 1930s, IBM had an interest in research being done by others on mechanical computational devices. One of these areas of concern to IBM involved what would ultimately become the electronic digital computer* of the 1940s. An early step in that direction was represented by Aiken's work at Harvard while a graduate student. Theodore H. Brown,** who taught there and consulted for IBM, encouraged the company to work with Aiken on a joint project. IBM agreed to help sponsor Aiken's research and assigned Lake as its contact with him.

On the eve of World War II, Lake promptly set engineers to work with Aiken at Endicott and made components from IBM equipment available to him. Meanwhile, Lake directed his own energies to building a fast calculator for the Aberdeen Proving Ground and, later, card reading and punching equipment for the U.S. Army. Lake's engineers built parts for the Mark I during 1941 and 1942, and, in January 1943, its first tests were conducted. Before the end of the war, the machine was fully operational.

Considerable historical controversy exists over who should get credit for building the Mark I: Aiken or IBM. At the time of the Mark I's unveiling, Aiken clearly implied to the press that he had built the machine. As a result, Watson refused to have anything further to do with him, and some people incorrectly assumed that IBM's head had washed his hands entirely of computers. Engineers at IBM's Endicott plant argued that they had built the machine using Aiken's ideas and many of their own. The latest historical evidence suggests that (1) Aiken was sloppy in the way the machine received initial publicity, thereby cutting himself off from any future aid from IBM; (2) he did create the overall design/architecture of the machine; (3) but Lake's staff and others at IBM bolted the pieces together, invented components, and made innumerable enhancements to his designs; (4) finally, IBM expended the dollars involved. One consequence of the entire project was a set of experiences with advanced computational components among IBM engineers at Endicott.

In the late 1940s, Lake, taking advantage of that newfound knowledge, helped build high-speed counters and calculators, relays, and other components which appeared in IBM's products in the late 1940s.

Lake was an effective, "results-oriented" inventor who was able both to create new components and equipment and improve existing ones. He was an early model of the technical manager who was responsible for the efforts of many engineers working on complex portions of larger projects. The next great model of a manager of large data processing projects was Jay W. Forrester** who developed WHIRLWIND* at the Massachusetts Institute of Technology between the waning years of Lake's professional career (late 1940s) and the mid-1950s.

For further information, see: Charles J. Bashe et al., *IBM's Early Computers* (Cambridge, Mass.: MIT Press, 1986); Paul E. Ceruzzi, *Reckoners: The Prehistory of the Digital Computer, From Relays to the Stored Program Concept, 1935–1945* (Westport, Conn.: Greenwood Press, 1983).

LANGEFORS, BÖRJE (1915–). This Swedish computer scientist has designed and built analog and digital computers* and is a leader in the Swedish data processing community.

Langefors was born in Ystad, Malmöhus, Sweden, in 1915 and finished his studies in engineering at the School of Engineering in Malmö in 1937. He attended the Royal Institute of Technology in Stockholm in 1944; earlier, from 1937 to 1944, he was an engineer at a shipyard designing turbines. Between 1940 and 1942, he also conducted research on aerodynamic windtunnels at the Swedish Aeronautical Research Institute. He developed computational methods for the design of turbines for Ljungströms Angturbin in 1942. Between 1943 and 1946, he was head of the Aircraft Instruments Laboratory where he helped to design an automated dynamic balancing machine. This was an important experience for Langefors because it had analog computing qualities. Between 1946 and 1949, he ran the Hydraulic Laboratory. His most dramatic entry into the field of computing came in 1949 when he was appointed head of the Computation Laboratory and Engineering Data Processing Center, charged with designing and building analog computers.

In addition, Langefors constructed a digital computer called the SARA (1955–1957) and an error-correcting magnetic tape system, completed in 1957. His staff also worked on a numerically controlled milling machine in the 1950s when similar work was being done in the United States. They also worked on a variety of computer-based mechanical design systems, including matrix methods for structural mechanical analysis and systems analysis, design tools for use with aircraft and ships, and, finally, computer graphics in general.

Between 1962 and 1965, Langefors was manager of the Systems Department at Data Saab, a computer manufacturer. He headed up the project to develop a computer architecture which later appeared in the Saab D21. In his first year he also acquired responsibility for developing software,* including the ALGOL-Genius general-purpose programming language and its compiler, which became operational in April 1964. He did systems analysis and designed computer-aided data systems while developing ideas about the theory of information systems.

In 1965, Langefors became professor of business information systems at the Royal Institute of Technology and at Stockholm University. From 1960 to 1965, he served as president of the Swedish Society for Information Processing and as his nation's representative on the International Federation for Information Processing (IFIP)† Council. In 1965, he also served as chairman of the Program Committee for IFIP's congress, and in 1966, he was named president of the Swedish Society of Engineers, a position he held into the 1980s.

For further information, see: Börje Langefors, "The D21 Data Processing System by Svenska Aeroplan Aktiebolaget, Sweden," *IEEE Transactions on Electronic Computers,* EC–12, No. 5 (December 1963): 650–662, and "ALGOL-GENIUS, A Programming Language for General Data Processing," *BIT* 4, No. 3 (1964): 162–176.

LEHMER, DERRICK HENRY (1905–). This American mathematician designed a numbers theory machine in the 1930s and used the ENIAC,* the world's first electronic digital computer,* in the 1940s. As a consequence, early in the era of computing he established a reputation as a user of such technology in order to probe complex theoretical number problems. At the Ballistic Research Laboratory in the United States, he was instrumental in determining how the ENIAC would be used.

Lehmer was born in Berkeley, California, on February 23, 1905, and attended the University of California, completing his A.B. in 1927, and his Sc.M. and Ph.D. in mathematics at Brown University in 1929 and 1930, respectively. He became a mathematician for the National Research Council at the California Institute of Technology (1930–1932), and a researcher at the Institute for Advanced Study at Princeton, New Jersey (1933–1934), and moved to Lehigh University for the rest of the decade, serving as instructor and assistant professor in mathematics (1934–1940). During the 1938–1939 academic year, he was a Guggenheim Fellow at Cambridge University. He returned to Berkeley in 1940, rising to full professor; he retired in 1972. In 1945–1946, he was a mathematician for the Aberdeen Proving Ground and, between 1951 and 1953, director of the Institute for Numerical Analysis. Both organizations were early and important supporters of computing in the United States.

Lehmer showed his faith in data processing not only through his research but also through his leadership role. He was vice-president of the American Mathematical Society in 1953 and vice-president of the Association for Computing Machinery (ACM)† between 1954 and 1957. Beginning in 1952, he served as a member of the advisory panel of the National Science Foundation (NSF) which funded many research projects in the general field of computing.

Lehmer's most productive period came in the 1930s and 1940s, particularly during World War II when he was associated with the Ballistic Research Laboratory along with a distinguished set of colleagues. On the staff with him, for example, was Herman H. Goldstine** who obtained U.S. Army support for the ENIAC and later had a highly successful career as a scientist with International Business Machines Corporation (IBM).† Lehmer was therefore very familiar with the development of the ENIAC and other computer-related projects. He encouraged the dissemination of material on the ENIAC and on computing in general. In 1945, he was a member of a committee created by the National Research Council to hold a conference on computing equipment at the Massachusetts Institute of Technology (MIT) and Harvard University, one of the earliest such gatherings to discuss computers. Always interested in numbers theories, Lehmer also employed the ENIAC in his own research. His most famous use of the ENIAC involved a study, together with his wife Emma in 1946, of certain primes done with the machine. This project was part of his larger interest in theoretical number problems. When the Moore School of Electrical Engineering,† which had built the ENIAC, turned the system over to the Ballistics Laboratory in 1946, he played a leadership role in determining how it would be used.

Lehmer's earlier and initial contact with the concept of computing with hardware is not as well documented as his relationship to the ENIAC. He designed and built a primitive processor in the 1930s. His father, also a mathematician, had developed a set of card stencils which one could use to find quadratic residues. These cards were first published in 1929 and proved relatively easy to use. Thus, a mathematician now had a tool to find out quickly if a specific number was prime or composite. Ten years after its initial publication, another American, John Elder, converted this set of stencils to cards which operated on IBM punch cards. Using data processing equipment made it possible for a mathematician to determine whether any number smaller than 3.3 billion was prime.

It was against this background that the younger Lehmer studied numbers. In 1933, he had worked out a description of a fast-operating photoelectric number sieve which he then constructed. Its purpose was to solve problems in number theory. In effect, that device partially mechanized his father's set of stencils. Lehmer could now generate the "Sieve of Eratosthenes," named after its creator who lived from 275 to 195 B.C. Lehmer's device comprised a collection of gears sharing a shaft, all of which rotated quickly. The gears had holes at their peripheries. The solution to a problem was to search for a particular sequence of holes. These holes were sought out by the photocell. The gadget worked, and up to 300,000 numbers could be handled in one minute.

This project gave Lehmer the necessary experience and confidence to enthusiastically support mechanical computing when, a decade later, he influenced the use of digital computers.* Thus, for instance, when the Moore School was entertaining the notion of building a follow-on machine to the ENIAC (known as the EDVAC*), Lehmer blessed the project. He even pointed out how a new computer might appear employing a bank of acoustic delay lines to do the kind of work originally executed on his gear and shaft machine.

Lehmer had already modified his own machine, transforming it from a strictly mechanical to an electronic device. He experimented with it for years, and it eventually acquired the formal name of the Delay Line Sieve (DLS–127). Constructed at Berkeley, it was a solid-state machine built with the help of students and faculty as a joint project shared by the departments of mathematics and electrical engineering. The DLS–127 became operational in 1965.

Lehmer also did research using yet another early, famous computer, the SWAC,* which was located at the Institute for Numerical Analysis at the University of California at Los Angeles in the early 1950s. The jobs he ran on this computer involved the study of a Fermat theorem in number theoretical conjectures.

For further information, see: Herman H. Goldstine, *The Computer from Pascal to von Neumann* (Princeton, N.J.: Princeton University Press, 1972); Derrick H. Lehmer et al., "An Application of High-Speed Computing to Fermat's Last Theorem," *Proceedings, National Academy of Science, USA* 40 (1954): 25–33; D. H. Lehmer, *Factor Stencils* (1929), as published by John D. Elder (Washington, D.C.: Carnegie Institution of Washington, 1939).

LEIBNIZ, GOTTFRIED WILHELM VON (1646–1716). This German mathematician built mechanical calculators and produced important findings on symbolic logic, the basis of computer technology today. His accomplishments are remarkable given the paucity of mathematical training in Europe in his day. It was not uncommon, for example, for a university graduate not to know how to multiply well. Yet Leibniz lived in an age when scientific activities in astronomy, mathematics, navigation, and physics were expanding. In this awakening environment he was a star: he flirted with binary numbers, developed the basis of differential calculus, launched the study of formal logic, recognized the waste of human talent slaving over arithmetic calculations that should be done by mechanical means, and even argued the case for testing mathematical hypotheses mechanically. With regard to his impact on computer technology, he is important for his work on calculus and logic, and his input into the development of languages and language/machine relations.

Leibniz was born in Saxony. A child prodigy educated at home, by the age of ten he had a working knowledge of Greek and Latin, by fifteen he matriculated at the University of Leipzig, by seventeen he completed his studies, and by twenty he had his doctorate. He spent the next six years in law and diplomacy. While on a diplomatic mission to Paris during 1672–1676, he studied mathematics, particularly the work of Blaise Pascal** and René Descartes.

Leibniz's contact with Pascal's calculator led him to study the problem of making useful calculators. He shared Pascal's perception that certain laborious mathematical calculations should be automated. Also like the French scientist, his father had also been a tax collector and thus had personally witnessed the long hours required to prepare sets of numbers (tax figures in their cases). After studying Pascal's adding machine, called the Pascaline, Leibniz saw the need for a device that could also multiply and divide—functions Pascal's calculator was not able to perform. He envisioned a calculator that would add and subtract just like Pascal's and another part that conducted multiplication and subtraction, all four pieces working compatibly rather than competitively to the Frenchman's creation.

In order to make the two more complicated mathematical functions possible, Leibniz developed a multiplier wheel with digits of 9 to 0 and with teeth of varying sizes. This arrangement allowed him to build a machine with a movable carriage. It would shift by wheels employing the principle of an active-inactive pin arrangement. He also used a delay carry mechanism, the design for which was still employed in desktop calculators as recently as the post–World War II period. The net effect of his entire design was a machine that could multiply and divide without requiring an operator to use an algorithm.

Leibniz's original idea was to use a rachet-carry mechanism, which he later replaced with a Geneva gear system. This modification made it possible to build a machine that could carry digits by a series of star-shaped gears, each with five points in a discontinuous manner. Division and subtraction could be performed by reversing the rotation of the addition and multiplication parts of the machine. The device worked very well and in varying forms inspired other inventors to adopt his mechanism in calculators for the next 300 years. And for his day, of all those working on calculators, his was the most advanced and well received. A copy of one of his machines, built in the 1670s, was sent to Peter the Great to pass on to China as proof of the sophistication of Western arts and industries.

Leibniz's work in mathematics proved equally important in the long history of the computer. Mathematics captured his interest in Paris and, as the 1670s progressed, caused him to spend less time with calculators and more with the pure science itself. Yet as early as 1666, this genius had published his first paper on mathematics. It concerned what he called Universal Combinatorics, entitled *De Arte Combinatorica*. It was his first attempt to reduce the logic of thought to mathematical calculations. Although neglected in his lifetime, George Boole* and others in the nineteenth century used his ideas to develop Boolean algebra. This branch of mathematics led to theory of probability, the theory of lattices, the geometry of sets, and information theory. Without this foundation, the design of the digital computer* would not have taken the form it did.

Leibniz's work on symbolic logic was therefore critical. In the 1670s, he also worked on differential calculus, and in 1675, he publicized the fundamental theorem of calculus. To paraphrase his idea very simply, he observed that the process of summation of integration was the same as reversing the operation of differentiation. Although Newton had already reached this conclusion, there seems to be no evidence that Leibniz was aware of the Englishman's work. Hence, his work on calculus is as significant as Newton's for the purpose of computer history.

Leibniz left Paris in 1676 after a brilliant period in his investigations. He moved to Hanover where for the next forty years he served as both librarian and historian. In these decades he wrote history and philosophy, corresponded with other scientists, and held several diplomatic positions. In 1700, he founded the Berlin Academy of Science. At the time of his death, he was studying the theological ramifications of Newton's *Principia* and *Optiks*. Leibniz's work in Paris was appreciated by other scientists, and in 1673 he was made a Fellow of the Royal Society in London. In that same year he exhibited his calculator. Although at the time of his death many had already forgotten him, future centuries would not neglect him. Copies of his machines remained in existence in various countries and he left behind an impressive list of publications in mathematics.

For further information, see: J. E. Hoffman, *Leibniz in Paris, 1672–1676* (London: Cambridge University Press, 1974); L. L. Locke, "The Contributions of Leibniz to the Art of Mechanical Calculation," *Scripta Mathematica* 1 (1933): 315–321; R. Taton, *Histoire du Calcul* (Paris: Presses Universitaires du France, 1969 [5th ed.]).

LOVELACE, AUGUSTA ADA, COUNTESS OF (1816–1852). Augusta Ada, daughter of the poet Lord Byron, provided the clearest expression of Charles Babbage's** difference engine* and analytical engine* written in the 1800s. She and Babbage were close friends, and Lovelace frequently explained to his visitors how his machines were intended to work.

Lovelace was the only child of Lord Byron who separated from his wife only one month after Augusta's birth. Although Lovelace never saw him again, she was always emotionally tied to him and even expressed the desire to be buried with him after she died. Her wish was granted, and, ironically, she died at the same age as her father had. Augusta became the Countess of Lovelace and Baroness Wentworth by marriage, providing historians with a variety of names for a woman who left few records of her activities. Most commonly, she is called Augusta Ada Lovelace, or simply Countess Lovelace. She was gifted intellectually and artistically, and by the time she was fifteen, she had already shown a remarkable ability in mathematics, an unusual talent for an English woman of her century. She had the good fortune of being a student of the French mathematician Augustus de Morgan, who was well known for his work in algebra.

In 1833, Lovelace attended a lecture concerning Babbage's engines. She became so interested in his work that she visited his workshop. She formed a strong friendship with Babbage that lasted until her death in 1852. According to de Morgan's wife, when Lovelace first saw Babbage's pieces for an analytical engine, the young woman reacted in the following manner: "While the rest of the party gazed at this beautiful instrument with the same sort of expression and feeling that some savages are said to have shown on first seeing a looking glass or hearing a gun, Miss Byron, young as she was, understood its working and saw the great beauty of the invention."

Lovelace imparted her understanding of Babbage's machine to others by undertaking to explain his work in detail. For example, General Luigi F. Menabrea, later a prime minister of Italy, became fascinated with Babbage's work while the English inventor visited Italy and wrote an article on the device which the Bibliothèque Universelle de Genèva published in 1842. Lovelace translated the article into English, added explanatory notes that were twice the length of the original paper, using help from Babbage, and published the account. She followed this up with other publications on Babbage's ideas. Simultaneously, she used her own mathematical talents to encourage him, even suggesting improvements of the design of the analytical engine. Her ideas were supportive of his work.

Augusta Ada Lovelace died of cancer in late November 1852. Her last years had been years of bored existence with her husband, of almost fanatical gambling and possibly of a too-close relationship with a man named John Crosse. Her death ended a sad period in her life and depressed Babbage greatly. However, by the time of her death, Babbage's greatest contributions to the development of computing devices were nearly over.

For further information, see: V. R. Huskey and H. D. Huskey, ''Lady Lovelace and Charles Babbage,'' *Annals of the History of Computing* 2, No. 4 (October 1980): 299–329; A. Hyman, *Charles Babbage: Pioneer of the Computer* (Princeton, N.J.: Princeton University, 1982); Doris Langley Moore, *Ada, Countess of Lovelace* (London: John Murray, 1977).

LUDGATE, PERCY E. (1883–1922). Little is known about this Dublin accountant except that he took a great interest in calculating devices, designing one which on paper at least appeared to be a significant improvement over the work done by Charles Babbage**. He was one of the last examples common in the nineteenth century of the inventor working privately and quietly in his spare time. Within a decade of his death, it became the norm for major scientific work to be done by teams spending great sums of money in well-stocked laboratories.

Ludgate was not attached to any university or research laboratory. With a full-time job as an accountant, his work on an analytical engine* had to be done in the evenings and on weekends. Indeed, he conducted his work in such isolation that he may not have been aware of previous research projects on automated calculators other than for what was commercially available to accountants of his day. Two of his publications on his design reveal that he eventually learned about Babbage's efforts. Perhaps his sparse knowledge of previous attempts made it possible for him to design a highly original machine. The only other known contribution he made was organizing the supply of food for horses of the cavalry during World War I.

In 1909, Ludgate published a paper in which he described the design of a program-controlled general-purpose computational device. He used the term *analytical engine* to describe his machine. It had three parts: a store function, an arithmetic feature, and a sequencing mechanism. Multi-digit numbers were represented by sets of rods that could slide within a shuttle box. They were rotated in order to send the correct number to the arithmetic unit. For multiplication, he proposed converting digits of the multiplicand and one of the multiplier into index numbers. Each index number representing the multiplicand digit would be added to the index number of the multiplier, employing additive linear motion. He then suggested that these results could be converted to sets of two-digit partial products. To facilitate the process, he developed tables which one author dubbed Ludgate's ''Irish logarithms.'' Division would be handled by a method using a series of approximations. By this means he hoped to avoid using the then current method of trial and error with repeated subtractions. He also envisioned that the subtraction process would be incorporated into the machine as a built-in subroutine. He thought that the best way to handle the process would be to use a perforated surface of a metal cylinder that could rotate within the box. Brian Randell** has likened the process to read-only memory on a computer.

Another aspect of Ludgate's machine was sequence control, that is, that portion of the design which told his machine how to perform and when. He thought of using perforated tape which would have specified operation code, two operands and addresses, and finally an address for the result (one or two depending on the requirement). Holes in the tape would provide the means of detailing these functions. Randell thought the whole mechanism was far simpler than and superior to Babbage's. If he had ever built the machine, it would have been much smaller than Babbage's because his specifications called for a device that would take up only 8 cubic feet and that would probably be operated by electricity rather than steam. Ludgate's description suggested that this machine would have been able to multiply two two-decimal digit numbers in approximately six seconds.

Like Babbage's analytical engine, Ludgate's was never built. Furthermore, except for his two publications and Randell's research during the 1970s, we have little information on either his life or work. His direct impact on the history of computing remains an open issue.

For further information, see: P. E. Ludgate, "On a Proposed Analytical Machine," *Scientific Proceedings of the Royal Dublin Society* 12, No. 9 (1909): 77–91 and "Automatic Calculating Machines," in E. M. Horsburgh, ed., *Napier Tercentenary Celebration: Handbook of the Exhibition* (Edinburgh: Royal Society of Edinburgh, 1914), pp. 124–127. His 1909 article is reprinted in Brian Randell, ed., *The Origins of Digital Computers* (Berlin: Springer-Verlag, 1982), pp. 73–87. For a biographical treatment, see Brian Randell, "From Analytical Engine to Electronic Digital Computer: The Contributions of Ludgate, Torres, and Bush," *Annals of the History of Computing* 4, No. 4 (October 1982): 327–341 and his earlier article, "Ludgate's Analytical Machine of 1909," *Computing Journal* 14, No. 3 (1971): 317–326.

LUKOFF, HERMAN (1923–1979). This computer scientist was the single most important manager at Sperry Rand Corporation† responsible for the manufacture of the UNIVAC* computers. He also helped develop the ENIAC* during the 1940s, the first electronic digital computer.*

Lukoff was born in May 1923 in Philadelphia, the electronics capital of the United States during the two decades before World War II. As a child, he was interested in radios and built several before attending the University of Pennsylvania. There he studied electrical engineering at the Moore School of Electrical Engineering† and worked on the ENIAC. While a student, he designed electronic devices to test various components of the ENIAC. When the inventors of the ENIAC left the University of Pennsylvania in 1946 to form what would be first known as the Eckert-Mauchly Computer Corporation and later a division of Remington Rand,† he went with them. When Remington Rand merged with

Sperry, he continued with the firm as one of the key engineering managers on the UNIVAC project.

Beginning with the Eckert-Mauchly Computer Corporation, Lukoff was an important figure in the design of what became known as UNIVAC I—the first commercially constructed and widely used digital computer. This computer launched the data processing industry. Lukoff was specifically responsible for the design of its input/output features (I/O channels and magnetic tape control unit in today's parlance). When the machine went into production, he promoted its mass manufacture, the first time such a technique was applied to computers. His techniques were perfected by the time the first dozen UNIVACs were built. His manufacturing techniques proved that computers could be replicated exactly alike in a cost-effective manner. He continued to improve his manufacturing methods during the construction of other versions of the UNIVAC throughout the 1950s.

During the late 1950s, Lukoff headed Sperry Rand's design team working on the UNIVAC-LARC, which resulted in operational equipment during 1960. Thus, he had managed first-and second-generation computer manufacturing before being named head of the Engineering Department at Sperry in 1960. He held that post until 1968. During the 1960s, he supervised the design of UNIVAC III, UNIVAC 1050, and UNIVAC 9200. He established semiconductor manufacturing capability within the company and was an early user of computer-aided design software (today called CAD).

In the mid-1970s, Lukoff contracted leukemia; he died on September 24, 1979.

For further information, see: Herman Lukoff, *From Dits to Bits. A Personal History of the Electronic Computer* (Forest Grove, Oreg.: Robotics Press, 1979); Albert B. Tonik, "Eloge: Herman Lukoff, 1923–1979," *Annals of the History of Computing* 2, No. 3 (July 1980): 196–197.

LULL, RAMON (1235–1315). Lull, a Spanish mystic and one of Spain's greatest poets, was a missionary who designed a logic machine that later influenced Gottried Wilhelm von Leibniz** and impressed Georg Wilhelm Friedrich Hegel. Some historians regard him as an early figure in the history of artificial intelligence.*

Lull was born in approximately 1235 at Palma, in the offshore Catalan (Spanish) province of Mallorca. As a tutor at the Aragon court of James I, he led a dissipated existence. In 1266, he claimed to have had five visions of the crucified Christ, as a result of which he renounced his way of life and resolved to convert the Moors to Christianity—a goal he pursued as a missionary and writer throughout the Mediterranean world for the rest of his long life. As part of his effort, he studied Arabic and wrote a number of treatises on philosophy, religion, and logic. He also wrote some of the most beautiful poetry to emerge from Iberia, the best of which included *El Desconort* (1285) and *Lo Can de Ramon* (1299), both composed in Catalan, a dialect spoken in northeastern Spain.

In the course of his studies of Arabic, Lull may have learned about a group of Arab astrologers who had designed a thinking machine called the *zairja*. It relied on the use of the Arabic alphabet of twenty-eight letters to represent the twenty-eight classes of ideas found in Arab philosophy. Combinations of numerical values associated with each class and letter supposedly gave insight. Lull devised his own version of the *zairja,* which he named the *Ars Magna*. It was a primitive logic machine which he claimed was infallible, a true inspiration from God. He described a collection or series of information (knowledge), and even faculties that were all set in a pattern of concentric circles. By matching these in different configurations, one could obtain answers to questions on religion, metaphysics, morality, and science. Although it is not known whether he actually constructed the machine (he wrote that he had), it would have been composed of discs made of metal and possibly pasteboard. These were spun and subsequently matched to arrive at answers.

Various writers and scientists referred to his contraption in subsequent centuries. Both Leibniz and Hegel acknowledged his device in their writings, whereas Jonathan Swift satirized it in *Gulliver's Travels*. Historians of artificial intelligence mention Lull at least in passing. The important point is that he was one of the last "scientists" to claim successfully (in his day) that an idea drawn purely from imagination could be called science. After him, experimentation and observation increasingly became the mode of operation for researchers and thinkers. Although his concept of "divine wheels" was intriguing and even inspirational, Leibniz and others focused on the realities of mechanical calculation. It was not until the Industrial Revolution of the eighteenth century that serious attempts were made to develop logic machines.

From the point of view of the historian of data processing, Lull is important because his *Ars Magna* was conceived on the premise that a person's thoughts could be mechanized—the fundamental basis of much twentieth-century research related to artificial intelligence and computers (with their "expert systems"). It is also very probable that Lull did not consciously recognize the premise.

Lull is more properly regarded as a literary and religious rather than a scientific figure. Clearly, the church was of greatest concern to him, which was true even in his death. When he went back to Bougie in North Africa to preach Christianity in an Islamic community he was stoned to death. He died of his wounds on June 29, 1315.

For further information, see: Margaret Harmon, *Stretching Man's Mind: A History of Data Processing* (New York: Mason/Charter, 1975); Pamela McCorduck, *Machines Who Think* (New York: W. H. Freeman and Co., 1979).

M

MARQUAND, ALLAN (1853–1924). Marquand, a professor at Princeton University, constructed a device in the 1880s to solve problems in formal logic. It was a wooden box with rods and levers connected inside with catgut strings and spiral springs. Through the manipulation of these levers and rods, the machine operated so as to exhibit the valid implications of simple logical propositions. Marquand later expanded on this manual device by designing an electrical version which historians believed was the first electrically designed logic machine, although whether he ever built it is not known.

Marquand, who had been a tutor of logic at Princeton, resigned that position in 1883 to become a professor of art and archaeology. He continued to design various logic machines, and the device he built (he claimed) came from a red cedar post taken from an old building in Princeton dating back to the eighteenth century. His logic machine worked. An often cited example of how it could provide the correct answer came from Lewis Carroll's *Symbolic Logic,* in which the following problem was presented:

No birds, except ostriches, are 9 feet high;
There are no birds in this aviary that belong to anyone but me;
No ostrich lives on mince pies;
I have no birds less than 9 feet high; . . .

The conclusion generated by the machine was "No bird in this aviary lives on mince pies."

Logic machines represented a whole area of activity dating back to antiquity. In the nineteenth century, various machines such as Marquand's were constructed to mimic logic. These machines, in conjunction with advances in hardware and the mathematical expression of logic questions, formed an early yet important intellectual heritage which, by the late 1930s, led to basic concepts about the relationship of computers to problem-solving. Logicians sought mechanical ways to do their work, and Marquand's effective, if simple, machine worked. He had once been a student of Charles S. Peirce, the great logician of nineteenth-century America best known as the father of pragmatism. Peirceknew of Marquand's machine and at one point even suggested how one might use switching circuits to execute problems in logic. Peirce even suggested that, properly developed, such a machine might also be employed in solving problems in algebra and geometry.

For further information, see: Allan Marquand, "A New Logical Machine," *Proceedings of the American Academy of Arts and Sciences,* N.S. 13 (1886): 303.

MAUCHLY, JOHN WILLIAM (1907–1980). One of the giants of the modern computer era, Mauchly was the co-builder, with John Presper Eckert, Jr.,** of the first electronic digital computer,* called the ENIAC.* But more than that, because of his efforts in making computers commercially viable, he ushered in the modern era of computer science. As a consequence of his direct efforts, several successful types of computers were designed and built: ENIAC, EDVAC,* BINAC,* and, finally, the UNIVAC.* The UNIVAC launched several computers by the same name, as well as the UNIVAC Division of Remington Rand* which eventually became Sperry Rand,† a giant of the data processing industry during the 1950s. By the mid-1950s, the term *UNIVAC* had become almost synonymous with the word *computer*. The UNIVAC had become the first widely used commercial computer. In many articles and books on the history of data processing of the 1940s and 1950s, Mauchly's contributions have accurately been likened to those of Thomas A. Edison, Alexander Graham Bell, and the Wright Brothers.

Mauchly was born on August 30, 1907, in Cincinnati, Ohio. He had a brilliant mind and early in life took an interest in science. In 1925, armed with a scholarship, he matriculated at Johns Hopkins University to study engineering. After two years he changed his major to physics. Mauchly continued his studies, completing the Ph.D. in 1932 after writing a dissertation on the carbon monoxide molecule. He spent the following year at Johns Hopkins as a research assistant, calculating, among other things, the energy levels of the formaldehyde spectrum. This work proved enormously important for the future of data processing because it was during this year of tedious calculations that he first thought about the benefits of automating such functions. With that growing concern, Mauchly took his first steps toward becoming involved with computers. In the 1934 school year, Mauchly became professor at Ursinus College in Collegeville, Pennsylvania, teaching physics. There, his interest in calculations broadened. Although Mauchly's primary responsibilities were teaching, not research, he still managed to do both quite well. On the one hand, his students considered him well prepared and dynamic, and on the other, he began working on his major research interest: weather prediction. He, together with many in the scientific community, began to believe that weather forecasting could be accomplished through the heavy use of mathematics. This line of research meant that meteorology would have to rely more on mathematics than in the past, and in turn this new emphasis would require greater amounts of calculations. Mechanical tabulating equipment was simply too slow. What few formulas were available in the 1930s took far too much time to calculate weather. Thus, the problems were twofold: (1) to develop formulas helpful in forecasting weather and (2) to find methods for faster calculation. Mauchly soon came to the conclusion that electronic methods had to replace mechanical procedures, and that meant creating new equipment. He believed that the same functions could be performed on devices based on vacuum tube technology. He even built a small analog computational machine that could do limited harmonic analysis of weather data. This exercise led to a paper in 1940 on the subject of quasi-periodicity of precipitation. That summer he presented other papers on metereological subjects.

Mauchly's interest in weather forecasting led him to seek additional training in electronics to support his own work with electrical calculators. During the summer of 1941, he attended an introductory course at the Moore School of Electrical Engineering† of the University of Pennsylvania. He so impressed the staff that at the end of the course the Moore School offered him a job, which he accepted. Over the next six years, he would focus all of his attention on computational devices which resulted in the creation of the electronic digital computer.

When Mauchly joined the faculty, the school was already working closely with U.S. defense agencies on war-related research. His first project was for the Ballistics Research Laboratory (BRL) which was expanding its work on ballistics by preparing new firing tables—all of which required enormous quantities of calculations. During most of the war mathematical calculations were done by hand using mechanical calculators operated by female graduates recruited from colleges up and down the East Coast. As new artillery was developed, the BRL had to prepare range tables that showed how far shells would go at what angle at which wind velocities, and so on, and for this the Moore School was put under contract. The pressures of war increased the need to find faster, more efficient ways of preparing these tables.

Thus, work on calculating tables and Mauchly's use of similar mathematics for weather forecasting pushed him even more into working on electronically automating much of the tedium involved. He went to John Grist Brainerd,** who had responsibility at the Moore School for working with the BRL, to suggest how some of these functions could be handled using a computational device that relied on vacuum tubes. Mauchly was sufficiently encouraged to prepare a document on the subject in 1942. His paper, "The Use of High Speed Vacuum Tube Devices for Calculating," outlined his concept, and is now recognized as one of the most important early papers on computers. The management of the Moore School initially rejected his paper, which outlined how such a machine might be built and used. Mauchly detailed how such a machine could handle specific problems related to the BRL's project, and he even described how an automatic device could be constructed. Mauchly continued to discuss his ideas with other members of the Moore School, including a young engineer, J. Presper Eckert, Jr. In 1943, Herman H. Goldstine,** then an Army officer with the Aberdeen Proving Ground and responsible for speeding up the production of firing tables, revived the proposal. Years later, he recalled that Mauchly was persuasive and enthusiastic, and had clear ideas of what needed to be done.

As a result of Mauchly's views and actions and Goldstine's ability to marshal government support, interest in an automated calculator revived, leading to a government contract with the Moore School for the construction of the ENIAC. Eckert became the chief engineer on the project and has been credited as the technical genius behind its design. Mauchly was the visionary, however, who conceived the idea and subsequently managed many of its developmental activities. Mauchly and Eckert were ably assisted by such outstanding members of the Moore School as Arthur W. Burks,** T. Kite Sharpless, and Robert Shaw.

The ENIAC was completed and demonstrated in February 1946. Compared to mechanical calculators of the day, the ENIAC could perform similar calculations 1,000 times faster. Between April 1943 when work began on the machine and February 1946 when it ended, Eckert and Mauchly designed new technology, while also employing existing electronic components in order to build the computer quickly. The end result was the Electronic Numerical Integrator and Computer (ENIAC). It had some 18,000 vacuum tubes, approximately 70,000 resistors, 6,000 switches, and 10,000 capasitors. This huge device had twenty accumulators capable of handling twenty numbers. They served as registers and multipliers, and could do division—all for a total cost of $400,000.

Because of the speed of this computer (it could do 5,000 additions per second), the Moore School and its staff received considerable publicity. That attention helped lay the groundwork for the public's acceptance of the practicality of computers and aroused interest in such devices within the business and scientific communities. While the ENIAC was in the final stages of construction, the Moore School's staff was seeking to improve on it and won another contract to build the follow-on device, the EDVAC. The contract was signed in October 1944 and called for a new device that was faster, smaller, and more reliable than the ENIAC. The team of Eckert and Mauchly wanted to reduce the number of vacuum tubes used in the EDVAC over the ENIAC, which had been sufficiently unreliable to cause the first machine to malfunction frequently. They also wanted to apply the stored-program concept, that is, the idea that machines can be more efficient and faster if all programs used by a computer are housed within it at the time they are used rather than be fed in piecemeal by humans when needed. Such a basic change in the architecture of a computer would allow sets of calculations to run even faster. Eighteen months after starting on the new project, Eckert and Mauchly completed the machine, and they were able to partially make this computer an electronic digital stored-program device. That was the first machine to do so. (Today, all computers have the capability of keeping programs in storage which they will use.)

In the final days of World War II, Mauchly (before Eckert) saw the potential for building machines that could be used by businesses and laboratories. He gave considerable thought to the best way to take advantage of the new technology being developed at the Moore School. Following World War II, he and Eckert, encouraged by friends in the military establishment, applied for patents on the ENIAC. The University of Pennsylvania reluctantly let the two apply for patents. Yet in early 1946, the university established a new policy requiring employees to sign a document giving the school all rights for any patents acquired as a result of work done at the University of Pennsylvania. Fearing the loss of potential commercial advantages, Mauchly decided to leave the Moore School. He talked Eckert into doing the same thing and into forming their own company for the purpose of building commercial versions of their computer. They both resigned effective March 31, 1946, thereby ending the Moore School's important contribution to the data processing industry. After they left, other engineers who had worked on the ENIAC also departed. In the whole crisis Mauchly's almost clairvoyant vision of the future was evident. His belief that computers had a potential major role in business applications was even more remarkable since in 1946 most government, business, and scientific leaders did not yet envision a future for such a device. Today many historians find it difficult even to imagine how knowledgeable individuals in the late 1940s could not see that data processing was the most important industry of the second half of the twentieth century. Mauchly's views were therefore that much more important for the unborn industry.

Mauchly and Eckert formed the Electronic Control Company which they renamed the Eckert Mauchly Corporation in December 1947. They gave talks on computer science, while also signing up government agencies for their next machine. In order to satisfy severe cash-flow problems in their undercapitalized company caused by design expenses on the UNIVAC, they negotiated a contract with the Northrop Aircraft Corporation to build quickly a small, airborne computer, called the BINAC, to calculate missile trajectories. They built this machine relying primarily on existing technologies with some refinements. Meanwhile, they focused considerable attention on the next major machine, what would become the UNIVAC.

The role of both scientists in this business venture was clearly discernible by their actions. Eckert had primary responsibility for the technical developmental efforts. He served as vice-president of the company, preferring to function as chief engineer. Mauchly, as president, contributed to the logic design of the machines, but was primarily concerned with general management of the company, including sales. He won contracts from the A. C. Nielsen Company and the Prudential Insurance Company, along with others from government agencies. Yet the company remained woefully undercapitalized. Therefore, Mauchly had to spend a considerable amount of time and energy looking for financial support. In addition to help from the companies and agencies already under contract, he obtained assistance at the American Totalisator Company, a manufacturer of race track equipment which calculated betting odds and totals.

Eckert was an engineer with a demanding and even an unfriendly dispositionat times. He was always concerned with detail and often lacked sound business judgment. Mauchly was the more jovial, friendlier of the two, capable of negotiating and persuading and willing to manage. Although engineering decisions dominated their company in 1947 and 1948 when business issues (such as how to raise money through building additional BINACs) should have, progress was made in the development of the UNIVAC. Mauchly also spent time building demand for information about computers at a period when little was known. In addition to academic presentations, particularly at the University of Pennsylvania, in 1947 he helped establish the Eastern Association for Computing Machinery, later named the Association for Computing Machinery (ACM),† the largest computer-related organization in existence today. In 1948 he served as its second president.

Despite the success in building the BINAC, which was the first *fully* operational stored-program electronic digital computer, lack of funding continued to slow progress on the UNIVAC. In order to alleviate this problem, Mauchly searched for a buyer of their company. Remington Rand† agreed to acquire the firm in February 1950, making the Eckert Mauchly Corporation one of its divisions. In 1955, Remington Rand merged with the Sperry Corporation to form Sperry Rand.† Mauchly and Eckert remained with their project and the company during these mergers and name changes. Eckert concentrated on the design of the UNIVAC and its follow-on products, whereas Mauchly provided guidance in the UNIVAC's logic design and for some of its software,* including the development of the computer's C–10 programming code. The UNIVAC I appeared in 1951 and became a successful product for the company. UNIVAC II was announced in 1957. Mauchly retained the title of director of Univac Applications Research. In 1958, he was asked to move out of the laboratory and into the sales department, which meant a move to New York. He elected not to go.

In 1959, Mauchly left the company and formed another, Mauchly Associates, in order to continue developing computers. Eckert remained with Sperry. While with his new firm, Mauchly developed the critical path method, called CPM today, for scheduling the use of computers. In 1967, he formed a consulting firm called Dynatrend.

Throughout his career and particularly toward the end of his life, Mauchly was never widely appreciated by the public at large but he was honored by his profession. As early as 1949, the Franklin Institute bestowed on him its Howard N. Potts medal. He received a number of other awards: the John Scott award, 1961; the Modern Pioneer award, 1965; the Harry Goode Memorial award from the American Federation of Information Processing Societies, (AFIPS),† 1966;

Pennsylvania Award for Excellence, 1968; and the Emanual R. Piore award by the Institute of Electrical and Electronics Engineers (IEEE),† 1978.

Mauchly married Kathleen McNulty, a one-time programmer on the ENIAC, one of the first female programmers in history. They had seven children. He invented snap caps for bottletops and had a lifelong interest in classical music. Friends considered him bright, accessible, personable, and amiable. An honest scientist who made brilliant contributions, he was also obviously less effective as a business manager. He died on January 8, 1980.

Mauchly's last years were marked by bitter strife over a controversy surrounding the issue of who actually invented the first computer and by battles with other scientists and lawyers concerning patent issues. Most upsetting to him was a court ruling on the origins of the ENIAC. The controversy came to a head in the 1970s, after years of frustration and negative publicity for Mauchly. During the early 1970s, Honeywell† and Sperry Rand battled in court over patent rights to technology inherent in the UNIVAC and even in the ENIAC. In 1973, Honeywell won its suit, and the patent rights originally signed over to Sperry, for work done on the UNIVAC by Eckert and Mauchly, were upheld. Mauchly became particularly upset by a finding on the origins of digital electronic computers: the federal court ruled that the first inventor of the electronic digital computer was not the team of Eckert and Mauchly, but rather a professor at Iowa State University, John Vincent Atanasoff.* For the rest of his life, Mauchly argued to the contrary, attending historical conferences, writing papers, and granting interviews in an attempt to clear the record.

Today historians credit Mauchly and Eckert with being the first to develop digital electronic computers because Atanasoff only partially conceived and constructed one machine, whereas Eckert and Mauchly designed and fully built various machines which they put into productive use, thereby ushering in the age of the computer. For these reasons, Mauchly is considered a giant in the history of computers and data processing, and a major American scientist of the twentieth century.

For further information, see: John Vincent Atanasoff, "Advent of Electronic Digital Computing," *Annals of the History of Computing* 6, No. 3 (July 1984): 229–282; J. G. Brainerd, "Genesis of the ENIAC," *Technology and Culture* 17, No. 3 (July 1976): 482–488; A. W. Burks and A. R. Burks, "The ENIAC: First General-Purpose Electronic Computer," *Annals of the History of Computing* 3, No. 4 (October 1981): 310–399; J. P. Eckert, Jr., "The ENIAC," in N. Metropolis et al., eds., *A History of Computing in the Twentieth Century* (New York: Academic Press, 1980), pp. 525–539; H. H. Goldstine, *The Computer from Pascal to von Neumann* (Princeton, N.J.: Princeton University Press, 1972); J. W. Mauchly, "Amending the ENIAC Story," *Datamation* 25, No. 11 (1979): 217–219, "The ENIAC," in N. Metropolis et al., eds., *A History of Computing in the Twentieth Century* (New York: Academic Press, 1980): 541–550, "Mauchly on the Trials of Building the ENIAC," *IEEE Spectrum* 12, No. 4 (April 1975): 70–76, and his "Mauchly: Unpublished Remarks," *Annals of the History of Computing* 4, No. 3 (July

1982): 245–256; Joel Shurkin, *Engines of the Mind: A History of the Computer* (New York: W. W. Norton and Co., 1984); Nancy Stern, *From ENIAC to UNIVAC: An Appraisal of the Eckert-Mauchly Computers* (Bedford, Mass.: Digital Press, 1981).

MCCARTHY, JOHN (1927–). This scientist was one of the fathers of artificial intelligence (AI),* a major branch of scientific inquiry related to computers. McCarthy first coined the term *artificial intelligence* in the title of a conference he helped put together in 1956 at Dartmouth College where he was a young professor. His major contribution to the study of artificial intelligence came with his development of LISP,* one of the most widely used programming languages* used in the study of AI during the 1960s and especially in the 1970s and 1980s. He also helped develop two ALGOL* languages, ALGOL 58 and ALGOL 60, in the early 1960s and encouraged a large number of students and scientists to study AI over the past thirty-five years.

McCarthy was born on September 4, 1927, in Boston and completed his B.S. in mathematics at the California Institute of Technology. He did his graduate work at Princeton University, completing his Ph.D. in 1951. While there, a fellow student, Marvin L. Minsky,* also became a giant in the field of computer science and one of the best known specialists in artificial intelligence. Between 1951 and 1953, McCarthy was an instructor at Princeton. In 1953, he moved to Stanford University as an acting assistant professor of mathematics. In 1955, he joined the faculty at Dartmouth College as an assistant professor, staying until 1958. He next taught at the Massachusetts Institute of Technology (MIT), first as an assistant and then associate professor in communications science. He returned to Stanford in 1962 as a full professor, remaining there to the present.

McCarthy's mathematics ability became evident early in his academic career. While at Princeton he was named a Proctor Fellow and later the Higgins Research Instructor in the field of mathematics. Years later, he taught computer science at Stanford and was director of its Artificial Intelligence Laboratory.

McCarthy initially began his research wedded rigorously to mathematics. Yet during the summer of 1955 he worked at an International Business Machines Corporation (IBM)† laboratory where he was introduced to computers and developed his lifelong interest in them, becoming convinced that they had intelligence. The following year he arranged a conference at Dartmouth that brought together scientists working in the field of computer science and particularly in what eventually would become AI. Many of the important researchers in AI were present, including Marvin Minsky and Claude E. Shannon.** One result of the conference was a series of papers on the subject edited by him and Shannon called *Automata Studies* (1956), published by Princeton University Press. One of the first collections to deal with the subject, the book became a minor classic and a basic reference on the subject of AI for decades.

Following the conference, McCarthy thought more about intelligence and its characteristics while searching for tools to help in the study of this subject. He wanted to rely as much as possible on the precision of mathematics, and yet he also saw computers as a tool furthering that research. In 1958, he argued that human knowledge could be given some formal representation, heavily relying on mathematical discipline. McCarthy wanted scientists to build "theorem provers" that could directly link symbolic expressions to reason. His first effort in this direction was the initial creation of LISP between 1958 and 1962. By the 1980s, it had continued to evolve and became the most widely used programming language to do research on artificial intelligence.

LISP could compute using symbolic expressions rather than simply numbers. It used list structures to represent symbolic expressions, and it had conditional expressions and the ability to do recursive transactions. Boolean logic played a critical role in the development of this language. The language gained a foothold while McCarthy worked with Marvin Minsky on their new artificial intelligence project at MIT. Throughout the 1960s, new releases of the language appeared, operating on IBM's computers and on Digital Equipment Corporation's† PDP* minis in the beginning.

McCarthy's interest in LISP was part of his general analysis of issues associated with artificial intelligence. Like many other scientists, he saw chess as yet another means of understanding intelligence, and he thought that building chess-playing devices would offer insight into mechanical intelligence. Throughout his entire career, he never lost interest in such equipment.

McCarthy also studied how a program could take advice to improve its performance. To address this issue, he proposed a system (which was never built) called the Advice Taker in 1957. Part of the work on this proposal went into LISP as an attempt to create software* that had common sense, used what it knew and what it was told, and appreciated the consequences of contemplated actions. Advice Taker gave McCarthy an opportunity to propose that such devices also do time-sharing, a concept that may be common today but was alien in the 1950s. He believed that computers would be useful to many if they were designed so that one processor could deal with several users simultaneously, giving each the impression that the machine was dedicated to them. During the early 1960s, he encouraged architects of computer systems to design that capability into their machines. It is difficult at this time to tell whether he was the only voice or just one of many advocates of that concept at the same time. Nevertheless, today time-sharing capability is normal in all digital computers* and in many analog devices.

McCarthy's Advice Taker became a convenient anthology of his ideas regarding intelligence and hardware as of the late 1950s. He envisioned a device that operated in real time and took in information about the environment it operated in, while giving it instructions about what to do. He assumed that it would be able to then understand what to do next. In the 1980s, when asked why it was still only a proposal, McCarthy commented that "in order to do it, you have to be able to express formally that information that is normally expressed in ordinary language. As far as I'm concerned, this is *the* [McCarthy's emphasis] key unsolved problem in AI.''

From the 1950s to the present, McCarthy developed ideas on the issues involved in artificial intelligence, launched a series of projects to deal with them (e.g., LISP), and, in the 1960s, attempted to expand the community of students and scientists dealing with AI. At Stanford he sought to escape the politics which he perceived at MIT and to establish a research program in AI. By 1970, the question concerning work with robotics began sweeping the artificial intelligence and computer science communities. Not surprisingly, robotics tempted these scientists because robots represented an opportunity to apply their skills and theories on useful tools.

In 1969, McCarthy and a colleague, P. J. Hayes, published one of the earliest papers available outlining what they saw as many of the criticial issues concerning robotics—a subject of considerable research during the 1970s. They had already started to build robots (four by 1969) and thus could speak about both practical and theoretical issues. McCarthy's fundamental concern was to define intelligence, an extension of his lifelong work. He was nervous about using the classical Alan M. Turing* approach that related intelligence to human behavior, suggesting instead that people might not always be a good model. Although McCarthy saw early steps being taken to develop machines that could manipulate facts, he insisted that a machine was intelligent only if it could construct a model of its world, answer a great number of questions concerning that universe, and then perform tasks required of it within that world. The machine's physical capabilities represented the only limitation in his mind. His ideas on robots came at the dawn of a new era in the development of robots, much like his thoughts in the mid-1950s came at the beginning of new era in artificial intelligence and the growth in sophistication of computers in general. Throughout the 1970s, McCarthy witnessed many new projects that led to the manufacture of robots utilizing computerized technology, were intelligent (that is, could be programmed), and were cost-effective.

In 1971, McCarthy was honored with one of data processing's most prestigious awards, the A. M. Turing award, given to those who make an outstanding contribution to the field of data processing. This recognition called for a synthesis of his contributions up to that time and for analysis of his position within the data processing community. Clearly, many felt he was a leading figure in a brilliant circle of scientists working on AI. More so than many others, he approached the subject from theoretical grounds, focusing on the definition of intelligence while relying heavily on his mathematical background for answers. In the mid-1980s, he still remained one of the key scientists in the field of AI and the leading proponent of LISP.

For further information, see: John McCarthy, ''History of LISP,'' in Richard L. Wexelblat, ed., *History of Programming Languages* (New York: Academic Press, 1981), pp. 173–185; Pamela McCorduck, *Machines Who Think* (New York: W. H. Freeman and Co., 1979); David Ritchie, *The Binary Brain: Artificial Intelligence in the Age of Electronics* (Boston: Little, Brown and Co., 1984); Patrick Winston, *Artificial Intelligence* (Reading, Mass.: Addison-Wesley, 1977).

MEAGHER, RALPH ERNEST (1917–). This computer scientist was the chief engineer on the ORDVAC* and the ILLIAC* projects, two early electronic digital computers.* He extended computational activities at the University of Illinois, the first major American university to own a relatively large digital computer and one of the earliest to establish a Department of Computer Science leading to degrees in the subject from bachelor to Ph.D.

Meagher was born on September 22, 1917, in Chicago. He completed his B.S. at the University of Chicago in 1938, his M.S. at the Massachusetts Institute of Technology (MIT) the following year, and his Ph.D. in physics at the University of Illinois in 1949. He was a member of the staff at the Radiation Laboratory at MIT from 1941 to 1945. He was brought to the University of Illinois in 1948 as an assistant professor of physics as part of Dean Louis Ridenour's effort to build a program in data processing and computer science. From 1950 to 1951, Meagher was research associate professor, then research professor (1951–1957), research professor of physics and electrical engineering (1957–1958), head of the digital computer laboratory between 1957 and 1958, and, subsequently, a member of the Computer Science Department. He also consulted on data processing for American industry.

The Digital Computer Laboratory, where Meagher spent his early years at Illinois, built the ORDVAC and the ILLIAC. The ORDVAC was constructed for the Aberdeen Proving Ground, while the ILLIAC went to the University of Illinois. The ORDVAC completed its acceptance tests on March 10, 1952, and the similar ILLIAC followed on September 1, 1952. Four other ILLIACS were built between the 1950s and 1970s, but each was different from its predecessor, reflecting improvements in technology evident throughout the data processing community. In his earliest days at the University of Illinois, Meagher focused on improving memory* based on the Williams tube. In effect, he developed techniques that reduced the requirement to refresh or "tune" this kind of memory. Data, stored on a cathode ray tube's surface much like a picture on a television screen, faded or "leaked" from their locations and thus had to be constantly refreshed. That onerous task had such negative consequences as introducing errors, lowering reliability, and reducing speed of operation. Fixing such problems thus made it possible to use memory in computers more effectively.

Meagher's management of the construction of the ORDVAC was important because the machine was an early example of what became known as the von Neumann computer, a design that has influenced processors down to the present. It was one of the first to store the programs it would use within it as conceived by John von Neumann.** The ILLIAC, designed much along the same lines, was intended as a tool with which to train computer scientists and programmers while facilitating research in the field of computer science.

For further information, see: Herman H. Goldstine, *The Computer from Pascal to von Neumann* (Princeton, n.g.: Princeton University Press, 1972); Michael R. Williams, *A History of Computing Technology* (Englewood Cliffs, N.J.: Prentice-Hall, 1985).

METROPOLIS, NICHOLAS C. (1915–). This mathematical physicist and computer scientist was instrumental in constructing the MANIAC* at the Los Alamos Laboratory in the 1940s, completing the machine in 1952.

Metropolis was born in Chicago on June 11, 1915. He completed his B.S. degree at the University of Chicago in 1936 and his Ph.D. at the same institution in 1941. He remained at the campus to work in the Metallurgical Laboratory (1942–1943), and then moved to Los Alamos as a group leader (1943–1946). He came back to the University of Chicago as assistant professor of physics (1946–1948) but returned to the Los Alamos National Laboratory (1948–1957) as a group leader. During this period he worked with MANIAC and its successors. He was named a senior fellow at the laboratory in 1965; however, he was back at the University of Chicago between 1957 and 1965 as professor of physics. He also served as the founding director of the Institute for Computer Research between 1957 and 1963. During the 1970s, he was active as a consultant to the National Science Foundation, in addition to his normal academic duties. In the late 1970s and early 1980s, he helped generate interest in the history of data processing and published on the subject.

Metropolis' most important years for historians of computers were the years he worked on the MANIAC. As at most government laboratories after World War II, scientists at Los Alamos wanted to build computers to support their research. Metropolis gained some early experience by running problems on the ENIAC,* the first electronic digital computer* in the United States. In addition to constructing this machine as part of a team, he was involved in others: the MANIAC II (started in 1955) and the MANIAC III (built at the University of Chicago). These machines largely followed the basic design established for the IAS Computer* built in Princeton, New Jersey, in the late 1940s and early 1950s under the direction of John von Neumann.**

For further information, see: Nicholas C. Metropolis, "The MANIAC," in Metropolis et al. eds., *A History of Computing in the Twentieth Century* (New York: Academic Press, 1980): 457–464 and with E. C. Nelson, "Early Computing at Los Alamos," *Annals of the History of Computing* 4, No. 4 (October 1982): 348–357.

MINSKY, MARVIN LEE (1927–). Minsky was one of the most important scientists in the field of artificial intelligence (AI)* during the 1950s and 1960s. He was also one of the most visible members of this small community of scientists, representing their views and the relevance of their work to the public and to the rest of the academic community in the United States and in Europe. Minsky is best known for his work on how the mind perceives and learns. His work at the Massachusetts Institute of Technology (MIT) ranged from theoretical considerations about AI to the construction of robotic devices while training a generation of scientists during the 1950s and 1960s who went on to do their own research on AI. Like his students, he saw computer technology as a tool to further the study of intelligence and its functions. Ultimately, his greatest contribution may be his encouragement of new definitions of AI and their relationship to the development of robotic devices of considerable power among the scientists he trained.

Minsky was born on August 9, 1927, in New York and was raised there. His father was a medical doctor. The younger Minsky was a brilliant student. While an undergraduate at Harvard University, he majored in physics but also expressed an interest in biology and the psychology of the mind. After graduating with a B.A. in 1950, he continued his studies at Princeton University, completing his Ph.D. in 1954. He was a participant in the first important conference on artificial intelligence in 1956 at Dartmouth College. Two years earlier he had a summer job at Bell Laboratories,† working for Claude E. Shannon,** another early specialist in AI. Thus, by the mid-1950s Minsky was deeply interested in pursuing the workings of the mind through the study of neurons and related subjects. This effort came at a time when artificial intelligence was just emerging as its own field of study, ill-defined, with little direction, and confusing for many throughout the 1950s and 1960s. The subject did not fit into any simple categorization such as physics, biology, electronics, or psychology but instead seemed to be an amalgm of each. Yet his academic career stabilized on more precise terms when in 1957 he left a position at Harvard to work at the Lincoln Laboratory† at MIT. Between 1958 and 1961, he was a professor of mathematics at MIT and then switched to electrical engineering at the same institution. Beginning in 1958, he was also director of the artificial intelligence group Project MAC* (MIT's first effort at an AI lab), and in 1970, he was named director of MIT's Artificial Intelligence Laboratory. He was in the U.S. Naval Reserve between 1945 and 1946. In 1970, he was recognized for his contributions to AI with one of data processing's most prestigious honors, the Turing award. It was named after Alan M. Turing,** a British scientist who was also interested in computers and intelligence, before World War II and throughout the 1940s.

Like many of his peers who would eventually be identified as researchers in the field of artificial intelligence (a term coined only in 1956), Minsky sought to understand the functioning of the mind. One of his most widely quoted statements—"the brain happens to be a meat machine"—suggests an analogy of his thinking. Many of his contemporaries, himself included, dabbled in research involving biology, psychology, engineering, and mathematics to better understand the general theme of intelligence. Minsky's first efforts (actually begun at Harvard) involved neurological analysis. He thought that certain electrical impulses on nerves could cause the muscles to do specific tasks. In the early 1950s, Minsky had already stumbled on the logic of "on-off" cells in nerves, much as the "on-off" functions in an electronic digital computer.* He was later to observe that in the 1950s he had to "switch from trying to understand how the brain works to understanding what it does." Like others, he searched for theories on how a "thinking" machine could function. As the 1950s passed, he increasingly saw the need to distance himself from the neural or purely biological approach, even though he had grown up in a home with a medical doctor, was married to one, and had studied biology. Thus, turning away from a total dependence on the physiological caused him some anxiety but moved him in the direction of understanding information processing. The result was a series of models of such behavior. Graduate students wrote models of information processing, while he looked in different directions himself.

By the 1960s, Minsky and his colleagues were looking at robotic behavior as a fruitful area for research. Robots were seen as devices that could perform human-like actions. The study of robots was a logical extension of AI, which was and remains an effort to create things that are mechanical (machine-like) and yet function in ways humans consider "intelligent." An early step in this direction involved the concept of perceptrons. Originally the result of work in the 1950s by Professor Frank Rosenblatt and his students at Cornell University, it was the idea of a machine, actually constructed, using photocells and random wiring. It could recognize specific shapes (such as a letter). In the 1960s, Minsky felt that work on perceptrons was too unclear. No precise theories of behavior had emerged from such experimentations by others. Along with another scientist, Seymour Papert, Minsky wrote a book defining perceptrons more precisely, while describing relevant theorems governing their behavior and functions. This book, published in 1968, represented the highwater mark of research on perceptrons which did not lead to a better understanding of how a thinking machine could be built to mimic the brain. Devices relying on similar functions, however, were once again under construction during the 1980s but by different scientists.

As part of his research efforts in the 1950s and 1960s, Minsky developed a concept useful in the study of artificial intelligence. Beginning almost at the point of trying to find mathematical ways to look at AI, including the use of recursive function theory, discrete mathematics, and later theories of formal languages, he emerged with ideas concerning pattern recognition. Minsky argued that knowledge could be catalogued into "frames." A certain quantity of knowledge would fill a frame. Such a body of knowledge might be car-driving or piano-playing. A frame might contain subframes. Using our analogy of the car, a subframe might be the knowledge required to do a left turn or to fill a gas tank. Frames in turn were part of larger frames. Thus, car-driving might be a part of a bigger frame we might call "daily-routine" or "transportation." Minsky's idea was crucial because it led to the development of a specialty called "knowledge engineering." It was a concept that forced research on how to move information to where it was needed in a timely fashion. Thus, a knowledge engineer would have the role of collecting information, organizing it in a relevant way, and then moving it to where it was needed. The thought raised the question as to whether such people could then build intelligence into a machine.

This work led Minsky and his students to study robotic behavior and to develop their own machines in the 1960s. In an important summary of this work done at MIT and published in 1971, Minsky argued that robots were a good way to test concepts concerning intelligence. He maintained that any idea concerning the nature of intelligence must work empirically if it was to be useful. Thus, in his own way, he contributed some answers to the central issues of what is intelligence and gave some descriptions of its characteristics.

In the late 1960s and early 1970s, Minsky concentrated on vision, a complicated set of processes involving intelligence. He had come to the conclusion that people did not do a very good job in describing their mental actions, nor did they agree with each other on what they thought they saw or how it happened. Therefore, scientists had to develop better approaches to describe learning and observation, much as they have described scientific theories in general. These thoughts emerged as a result of his attempt to construct a robotic eye. This exercise encouraged him to comment on thinking in general terms and more precisely on vision and imagery. He maintained that any theory concerning these subjects would not be relevant unless they could be used to make a seeing machine or one that thought.

By 1970, this brilliant, soft-spoken scientist had arrived at the position that learning required an understanding of "how something might be done at all," followed by defining "how a task can be learned." He did not hesitate to engage in intellectual combat with others in the field because much had yet to be defined and agreed on. By the early 1980s, he was arguing that the new problem with learning was to answer the question, "how do you decide what you want to have in your memory?" Using the language of a systems analyst designing a software* program, Minsky believed that good descriptions, the differences between them, and finally stating both, represented problems for future scientists. His point was that "the difference isn't between the things, it's the difference between their descriptions." His ideas had come full circle to an earlier idea about wanting to understand how the brain or intelligence in general carried out a function. The biological metaphor was still hard to desert.

Minsky made yet another important contribution to the study of AI and robotics, which also related to his earliest interest in biological relationships. Still fascinated with the idea that biology was a key to understanding some of the components and behavior (characteristics) of intelligence, he remained enamored of the idea of a biochip as possible future artificial intelligence. He coined a word to describe his interest: telepresence. The word represented a device or circumstance that allowed a person to experience something without being physically involved in experiencing the event. Thus, for example, one would be able to experience the sensations of flying an airplane, fighting in a war, or doing underwater swimming without actually doing these things through the projection of a body in an airplane or onto a battlefield or into a wet suit and into the water. He proposed that a device (e.g., wingtip cameras and motion sensors on an airplane or similar equipment on a robot on the bottom of the ocean) could feed back sensations (images and other input) through the medium of telecommunications to an individual. That person could then go through the motions of flying an aircraft, firing a rifle, or swimming but with the difference (using the idea of the airplane) that the individual would experience the sensation of the airplane being that person's own body. If these ideas were applied to robotics, it could mean that dangerous tasks or work done in hostile environments would be possible without the risk of physical damage to people. By the late 1970s, practical applications were already present, with more on the way. For example, the city of New York had a robot that could dismantle and disarm bombs under the control of a police officer far away from the explosive and its dangers.

Yet in the 1960s, the idea of telepresence was new and provided direction for research on robots of some intelligence. But Minsky's ideas went further. He argued that devices (robots) could be developed to magnify or shrink movements and force human fingers doing delicate work. He cited the example of a surgeon who would be able to do microsurgery using his own hands to guide a device, which would do the actual incisions. Such a robot might be microscopic in size with its movements scaled up or down in scope and size, depending on what the human operator desired. A doctor, receiving ''feedback'' from such a machine, would be able to cut or scrape tissue, selecting almost at the cell level how much to do and exactly where, thereby reducing the size of the wound created by surgery and its pain while increasing the safety of the procedure.

Minsky's ideas received considerable attention, and his publications profoundly influenced research on robotics in the 1970s. For over twenty years he produced a stream of students working in the general field of artificial intelligence, focusing their dissertations, for example, on problems of interest to him. Ultimately, historians might conclude that his development of students might be his lasting contribution to the field of AI and his most pervasive influence on computers and their robotic cousins. He has been described as one of the giants of artificial intelligence, frequently serving as their spokesman and representing an American community of 250 to 300 scientists. In mass publications he has commented on the nature of AI and the role of computer science in general. During the 1970s and early 1980s, Professor Minsky increasingly turned his talents to the composition and the playing of music.

For further information, see: Tom Logsdon, *The Robot Revolution* (New York: Simon and Schuster, 1984); Pamela McCorduck, *Machines Who Think: A Personal Inquiry into the History and Prospects of Artificial Intelligence* (New York: W. H. Freeman and Co., 1979); Marvin L. Minsky, *Semantic Information Processing* (Cambridge, Mass.: MIT Press, 1968), with Seymour Papert, *Perceptrons* (Cambridge, Mass.: MIT Press, 1968), and *Artificial Intelligence,* Project MAC Report (Cambridge, Mass.: MIT University Artificial Intelligence Laboratory, 1971), but see Minsky's ''Steps Toward Artificial Intelligence,'' in E. A. Feigenbaum and J. Feldman, eds., *Computers and Thought* (New York: McGraw-Hill, 1971) which represented his thoughts as of 1961; David Ritchie, *The Binary Brain: Artificial Intelligence in the Age of Electronics* (Boston: Little, Brown and Co., 1984).

MOORE, GORDON E. (1929–). Moore was one of the creators of the modern computer chip,* the fundamental building block of the twentieth-century ''thinking'' machines. He worked with Robert N. Noyce** in pushing electronics forward from transistors to chips. Their creation drove computing costs down dramatically while improving efficiencies during the last forty years. Dr. Moore has most precisely described the cost element in what has become known as ''Moore's Law,'' a guiding principle behind the economics of manufacturing chips from the 1960s through the 1970s.

Moore was born on January 3, 1929, in San Francisco. He completed his B.S. in chemistry at the University of California in 1950 and continued his studies in both chemistry and physics at the California Institute of Technology, earning a Ph.D. there in 1954. He then joined the faculty at Johns Hopkins University and soon found himself in the center of research in the 1950s leading from transistors (invented in the late 1940s) to what would become the chip, or the wireless cluster of transistors. In 1956, he joined the staff of the Shockley Semiconductor Laboratory, and, when Fairchild Semiconductor Incorporated† was established in 1957, he was a co-founder with Noyce. As manager of engineering, this thoughtful and conservative physical chemist became a sounding board for Noyce who really came up with most of the early ideas at Fairchild leading to the development of the chip.

While at Fairchild, Moore did most of his work with Noyce. It was the most fruitful period of his life as a scientist. The two men began with the manufacture of individual transistors without wires and progressed to the monolithic chip which was an entire electrical circuit in a package (a semiconductor) without cabling. Using silicon wafers, they printed copper lines on them for circuits. By early 1959, Noyce and his staff had a design for the monolithic circuit, the basis of the modern computer chip, and applied for a patent. The application came at exactly the same time as one from Texas Instruments† where work on such a chip was also underway. In the early 1960s, Fairchild grew as it manufactured and sold chips. In 1964, International Business Machines Corporation (IBM)† announced its System/360* family of computers based on chips, not vacuum tubes or transistors, thereby ushering in a new era of computerized technology. During this period, Moore worked diligently to help develop efficient manufacturing methods for producing chips. The art of making chips efficiently and cost effectively presented some of the most difficult challenges of the entire computer manufacturing community throughout the next thirty years.

This kind of exercise led to what would soon be called Moore's Law. In 1964, Moore was asked to predict how far chip technologies would go in the next decade. His off-the-cuff response became his law: he said that the number of transistors on a chip seemed to double every three years, and thus, he thought that this rate of compaction would continue. Put another way, he discovered that there was a ''learning curve'' in semiconductors. Prices of chips declined 20 to 30 percent each time production doubled during the life cycle of a particular generation of chips. This sloping downward on a chart, so to speak, was his learning curve because prices dropped as a manufacturer learned to make better quality chips. Improved chips in turn could be sold in greater quantities, allowing a manufacturer to charge less, usually 20 to 30 percent from generation to generation of new chips. What he said so casually did in fact turn out to be the case throughout the 1960s almost to 1980. But because chips represented such a critical part of the equation of costs for any computer technology, the expense of chips has remained a guiding influence on the pricing of hardware in general. The 20 to 30 percent rule had become part of the industry's expectations such that it was evident elsewhere. For example, in the early 1980s IBM publicly stated that it intended to reduce the cost of computers by about 20 percent a year, an objective it frequently achieved.

Moore had been surprised that the press and, later, other chip manufacturers took his comment seriously. He was quoted as saying, ''At the time I had no idea that anybody would expect us to keep doubling (capacity) for ten more years. If you extrapolated out to 1975, that would mean we'd have 65,000 transistors on a single integrated circuit. It just seemed ridiculous.'' But in 1975 transistors were indeed packaged into chips at the rate of 65,536, and by 1985, there were experimental chips with a million transistors while 256,000 on a chip had become normal fare. Although Moore was at the center of much of the research and development leading to such miniaturization and compaction, it amazed him. In the early 1980s, he commented, ''I still have a tough time believing that we can make these things.''

His amazement notwithstanding, even in the early days Moore prospered as a result of this technology. In 1959, he became director of research and development at Fairchild. In 1968, he and Noyce left Fairchild to form Intel Corporation† where he served as executive vice-president (1968–1975). Intel became the most important manufacturer of memory chips in Silicon Valley during the 1970s. In 1975, Moore became president and chief executive officer of the firm and from 1979 served as its chairman and chief executive officer. During the 1960s and 1970s, he also received numerous industry awards and was made a Fellow of the Institute of Electrical and Electronics Engineers (IEEE).†

For further information, see: Dirk Hanson, *The New Alchemists* (Boston: Little, Brown and Co., 1982); T. R. Reid, *The Chip: How Two Americans Invented the Microchip and Launched a Revolution* (New York: Simon and Schuster, 1984); Everett M. Rogers and Judith K. Larsen, *Silicon Valley Fever: Growth of High-Technology Culture* (New York: Basic Books, 1984).

MORLAND, SAMUEL (1625–1695). Morland, an English diplomat-inventor, developed a series of calculators, one of which employed Napier's bones (logarithms). His first machine was designed for adding and subtracting and was one of the first to be commercially marketed in Europe. Other contemporary, as well as earlier devices, were built by other inventors for experimental purposes only. Morland's work is of interest for another reason: most work on calculators took place in Europe (especially in France) and not in England.

Morland was born into a family headed by a minister and at university studied mathematics rather than his family's traditional subject of theology. His serious introduction to mathematics took place during the tumultuous rule of Oliver Cromwell. In 1649, Morland was elected to the position of Fellow at Magdalene College. In 1653, he joined the diplomatic staff of Bulstrode Whitelocke, who had just been named ambassador to the court of Queen Christina of Sweden. While in Sweden, he may well have seen an adding machine given to the queen by Blaise Pascal.** Her court was the scene of numerous scientific debates concerning mathematics, astronomy, and possibly the adding machine which she frequently showed guests. Morland thrived in such an environment. The ambassador later wrote of him that he was "an excellent scholar," and "an ingenious mechanist." After his tour of duty in Sweden, Cromwell sent Morland out as envoy to the Duke of Savoy in Italy. Both en route and on his return from Italy, as well as on other occasions, Morland established contracts with the scientific community in France and at the French court.

Although he served Cromwell as a diplomat, Morland also spied for King Charles during the monarch's exile. After Charles regained the throne, in gratitude he pensioned Morland, making it possible for the diplomat-inventor to devote all of his time to the study of mathematics. Almost immediately, he focused on the construction of a calculator, of which three were designed and at least one was actually manufactured and sold commercially in the late 1660s. He also worked on barometers, water pumps, speaking trumpets, and other scientific devices.

Morland's first calculator was an analog machine that reflected an advance model of a popular instrument of the time called a sector. He quickly followed with two adding machines and then a device that manipulated Napier's bones. Fortunately for historians of data processing, he left a detailed description of his adding machines and their potential applications in a short book entitled *The Description and Use of Two Arithmetic Instruments* (London, 1672).

In his adding machine, Morland employed wheels rotating on individual styluses. An operator would put a stylus into the hole of the number to be added and so on with each being turned clockwise into the stylus, representing the added number at the top of his device on a dial. Peering through a window at the twelve o'clock position on the dial would show the results. Dials were in rows on a contraption that vaguely resembled a cash register. The major problem with this device was that it did not have a carry function. To resolve this requirement for adding, he provided auxiliary dials above the larger dials, so that every time a major dial was turned from 9 through 0 a tooth would advance an auxiliary dial by one mark. At the end of the addition, these auxiliary dials showed how many carries should have been added to the neighboring next major dial. However, that subsequent carry had to be done by hand. It was a simple procedure, even though reliability was often a function of an operator's attention to detail. It was an accurate instrument if the operator did not fail to add the carries forward at the conclusion of each set of adding operations.

Morland wanted the machine to be used primarily for counting English money (pounds, shillings, pence, and farthings) and designed the device to support this application. Nondecimal dials were positioned lower on the calculator than those employed in the addition of five-digit decimal numbers. The entire device measured only 4 by 3 by 0.25 inches thick and so could be carried in one's pocket. It was comparable to today's pocket calculator (minus the aggravation of having to replace the batteries).

Morland manufactured and sold a few copies of this adding machine. It was not a financial success; people probably thought it a novelty and not particularly useful. Nevertheless, it was an early commercial venture involving the use of mechanical devices to perform mathematical calculations in Europe. As early as April 16, 1668, Morland ran a commercial for the device in the London *Gazette*.

Morland's device illustrates the kind of technical issues with which he struggled. The use of wheels was the dominant concept with regard to calculators in the seventeenth and eighteenth centuries. Morland's design was in that mainstream and as such was not unusual. His wheels had digits etched on them from 9 to 0 and teeth to catch on to others as they were moved, indicating new combinations of numbers. In his book of 1672, Morland argued that additional wheels could be added to his machine to extend the capability of mechanized calculations. He fully understood the convenience of single teeth turning wheels. Also like other inventors, he had to contend with certain physical limits: only so many wheels could be strung out to handle multiple digit numbers. One can appreciate the strain imposed on the first several wheels that would have to force others to move to create the number 999999999. In his situation, a six-digit number seemed to be the limit. Even Charles Babbage,** working on his difference engine* in the 1820s, encountered similar problems caused either by the machinery being tooled with limited tolerances or the physical properties of metals and woods taking strain and jolts. These problems were only partially solved even in the nineteenth century (machining in particular). They were not completely eradicated until the mid-twentieth century when manufacturers of calculating devices employed electronics (electricity, transistors, and chips*).

Morland looked beyond the simple operations of addition and subtraction which could be done relatively easily with his wheels. Beginning in the 1660s and possibly even in the late 1650s, he considered the more complex issues of multiplication and division. He elected to use Napier's bones (logarithms) as a tool to make it easier to perform these functions mechanically. In the design of his second calculator, Morland developed a flat brass dish with a perforated hinged gate, circular disks engraved with numbers, and all mounted on semicircular pins. His disks carried logarithmic numbers in circular fashion in the same order as positioned by Napier. Products were on the disk edges, so that two digits of a number were at diagonal opposites. Thus, for instance, to use one author's example, if Napier's bones showed the number 3 and the calculation was $4 \times 3 = 12$, the product (12) would appear as the number 1 at one end of the disk and 2 at the opposite side of the same disk. Digits engraved in the center of a disk told the user what number from Napier's bones was being represented.

Rack-and-pinion machining allowed an operator to turn the disks as needed under the gate, thereby moving pointers to product indicators. This second machine was configured with thirty disks for routine multiplication and five additional ones for determining square and cube roots. Morland intended that both his adding and multiplication machines be used together to perform all four arithmetic functions of addition, subtraction, multiplication, and division while including the capability of calculating square and cube roots.

Today Morland's work is regarded mainly as a curiosity. By the end of the nineteenth century, calculators were rapidly changing in both function and technology. Nonetheless, he is still studied as an example of the seventeenth-century European mathematician and as representative of the art of mechanized mathematics in that period.

For further information, see: H. W. Dickinson, *Sir Samuel Morland, Diplomat and Inventor, 1625–1695* (Cambridge: Heffer and Sons, 1970); M. Gleisser, ''Samuel Morland: From Seals to Wheels,'' *Datamation* (April 1976): 54–57; M. R. Williams, ''From Napier to Lucas: The Use of Napier's Bones in Calculating Instruments,'' *Annals of the History of Computing* 5, No. 3 (July 1983): 279–296.

N

NAPIER, JOHN (1550–1617). Napier is known as the father of logarithms. In addition to developing these tables, he inspired later mathematicians to apply his logarithms (also called Napier's bones) to calculating devices, especially in the seventeenth and early eighteenth centuries. Logarithms revolutionized normal calculations, making mathematical procedures easier for heavy users of arithmetic procedures such as astronomers, navigators, and surveyors.

Napier, the Baron of Merchiston, was born in Scotland near Edinburgh in 1550. He considered his greatest hour to be his active participation in the Protestant movement, which was part of the Scottish Reformation of the sixteenth century. Although he published a book which he regarded as an important contribution to religious affairs—*A Plain Discovery of the Whole Revelation of St. John* (1593)—it is for his logarithms that he is remembered today. His work on logarithms was part of Napier's broad campaign to simplify mathematical procedures. Like those who developed mechanical calculators, particularly during the seventeenth century, he sought to remove the drudgery associated with arithmetic while improving its accuracy.

In addition to his work on logarithms, Napier was probably the first Western mathematician to appreciate the significance of the decimal point in arithmetic. By 1617, he had done enough study to describe the use of decimals. He built a box with rods to handle numbers and called the box Rabdologia. In addition, he built or designed other devices to handle various types of calculations. One such machine, a small box containing plates, he called the Promptuary of Multiplication, which he used to automate multiplication.

Napier's Rabdologia proved to be his most important device. It is most frequently called Napier's bones. In the better quality samples of the box still extant, parts are made from ivory or bone, hence the name. One historian of Napier uncovered other names applied to his invention: numbering rods, multiplying rulers, and speaking rods, but they are all the same thing.

Napier began work on logarithms in about 1600 and described the results in a small book called *Rabdologia* (1617). Logarithms are tables of numbers that simplify the process of division and multiplication. Napier discovered that each positive number has another (called its logarithm), so that multiplication of any two numbers can be accomplished by simply adding their logarithms. Division is achieved by subtracting their logarithms. In order to take advantage of this relationship, a number has to be converted to its logarithm by looking it up on a logarithmic table. The sum of the addition or subtraction is a logarithm and must be converted back to a normal number using an anti-log table. The time saved by this means exceeds the effort and time needed to do the multiplication or division, particularly for a large number.

In time, other mathematicians recognized that logarithms were variations of original numbers (logarithms are sometimes called compressed numbers) and that these could be transposed onto scales. Thus, multiplication or division could be done by adding or subtracting the lengths on a ruler or scale—the principle behind the slide rule.* Although Napier did not carry his thinking quite far enough to develop slide rules, he did put his multiplication tables on wooden cylinders, surfaces of which had numbers. By turning the correct cylinders, each of which represented a digit from 9 to 0, then adding or subtracting the numbers that appeared, one could get the results desired. This was the original version of Napier's bones.

Some scholars have suggested that Napier stumbled across this technique from an old method called gelosia. Since the concept behind gelosia is related and consequently served as another fine thread contributing to building a mathematics that could later support computers, it is important to understand the technique. It was probably first developed in India to do multiplications, although it was also known to exist in China, the Middle East, and Persia at least since the Middle Ages. During the 1300s, it was brought to Italy where it got its name gelosia because its pattern was similar to a popular window grating.

Gelosia is a matrix. In this grid the multiplicand appears at the top of each column selected and a digit (of the multiplier) beside each row. The product of each row and column digit comes from the appropriate box within the matrix. The tens digits appeared above the diagonal crossing boxes of numbers and all units digits below. One would start in the lower right-hand corner of a box and add digits from each diagonal, with carrying considered a part of the next diagonal.

Napier made a larger gelosia table with stripes of various possible columns. To multiply a number, one would select the numbers to be multiplied from those at the heads of columns. Each column had one digit at the top, so that a combination of stripes was needed. After placing these beside each other, one would read the partial products, adding all the parallelograms to determine each digit of the partial product. Then all the partial products were added to get the final answer. His bones could also be used to conduct division. Multipliers of the divisors could be determined, thereby saving time spent in trial multiplications. Napier also established a means for calculating square and cube roots by means of his columns of numbers (digits) on stripes.

The usefulness of his tables was quickly recognized. Usage multiplied all over Europe rapidly. Within a few years of his original publication on the subject, logarithms were seen in use as far away as China, primarily through the work of Jesuits communicating with others in Asia. The popularity of his tables is understandable since a user had only to jot down the additions and partial products of a multiplicand. It was easy to use stripes of wood or brass in a box to perform these calculations.

After Napier's bones became available, others tried to apply them to calculators. Wilhelm Schickard** successfully applied them to a calculator which he developed in the early 1620s in which logarithms were etched on cylinders that could be turned to select specific digits. Horizontal slides exposed numbers and their logarithms. Results could also be accumulated on his machine. Another positive application was by Samuel Morland** who used disks to reflect Napier's bones, showing digits through little windows by the 1660s. Gaspard Schott** and Athanasius Kircher also worked with Napier's bones at the same time. Schott invented a mathematical instrument that housed ten sets of tables to perform such diverse applications as arithmetic, geometry, tables for fortifications, calendaring, generation of tables for Gnomics (sundials), astronomy, and music. Data were organized much like Napier's bones. In one form or another, therefore, work on applying Napier's work had started almost immediately after his book of 1617 appeared, influencing the fundamental characteristics and operations of mechanical calculators, precursors of the computer.

For further information, see: D. S. Erskine and W. Minto, *An Account of the Life, Writings and Inventions of John Napier of Merchiston* (Perth: n.p., 1778); M. R. Williams, "From Napier to Lucas: The Use of Napier's Bones," *Annals of the History of Computing* 5, No. 3 (July 1983): 279–296.

NAUR, PETER (1928–). This important Danish computer scientist participated in the construction of electronic computers, including the EDSAC,* and in their use in scientific research. He is a prolific writer and a specialist on data.

Naur was born in Frederiksberg, Denmark, and in 1943 did some work in computational celestial mechanics. In 1949, he completed his M.A. in astronomy at Copenhagen University. Between 1950 and 1951, he used the EDSAC computer at Cambridge University. The EDSAC, one of the earliest electronic digital computers* built in Great Britain, taught him a great deal about computational equipment. He remained primarily interested in astronomy during the early 1950s, however, working between 1952 and early 1953 as a research assistant at the Yerkes and McDonald Observatories operated by the University of Chicago. He was exposed to more computer technology when he studied astronomy at the International Business Machines Corporation (IBM)† Thomas

Watson Laboratory at Yorktown Heights, New York. Between 1953 and 1959, he was employed by the Copenhagen Observatory.

During the same period Naur consulted for the designers of Denmark's first computer, the DASK. At the same time he worked on his Ph.D. in astronomy, which he completed for the University of Copenhagen in 1957. In 1959, he joined those who had worked on the DASK computer at the Regnecentralen and shared Europe's keen interest in ALGOL,* a universal programming language* of the early 1960s. He convened the very first meeting ever held on how to implement ALGOL (February 28, 1959) in Copenhagen, and soon after he became the first editor of the *ALGOL Bulletin*. He participated in the design of ALGOL 60 and in writing the final report on the language in 1960. He also made that language work on the DASK computer between 1959 and 1961. Between then and 1967 he produced other versions of ALGOL.

Beginning around 1964, Naur became increasingly interested in the general workings of data processing and computing. In 1966, he invented a new word—datalogy—to describe the study of data and data processes. By 1969, he persuaded the University of Copenhagen to create a professorship in the subject for him. During the 1960s, he lectured at various Danish schools and in the United States, namely, the University of North Carolina (1961) and the University of Pennsylvania (1962). Between 1962 and 1966, he represented his country on the International Federation for Information Processing Technical Committee 2 on Programming Languages. He was a member of several task forces on ALGOL.

For his contributions to data processing in his country, Naur was given the G. A. Hagemanns Gold Medal in 1963 and the Rosenhjaer Prize in 1966. Since 1966, he has been the first and (so far) only president of the Danish Society of Datalogy. Naur has over 150 publications to his credit.

For further information, see: Peter Naur, *A Course of ALGOL 60 Programming With Special Reference to the DASK ALGOL System* (Copenhagen: Regnecentralen, 1962), and with Alan J. Perlis, "ALGOL SESSION: "The European Side of the Last Phase of the Development of ALGOL 60," in Richard L. Wexelblat, ed., *History of Programming Languages* (New York: Academic Press, 1981), pp. 75–172.

NEWELL, ALLEN (1927–). Newell, the father of list processing languages, created IPL-V,* the earliest and most important processing language in the 1950s. In addition to influencing the technology of programming languages* in the 1960s, his work with IPL expanded the way programming was done. His language also encouraged the use of computers in research on artificial intelligence.*

Newell was born on March 19, 1927, in San Francisco. He completed his studies of physics at Stanford University with a B.S. degree in 1949 and did postgraduate work at Princeton University (1949–1950). He completed his Ph.D. in Industrial Administration at the Carnegie Institute of Technology in 1957. From 1950 to 1961, he was a research scientist at the RAND Corporation where he did most of his work on IPL. He joined the faculty at Carnegie-Mellon

University in 1961 and in 1976 became professor of computer science. He published a number of articles and books concerned with programming languages and their use.

For his work on data processing technology, Newell was awarded the Harry Goode award from the American Federation of Information Processing Societies† in 1971, and he and a research associate who helped him with IPL, H. A. Simon,** received the A. M. Turing award from the Association for Computing Machinery† in 1975. He became a member of the National Academy of Science, American Academy of Arts and Sciences, and was a fellow of the Institute of Electrical and Electronics Engineers, Inc.†

For further information, see: A. Newell and F. M. Tonge, "An Introduction to Information Processing Language-V," *Comunications, ACM* 3, No. 4 (April 1960): 205–211, his *Information Processing Language-V Manual,* 2d ed. (Englewood Cliffs, N.J.: Prentice-Hall, 1965), and with H. A. Simon, *Human Problem Solving* (Englewood Cliffs, N.J.: Prentice-Hall, 1972).

NOYCE, ROBERT NORTON (1927–). Noyce helped develop the integrated circuit, more commonly known as the chip,* which became the basic building block of all computers during the 1960s. He was also a founder of Intel Corporation†—a major manufacturer of chips and computer memory* during the 1970s and 1980s. The development of the silicon microchip has been the most important of his accomplishments because the creation of the microprocessor made possible significant reductions in the cost of computing, along with dramatic advances in the miniaturization of electronics (hence of computers) and increases in reliability of performance. No single technological development within the data processing/electronics community has been as significant as the chip. Some historians consider the chip to be one of the greatest achievements of twentieth-century technology, making the present pervasiveness of computing possible and affordable in the industrialized world.

Noyce was born in Burlington, Iowa, on December 12, 1927, and was raised in the Midwest. He graduated from Grinnell College in 1949 and completed his Ph.D. in physics at the Massachusetts Institute of Technology (MIT) during 1953. He then became a research engineer at Philco Corporation,† but in 1956 he joined William B. Shockley,** also credited with advances in chip technology, at the Shockley Semiconductor Laboratory at Mountain View, California, in the area later known as Silicon Valley. In 1957, Noyce became a founder and director of research at the Fairchild Semiconductor Corporation,† a venture established by Fairchild Camera and Instrument for the purpose of developing the electronics of computing. In 1959, he was promoted to vice-president and general manager at the company and, in 1965, group vice-president of the parent firm, serving in that capacity until 1968 when he left to form Intel Corporation, serving as its chairman until 1975 and vice-chairman from 1979 to the present.

For his work on semiconductor technology Noyce received many of the most important awards given to American scientists. In 1967, the Franklin Institute awarded him the Stuart Ballentine award. He also received the American Federation of Information Processing Societies'† Harry Goode award in 1978 and, in the following year, the National Medal of Science. Noyce received other forms of recognition throughout the 1970s for his work with chips.

This son of an Iowa minister had always expressed interest in machines and electronics and, while at Grinnell, constructed a solid-state transistor in 1948—at the same time similar efforts were going on at Bell Laboratories† in New Jersey. Semiconductors (substances that were partially conductors of electricity) fascinated him, and at Philco he worked within the transistor division. Noyce enjoyed running projects, and, when he left Philadelphia to join Shockley in 1955, he anticipated doing important work in the development of silicon semiconductors. This intent, however, was in opposition to the then fashionable study of germanium-based semiconductors. Thus, in 1957 he joined Fairchild's new venture partly in order to do research of his own choosing. This change came at a time when semiconductor manufacturing was on the verge of exploding into a major factor within the computer industry. In 1957, the Russians launched Sputnik, causing the U.S. government to dramatically increase its research support. The result was the miniaturization of electronics, which was an important part of its strategy to build rockets and satellites. Early successes involved the development of the concept of the integrated circuit, better known as the microchip.

Both Texas Instruments† and Fairchild Semiconductor claimed authorship of the chip. The idea was developed independently at the same time at each location. Both wanted to develop a package that had multiple transistors on a chip of silicon with minimum or no wiring. The ultimate objective was to put all the functions common to electronics (wiring, capacitors, resistors, etc.) as functions on a chip made out of properly treated silicon. The final product was a logical extension of the development of solid-state transistors which pointed the way to the miniaturization of electronics.

In February 1959, Texas Instruments filed for a patent, and Noyce did the same that July. In 1962, both firms started a legal battle over patent rights, a struggle that lasted throughout the decade and was never completely resolved. In 1969, the court ruled that Noyce was the first to develop what became known as the integrated circuit (chip). The two companies resolved their problems through cross-licensing of their technologies. By 1970, the chip had become a working product which Noyce had been producing for a number of years. As early as 1963, integrated circuits accounted for 10 percent of all electronic circuits manufactured in the United States.

Noyce drove that percentage up sharply during the 1960s as did Texas Instruments, which eventually became the largest chip manufacturer. Yet credit for spawning this industry went to Noyce.

Fairchild Semiconductor also contributed to the development and expansion of this industrial subset of electronics. It became the spawning ground for the largest number of electronics firms in Silicon Valley during the 1960s and 1970s. As the profits available from such technology became obvious, venture capitalists moved into Silicon Valley, gutted Fairchild's engineering talent, and launched dozens of computer-related companies, many of which were extremely successful for short periods of time. By the time Noyce left Fairchild in June 1968, much of the original talent in the company had already left.

Noyce linked up with Gordon E. Moore** and Andrew Grove—two ex-Fairchild employees—to form a company that would have an important influence on the data processing industry: Intel. Arthur Rock, a venture capitalist, helped put together a $2 million investment package to launch the company. Noyce and a dozen employees made up the firm of *In*tegrated *El*ectronics (also called *Intel*ligence), or more commonly Intel, to make integrated circuits, primarily to replace older, more expensive computer memory technologies (such as core memories). Noyce focused his attention on semiconductor memories, which resulted in success at a time when large-scale integration (LSI) meant very much larger computers from those of the past. These computers required massive increases in memory. All of these events were occurring when a new round of price performance improvements was influencing the design of computers. In 1970, Intel announced the 1103 random access chip (RAM), which allowed information to be stored on silicon at costs far below those of existing devices. Intel's 1103 also represented new levels of miniaturization previously unavailable. It killed magnetic core memory and became the basis for all computer developments in the 1970s. The microprocessor and the development of cost-effective microcomputers in the 1970s resulted and were as significant as had been the evolution toward the integrated circuit around 1960.

Throughout the 1970s and 1980s, Noyce continued to produce increasingly dense chips, such as the 4004 and later the 8080. The 8080 became the most widely used chip after 1974 and set the standard for the industry. It also became the basis for the technology that went into hand-held calculators and later for such microcomputers as the IBM PC. Developments continued, and in 1983, Intel announced the 256,000-character memory chip and hinted that a one million byte chip would be available by 1990. Thus, in less than a decade, his company went from a chip that could hold some 8,000 characters of information to one housing a quarter of a million. Such progress meant that many intelligent machines could be made using chips from microwave ovens to traffic lights to shop-floor process control equipment. Even gasoline consumption in an automobile could be controlled effectively with such technology. These technological innovations contributed to Intel's success, and on its tenth anniversary the company proudly boasted that its sales were $300 million per year in a highly competitive market and that the largest computer manufacturing firms were its customers.

Noyce entered the 1980s as the individual who had most contributed to the growth of Silicon Valley and could claim to be the founder of its flagship company. As the valley's elder statesman, he spoke frequently on its behalf, particularly in making recommendations on how the American semiconductor industry could stave off competitive threats from Japan.

For further information, see: Stan Augarten, *Bit by Bit: An Illustrated History of Computers* (New York: Ticknor and Fields, 1984); Dirk Hanson, *The New Alchemists: Silicon Valley and the Microelectronics Revolution* (Boston: Little, Brown and Co., 1982); T. R. Reid, *The Chip: How Two Americans Invented the Microchip and Launched a Revolution* (New York: Simon and Schuster, 1984).

NYGAARD, KRISTEN (1926–). This Norwegian computer scientist is best known for his work within the Norwegian Defense Research Establishment (NDRE), especially for the development of SIMULA,* a programming language* developed in the 1960s.

Nygaard was born in 1926 in Oslo, Norway, where he did his college studies. His master's thesis in 1956 was on abstract probability theory (''Theoretical Aspects of Monte Carlo Methods''). He went to work for the Norwegian Defense Research Establishment in 1948, remaining there until 1960. Between 1948 and 1954, his duties focused on computing and programming, and from 1952 to 1960, his interests shifted to operational research. Between 1957 and 1960, he managed all operational research groups within the NDRE. He was also a co-founder and the first chairman of the Norwegian Operational Research Society (1959–1964). In 1960; Nygaard went to work for the Norwegian Computing Center (NCC) and in 1962 was promoted to director of research. During the 1960s, he played an active role in the development of SIMULA I and SIMULA 67. After these projects were completed, he conducted research for trade unions in Norway on planning, control, and data processing (1971–1973). Between 1973 and 1975, he did research on the social impact of computer technology and wrote a general system description for DELTA, a new programming language. He did the same for BETA between 1976 and 1980. He served as professor in Aarhus, Denmark (1975–1976), and at Oslo (from 1977 to the present). While at both universities he did research on the social impact of computer technology. Nygaard also served as a member of the Research Committee of the Norwegian Federation of Trade Unions. Between 1961 and 1971 Nygaard worked with Ole-Johan Dahl** of the University of Oslo on SIMULA, a language that offered basic new concepts which later emerged in computer languages of the 1970s and 1980s. Their two languages were developed at the NCC, however. They sought to create a programming tool that would employ simulation but that could also serve as a system description. Their language was designed in the early 1960s, and, in January 1965, they completed a compiler that ran on a UNIVAC* 1107 computer. SIMULA was related to another European language called ALGOL*; both underwent almost simultaneous evolution during the 1960s, leading to fully developed tools. SIMULA I was Nygaard's primary project during the first half of the decade, and he worked on SIMULA 67 from 1965 to the end of 1967, the bulk of this project being done in ten months (1967). In the 1970s, the language gained considerable currency and support, particularly in the United States where one version ran on a DEC 10 between 1973 and 1974.

For further information, see: Kristen Nygaard and Ole-Johan Dahl, ''The Development of the SIMULA Languages,'' in Richard L. Wexelblat, ed., *History of Programming Languages* (New York: Academic Press, 1981), pp. 439–493.

O

ODHNER, WILLGODT THEOPHIL (1845–1905). In 1874, this Swedish inventor developed the Odhner, a calculating machine that represented a major improvement over earlier devices. It was based on the concept of the "pinwheel." By using a variable toothed gear, Odhner could change the number of teeth projecting from a wheel's surface. To "input" numbers, levers on a wheel's surface were adjusted to push pins in and out, and the combination of pins represented numbers. A number of disks were used around a commonly shared central shaft which, when turned, caused results to register digits. The increase in these results was a function of the proportion of the number of positions to the number of pins protruding beyond the edge of a wheel. The device was relatively easy to use and proved reliable. The overall size of the machine shrank, from covering a full surface of a table to simply one corner. It was a practical adding machine. Two years before Odhner, Frank Stephen Baldwin of St. Louis, Missouri, developed the same kind of device independently.

Both men enjoyed financial success with their machines, with Baldwin dominating the market in the United States and Odhner in Europe. The Swedish inventor lived in Russia, and by the start of the Russian Revolution, over 30,000 of his machines were in use just in that country. Various manufacturers built and marketed his device in Western Europe. The owners of his Russian factory left the country at the start of the revolution and reestablished the firm in Sweden under the name of Original-Odhner, a company that continues to operate today. The most widely used versions of his machine in Europe were produced by Brunsviga,† Marchant Calculators, and Friden.

For further information, see: An Illustrated Chronicle of "A Machine to Count On" (Goteborg, Sweden: Aktielbolaget Original Odhner, 1951).

OLSEN, KENNETH HARRY (1926–). This data processing executive was the founder of Digital Equipment Corporation (DEC)† and has often been called the most important creator of the minicomputer. That device made distributed processing a reality in the 1960s.

Olsen was born on February 20, 1926, in Bridgeport, Connecticut. Between 1944 and 1946, he served in the U.S. Naval Reserve. He completed work for a B.S. degree in electrical engineering at the Massachusetts Institute of Technology (MIT) in 1950 and his M.S. there in 1952. Between 1950 and 1957, he worked in the Lincoln Laboratory† at MIT as an electrical engineer. During his student days and in the years immediately following, Olsen worked with and saw the construction of WHIRLWIND,* the largest digital computer* built up to that time. He participated in the development of yet another military project, SAGE.* Thus, when he left MIT in 1957 to establish his own company (DEC) in Maynard, Massachusetts, he had considerable and important technical experience with computers.

Olsen's concept for a new computer in the late 1950s called for a low-cost, efficient machine for engineers and scientists. His first computer was announced in 1960 and sold for only $120,000—nearly $900,000 less than for large mainframes suitable for scientific applications. Additional products and technical innovations made it possible for him to sell computers in the mid-1960s at prices closer to $20,000. As a result, Olsen ran one of the most successful companies in the data processing industry. He developed products constantly, typically spending 8 to 11 percent of all revenues on development. That technological leadership made it possible for DEC to become one of the largest companies in the data processing industry.

Olsen's most successful products were the PDP* series of minis. The PDP 1 was introduced in 1960, and, by the end of 1970, the PDP family had grown by over sixteen models of various sizes. By the early 1970s, DEC was the third largest company in the industry and, by the early 1980s, was settling into second place after International Business Machines Corporation (IBM).† Sales in 1985 totaled $6.7 billion. The company was successful largely because it sold good products that were current, practical, and cost-effective.

Olsen was an engineer with good technical ideas who was also an effective executive. He allowed groups of engineers to concentrate on specific products, developing them as they deemed appropriate. He felt that this decentralized approach would spark creativity and initiative. With regard to this approach, in 1982 Olsen stated: "It is more and more important that the people with the knowledge of the technology are in the position to make the proposals, which is almost the same as making the decisions." Another of Olsen's strengths was his flexibility. For example, when he formed his company, Olsen did not put salesmen on commission; rather, he had engineers on salary selling to other engineers. By the mid-1980s, his was the only company in the industry with noncommissioned marketing representatives. At that point he modified his view, meeting market conditions by paying bonuses to his top performers.

Olsen's career is a successful model of the data processing entrepreneur. Many engineers who came into the industry were initially successful with a company but failed to survive more than a few years and typically had only one product. In contrast, Olsen has been running a profitable company for thirty years. In the beginning he personally developed products, and later he became an effective general manager, an evolution which few of his peers have experienced. By the 1980s, he was first and foremost a manager. As he told a reporter in April 1986: "My job is to make sure the company has a strategy and that everybody follows it." He pays attention to the concerns of his employees, does not automatically resort to layoffs in difficult times, and has imposed what some have termed an "egalitarian style" of management.

For further information, see: Marcia Blumenthal, "A Q&A with DEC's President Olsen," *Computerworld,* June 7, 1982, pp. 10–11; "Digital Equipment: A Step Ahead in Linking Computers," *Business Week,* April 21, 1986, pp. 64, 66; Stephen T. McClellan, *The Coming Computer Shakeout: Winners, Losers, and Survivors* (New York: John Wiley and Sons, 1984).

OPEL, JOHN R. (1925–). This American executive was president of International Business Machines Corporation (IBM)† during the 1970s and chairman of the board in the early to mid-1980s. During his tenure at corporate headquarters, IBM enjoyed a period of enormous growth and prosperity while fighting successful court battles involving private antitrust suits as well as the U.S. Department of Justice. That federal case lasted twelve years, hanging like a dark shadow over the company and the industry it served, but it was finally dropped by the government in the early 1980s for lack of evidence.

Opel, the son of a hardware store owner, was born on January 5, 1925, in Kansas City, Missouri, and was raised there. In 1948, he completed his A.B. at Westminister College and, in 1949, his M.B.A. at the University of Chicago. He joined IBM after school as a salesman in his home state. Between 1949 and 1966, he held a variety of marketing jobs with IBM: salesman, marketing and branch manager, and other executive positions, rising very quickly. Between 1966 and 1968, he served as a vice-president and, in 1967, on the management committee. He was vice-president of corporate finance and planning (1968–1969), senior vice-president of finance and planning (1968–1972), group executive for the data processing group (1972–1974), and a company director since 1972. He was named president of IBM in 1974, working for the chairman, Frank T. Cary,** and he was made chief executive officer in 1981, relieving Cary of many day-to-day responsibilities. In 1983, he became chairman of the board of IBM, then one of the ten largest companies in the world. Upon reaching the mandatory retirement age of sixty in 1986, he relinquished the top post and was replaced by John Ackers, an executive whose career was in marketing.

Opel was the best trained chairman IBM had ever had. In addition to his early academic training and career in marketing, he had worked in manufacturing, finance, and planning. He was also an administrative assistant to two other chairmen of the board, T. Vincent Learson and Thomas J. Watson, Jr.** In 1972, he was given responsibility for managing all of IBM's product development, which allowed him to influence decisions concerning the company's offerings of the late 1970s and early 1980s. Opel was brilliant and had a reputation for doing excellent staff work. He also enjoyed good relations with both the financial and press communities. During the years when the S/360* was being developed, he was working for Learson, serving as an effective bridge between marketing and engineering.

During his tenure as chairman, Opel called for IBM to participate in all aspects of the data processing industry and to be its low-cost producer—two objectives that remarkably influenced the company in the mid-1980s. IBM introduced the Personal Computer which, within three years, contributed nearly 4 percent of the company's revenues and accounted for approximately 20 percent of the microcomputer market in the United States. Five years earlier, it would have been inconceivable that this large mainframe manufacturer would have even entertained the notion of making such little machines. While president in the 1970s, though concerned about the government's lawsuit, Opel continued to broaden IBM's offerings.

Opel also participated in the company's reorganization during this period. The most notable change (effective January 1, 1982) was a major realignment of manufacturing divisions along integrated product lines and the marketing divisions along customer sets. In marketing, for example, the National Accounts Division (NAD) acquired responsibility for IBM's largest customers and the National Marketing Division (NMD), smaller ones. These two new divisions differed from the Data Processing Division (DPD) and the General Systems Division (GSD) (established in the early 1970s with Opel's involvement) in that each also marketed the entire product line, not just a set tailored to a particular customer size. In January 1986, yet another reorganization went into effect within the United States involving the marketing community at IBM. This change took a geographical form. NAD and NMD disappeared, replaced by the South West Marketing Division (SWMD) which had responsibility for marketing all products to all customers essentially in the Southern and Western United States, and the North Central Marketing Division (NCMD) which serviced the rest of the country. Other reorganizations involved splitting up IBM's overseas World Trade Corporations from two to three to defuse responsibility. Yet other changes in divisions and missions occurred within the manufacturing community throughout the early 1980s. No executive at IBM had managed as many reorganizations as Opel had, reflecting the growing and changing character of the company. From the time he became president to the year he stepped down as chairman (1974–1986), IBM's revenues nearly doubled.

After he resigned as chairman, Opel remained on IBM's board of directors. While chairman and earlier as president, he had been active outside of the company. In the 1980s, for instance, he served as vice-chairman of the Business Council and was on the Business Roundtable as co-chairman—two organizations that advised the President of the United States. He became a trustee for the Institute for Advanced Study, served on the boards of the United Way, the Wilson Council, the Council on Foreign Relations, the University of Chicago, Westminister College, and various American companies.

For further information, see: Katharine Davis Fishman, *The Computer Establishment* (New York: Harper and Row, 1981); Robert Sobel, *IBM: Colossus in Transition* (New York: Times Books, 1981).

OSBORNE, ADAM (1939–). Osborne was the founder of the Osborne Computer Company,† an early microcomputer firm and the manufacturer of the first portable computer. Osborne was also an early and prolific writer of books on computers and later became one of the first to sell software* for small manufacturing firms. The Osborne Computer Company is his most important claim to fame within the data processing industry because it helped open up a rapidly growing part of the industry: personal computing.

Osborne was born on February 6, 1939, in Bangkok, Thailand, of British parents and spent a portion of his childhood in India. At age eleven he moved to Great Britain; he graduated from Birmingham University in 1961 with a B.S. in chemical engineering. By this time, like his father and other members of his family before him, he developed an interest in writing. Following graduation he moved to the United States, worked for several companies, and completed a Ph.D. in chemical engineering in 1968 at the University of Delaware. Shell Oil hired him, and he moved to California near San Francisco. In 1971, finding corporate life unsuitable, Osborne resigned and did technical writing. In 1975, he published one of the first books to appear on microcomputers, *An Introduction to Microcomputers,* which he published himself. It quickly became the bible of microcomputers, and by the spring of 1976, he had sold 20,000 copies. Osborne established a publishing firm called Osborne and Associates which he sold to McGraw-Hill in 1979. During his tenure, the firm published twelve of his books and a total of forty on data processing.

In 1979, Osborne decided to make a low-cost, simple microcomputer that would be portable and have software packages associated with it—two firsts for data processing. His idea led to the establishment of the Osborne Computer Company and to the introduction of the Osborne 1 in March 1981. The company grew rapidly during the next two years. At its peak in late 1982, sales approached $70 million and the firm had 1,000 employees. But in September 1983 it declared bankruptcy. Despite his entrepreneurial successes, Osborne apparently could not manage an organization with the kind of growth that his was experiencing. The flamboyant, self-styled *enfant terrible* argued with the president of his company, Robert Jaunich, over why the firm had failed, and even wrote a book, *Hypergrowth* (1984), which told his version of why the company failed. He gained little hearing from an industry accustomed to the rapid turnover of companies.

After his computer company failed, Osborne established Paperback Software International which packages, distributes, and sells the products of twenty-five to thirty small software houses as one would books in a retail market. In fact, he aimed his products at bookstores. By the summer of 1984, the firm had fifteen employees, while the market for distributing software through book dealers had already developed across the United States.

For further information, see: Paul Freiberger and Michael Swaine, *Fire in the Valley: The Making of the Personal Computer* (Berkeley, Calif.: Osborne/McGraw-Hill, 1984); Robert Levering et al., *The Computer Entrepreneurs: Who's Making It Big and How in America's Upstart Industry* (New York: New American Library, 1984); Stephen T. McClellan, *The Coming Computer Industry Shakeout: Winners, Losers, and Survivors* (New York: John Wiley and Sons, 1984); Adam Osborne and John Dvorak, *Hypergrowth: The Rise and Fall of Osborne Computer Corporation* (Berkeley, Calif.: Idthekkethan Publishing Co., 1984) and see also his *An Introduction to Microcomputers* (Berkeley, Calif.: Osborne and Associates, 1975).

P

PACKARD, DAVE (1912–   ). Packard was one of the founders of the Hewlett-Packard Corporation (H-P),† a leading manufacturer of minicomputers and related products in the data processing industry, especially during the late 1970s and early 1980s. By the 1980s, according to surveys conducted by *Fortune* magazine, it was also one of the most admired electronics firms.

Packard was born on September 7, 1912, at Pueblo, Colorado. He graduated from Stanford University in 1934 and completed his E.E. in 1939. Between 1936 and 1938, he worked as an electrical engineer at General Electric's (GE)† vacuum tube engineering department located in Schenectady, New York. In 1938, he co-founded, with William R. Hewlett, the Hewlett-Packard Company in Palo Alto, California, the area which, by the mid-1960s, would be called Silicon Valley, home for hundreds of high technology-based companies. He served as president of the company, (1939–1946); chairman of the board (1947–1964); and chief executive officer (1964–1968), (1972–   ). Between 1969 and 1971, he was Deputy Secretary of the U.S. Department of Defense, an assignment he assumed at the height of the Vietnam War.

Packard's role in public affairs reflects a pattern evident with other top executives in the data processing industry. He was a member of the U.S. President's Commission on Personnel Interchange (1972–1974), the Trilateral Commission (1973–1981), the U.S.-USSR Trade and Economic Council (1974–1982), and the board of directors of various universities and public organizations. Among his numerous awards recognizing his work within the electronics industry were the Vermile medal from the Franklin Institute (1976), Institute of Electrical and Electronics Engineers, Inc., Fellow (1973), and membership in the National Academy of Engineering (1979).

With regard to the Hewlett-Packard Company, its beginnings go back to 1931, when, as a sophomore at Stanford, Packard befriended William R. Hewlett. Both played football, and they found that they shared another interest: ham radios. As a child, Packard had become interested in radios and had a ham radio operator's license. Both also studied engineering and were interested in establishing their own company after graduation. In 1934, Packard went to work for GE, while Hewlett continued his studies at the Massachusetts Institute of Technology (MIT). Four years later, they returned to Stanford for graduate study and rented an apartment together. In their spare time they built electronic gadgets. Hewlett elected to prepare a master's thesis on a variable frequency oscillator which performed over a broader band of conditions than existing devices and at a construction cost of nearly $55 as opposed to the then going expense of $500. Professor Fred Terman of the Electrical Engineering Department encouraged the two students to build a commercial version of the device and loaned them $538 to get started. They worked on this device in addition to other ad hoc electronics projects during the first year. Like others who went on to operate large, successful data processing companies, H-P started in a garage in Palo Alto in the late 1930s (the same as Apple Computers† some four decades later). The company expanded during the 1940s with defense contracts and in the 1950s with civilian electronics products as well.

During the 1970s H-P made small computers, the most famous of which was the HP 3000 used frequently in manufacturing plants. The company also marketed one of the first hand-held calculators, which it brought out during the early 1970s. By the mid-1980s, the company had sales of $4.4 billion and 68,000 employees who designed, manufactured, sold, and serviced some 5,000 different products. By the same time, Dave Packard's share of the company (18.5 percent of the stock) was valued at $2.115 billion.

Many companies within the data processing industry have been characterized as very "people-oriented," that is, they treat their employees as the firm's greatest asset. This pattern is reflected at H-P which maintains a democracy of ideas regardless of rank, encourages the use of first names between employees and their managers, allows casual clothes at work, has flexible working hours, and offers a large package of benefits including medical programs, day care centers, stock options, and so on, all of which are designed to link the employee closely and happily to the fundamental welfare of the firm with rewards and compensation in exchange. Many of these programs first appeared at International Business Machines Corporation (IBM)† and at H-P within the data processing industry. Packard introduced this management style to the company, believing that people were the key to his company's success over so many decades. Many companies in the Silicon Valley of California point directly to H-P as their model when defining their own corporate practices and beliefs. H-P has enforced these concepts even in difficult times. Thus, for example, in 1970, when the company was suffering severely because of the nation's economic recession, rather than lay off people, executives asked all employees to absorb a 10 percent cut in salary and work 10 percent fewer hours until the crisis was over. Such ideas came from the leaders of companies, such as Packard, and they worked.

For further information, see: Dirk Hanson, *The New Alchemists: Silicon Valley and the Microelectronics Revolution* (Boston: Little, Brown and Co., 1982); Thomas J. Peters and Robert H. Watterman, Jr., *In Search of Excellence: Lessons from America's Best-Run Companies* (New York: Harper and Row, 1982).

PASCAL, BLAISE (1623–1666). Pascal, a French mathematician and philosopher, constructed an adding machine between 1642 and 1644 of notable efficiency. His machine not only did addition and subtraction, but also drew attention to the general subject of mechanical aids to mathematics. His device inspired Gottfried Wilhelm von Leibniz** to construct his own.

Pascal, a child prodigy, discovered a proof to Euclid's Proposition 32 at the age of twelve, and four years later developed his now famous theorem of projective geometry. He became interested in mechanical calculators as a result of his desire to help his father, Etienne Pascal, do his work as a tax collector in Rouen. Troubled that many hours had to be spent doing simple calculations, he sought a simpler method, and mechanizing some of the process seemed the logical path to follow. Pascal finished his first machine in 1642, and in the next ten years built fifty others. In addition to his work on these machines during the 1640s, he studied and presented papers on vacuums, conics, and barometers. His interests subsequently shifted back to mathematics and in 1654, he presented two papers that became the basis of integral calculus and probability theory. In 1658 he challenged his fellow mathematicians to a mathematical contest; when he entered the contest himself under the pseudonym of Amos Dettonville and won, awarding himself a prize, he created quite a row.

Pascal converted to Jansenism in 1645 and wrote his *Provincial Letters* in this period, which many consider the start of classical French literature. He discontinued his research into mathematics and calculators at this time. His final service to society, completed at the time of his death, was the plan for the first system of public transportation in Paris.

Pascal's calculator had many of the characteristics of seventeenth- and eighteenth-century machines. He engraved upon wheels the numbers 0 through 9. The first wheel to the right of his device stored digits 0 through 9, the second tens, the third hundreds, the fourth thousands, and so forth, all in a row on the surface of a box. To save a number (for example, 239), one turned the first wheel until it read 9, displayed 3 on the second and 2 on the third, reading from left to right by adjusting wheels from right to left. To add numbers, one placed the first number on the first wheel (assuming it was one digit for our example) and cranked the wheel as many times as the second digit to be added. This cranking of wheels performed the "carrying" function, turning other wheels automatically as appropriate. The process was not very different from adding machines in common use as late as the 1930s.

Pascal's device was small and uncomplicated. Adding and subtraction were its only functions. It was not as sophisticated, for example, as Wilhelm Schickard's** machine, which was built in the same year as Pascal's birth and could do such nonlinear calculations as multiplication and division. Yet Pascal's device received more attention, first, because he built many copies of the device; second, it was seen by many contemporary writers and scientists; and third, Diderot described it in considerable detail in the *Enclyclopédie*.

Pascal conquered a significant technical problem that faced all those designing calculators: he figured out a mechanical way of performing the process of "carrying." Without solving the problem of carrying over to a second set of digits beyond 9, it would have been impossible to add numbers larger than ten. And because the concept he employed was publicized, it affected the work of other scientists who came after him.

Pascal's machine was not a financial success. While it worked, it proved expensive and its maintenance was complicated. When compared to the cost of accountants and other clerks of the time, it was less expensive to continue employing manual methods of operation in businesses. Only scientists without staffs could personally benefit from such devices. Ironically, foreshadowing a problem of the second half of the twentieth century, some accountants were also afraid that his devices might displace their jobs.

Copies of Pascal's machine were deposited in places where they were seen for many years, such as at the Conservatoire des Arts et Métiers in Paris and at the Swedish royal court. Today copies can be seen at the Science Museum in London, periodically at the Smithsonian Institution, and in various IBM exhibits. Even to this day, the Pascaline is often featured in histories of computers as a significant contribution, even though other seventeenth-century devices were far more advanced because they either used logarithms or attempted to compute multiplication and division.

For further information, see: M. Bishop, *Pascal: The Life of a Genius* (London: Bell and Sons, 1937); J. Fonsny, "Pascal et la Machine Arithmétique," *Les Études Classiques, Namur* 20 (1952): 181–191; J. Payen, "Les Exemplaires Conservés de la Machine de Pascal," *Revue d'Histoire des Sciences et de leurs Applications* 16, No. 2 (April–June 1963): 161–178.

PASTA, JOHN R. (1918–1981). Pasta was both an administrator and a researcher. Like many of his contemporaries, his formal education was in physics with a healthy background in mathematics and he moved into what would eventually be called computer science. Because he worked for a variety of institutions during the early stages of computer development, he helped establish administrative policies governing their evolution. He was a professor, a research scientist, an administrator, and an author. The institutions he worked for figure prominently in the early history of digital computers:* Los Alamos, the U.S. Atomic Energy Commission, the University of Illinois, and the National Science Foundation.

Pasta was born in New York City, attended public schools in Queens, and in 1935 started his studies at the City College of New York. Because of the Depression he dropped out to go to work after three years in school. In August

1941, he became a policeman in New York, and in 1942, he was drafted into the U.S. Army and subsequently served as an officer in the Signal Corps. While in the Army, he learned about electronics and radar, and he worked on cryptographical security. He also continued to pursue his reading in physics and mathematics. After World War II, he completed his studies at City College and went on to New York University for graduate work in physics. He earned his Ph.D. in 1951.

Pasta's first full-time job after graduate school was at the Los Alamos Scientific Laboratory where he worked on a variety of physics projects, doing considerable research on solution theory, mathematical physics, and hydrodynamics. He also came in contact with the MANIAC* computer, then under construction at Los Alamos. He developed a graphic display capability for this computer, one of the first instances of such an application on a computational machine. In 1956, John von Neumann* invited Pasta to join the Atomic Energy Commission (AEC) for the purpose of establishing a mathematics and computer branch. The AEC had been an early pioneering institution in the use of computers. After its organization in 1946, scientists within the AEC and outside built a computer, often with the help of the Office of Naval Research (ONR),† other government agencies, and the Institute of Advanced Studies (IAS) at Princeton. By 1956, the AEC was heavily involved in weapons research at several laboratories, and the number of computers being installed to support this work was rising. Pasta's function was to advise on computer matters for unclassified research projects. He was also brought in to contract out research on mathematics and computers. In subsequent years, while at the AEC, Pasta built up the agency's staff in computers and mathematics. He instituted the practice of buying existing hardware to form systems rather than the prevalent procedure of taking the time to build special-purpose, one-of-a-kind devices. After spending four years in the Division of Research at the AEC as manager of the Mathematics and Computer Branch, he accepted an offer as research professor in physics at the University of Illinois in 1961.

At the University of Illinois, Pasta's interests turned to such subjects as pattern recognition, nonlinear problems, data analysis, and high-energy particle events recognized in hydrogen bubble chambers. He also played a role in the development of the ILLIAC* computers, II through IV. Yet his primary contribution at the University of Illinois was not as a research scientist but as an administrator. He headed the Digital Computer Laboratory, which he renamed the Department of Computer Science and which he followed with the development of a curriculum of courses in the field of data processing that led to degrees in the field (the bachelor's and eventually the master's and Ph.D.). By the time he left the University of Illinois in 1970, over 400 students were enrolled in degree programs in computer science, followed by thousands more by the early 1980s, making this institution a major center of computer studies. In the decade he had been there, Pasta had expanded staffs and faculty, enlarged laboratories, and even acquired a building for computer science.

In 1970, Pasta accepted an offer to join the National Science Foundation (NSF) as head of the Office of Computing Activities. During his years at the NSF, Pasta relied heavily on his experience as an administrator and a political infighter. He advocated additional support for research on computing, mathematics, data security, and computer crime. He expanded the relations of the NSF and the National Security Agency (NSA), thereby influencing the kinds of research on computers which both his office and various government agencies would support and finance.

Pasta died of cancer on June 5, 1981. His death came after many years of influencing the course of government support for computer research in the United States.

For further information, see: Kent K. Curtis et al., ''John R. Pasta, 1918–1981: An Unusual Path Toward Computer Science,'' *Annals of the History of Computing* 5, No. 3 (July 1983): 224–238.

PASTORE, ANNIBALE (1868–1936). Pastore was a professor of philosophy at the University of Genoa, Italy, in the early decades of the twentieth century. He built a logic machine in 1903 to provide a mechanical representation of problems and logical solutions. It consisted of three pairs of vertical bars, each supporting perpendicular shafts that could be connected together with belts driven by differential gears. They could be combined in hundreds of combinations to represent up to 256 syllogistic configurations (structures). In his book on the subject of logic machines, Pastore gave an example of the kind of syllogism his device could work with:

Whatever is simple does not dissolve;
The soul does not dissolve;
Therefore, the soul is simple.

The machine would be belted up to correspond to each premise of a logic problem, and, in order to obtain the conclusion, the user simply turned a crank on wheel A. If the syllogism was not valid, the machine would not turn.

Pastore's work with logic wheels was part of a much broader movement in his era to devise mechanical ways to solve problems in logic. Charles S. Pierce, the American philosopher, described a switching circuit to perform logic while his former student Allan Marquand** built a machine in the 1880s to solve problems in formal logic. Thinking machines were believed to offer major improvements in the productivity of both philosophers and mathematicians.

For further information, see: Annibale Pastore, *Logica Formale: dedotta della considerazione di modelli meccanici* (Turin: Bocca, 1906).

PATTERSON, JOHN HENRY (1844–1922). Patterson dominated the National Cash Register Company (NCR)† as its chief executive from the late nineteenth century until his death in 1922. He helped mold the business practices of Thomas J. Watson, Sr.,** founder of International Business Machines Corporation (IBM),† later arch rival of NCR. Patterson was also one of America's most visible and important general executives in the late nineteenth century.

Patterson was born on December 13, 1844, near Dayton, Ohio, attended local public schools, spent one year at Miami University in Ohio, and in 1864 enlisted in the Union Army. He graduated from Dartmouth College in 1867. Unable to find employment, he returned to his father's farm, and, perhaps because of this experience, he concluded that a college education was not valuable. In 1868, he became a toll collector in Dayton, and then joined with his brothers in buying coal and selling it directly to consumers in town. In an attempt to control the loss of cash collected from customers, he installed a cash register. The new device had been invented in Dayton by James Ritty. It worked, and the coal business now began to show a profit. In 1881, he joined with his brother John H. Patterson and a group of investors from Boston to create the Southern Coal and Iron Company. Then in 1884, dissatisfied with his lack of complete control over the firm, he elected to market cash registers which were still a relatively new item. He bought controlling interest in the National Manufacturing Company of Dayton for $6,500; it made cash registers. In December 1894, Patterson renamed the company the National Cash Register Company and had thirteen employees. As Robert Sobel has noted, Patterson, now age fifty, had a product nobody knew about, wanted, or appreciated and one that users would object to using since it cut down on their pilfering of funds.

This executive was distinguished from many others entering on a new enterprise by his ability to sell and the manner in which he treated salesmen. The result was that within a few years no serious retailing operation could function without a cash register. He gathered around him a cadre of outstanding salesmen, and he personally boosted the efficiencies of the cash register as a modern, indispensable tool. He made sure that his company improved on earlier models of the machine, making them more marketable. He also devoted the bulk of his energy to selling. He trained salesmen, ran sales rallies, used advertising, and encouraged his salesmen to rely on knowledge of the product and appreciation for a customer's needs, not pressure tactics, as the basis for selling. Patterson insisted that his people dress well and service the products sold. He created exclusive territories for his salesmen, paid handsome commissions, and established quotas—practices which were new to American business. After Patterson, salesmanship developed along the line he had established.

One of Patterson's salesmen was Thomas J. Watson. The future head of IBM acquired and embraced nearly everything he was ever to learn about the art of salesmanship and the management of marketing representatives from Patterson. Watson would later employ the same principles and practices at IBM, and they continue to be the mainstay of the company's practices in the late 1980s. Both at NCR and at IBM, salesmen were crucial to the success of the business. Service was always an important element of work with customers and came to characterize the efforts of many within the data processing community who were marketing products after World War II which, as in Patterson's day, were not appreciated, wanted, or at first user friendly.

Patterson proved as innovative with factories as with the sales force. He enhanced employee benefits, provided entertainment, painted factories, used wooden floors to reduce fatigue, and made his facilities bright and cheerful, all in an attempt to improve his workers' productivity. His benefits package included educational programs, medical services, and improved working conditions. Patterson had trees and gardens planted around factory buildings. In exchange, he exacted efficiency and hard work, and he proved irascible with sloppy or incompetent workers.

Patterson's impact was reflected in NCR's success. In April 1885, he could only claim that sixty-four cash registers had been sold. During 1887, the number rose to 5,400 and in another three years climbed to 16,395. Through the use of direct mail campaigns (which did not become common in the data processing industry until the 1970s), he sold many machines. In one year alone, his company mailed out 18.5 million advertisements, a huge number by the standards of his day.

This American executive also implemented competitive marketing practices that created problems for him with the U.S. government. In 1910, the American Cash Register Company complained about NCR's predatory practices to the U.S. Department of Justice. In December 1911, the government filed suit against NCR for conspiring to restrain trade. Patterson was indicted with twenty-one officers of the company, including Watson who was a high-ranking sales manager. They were accused of violating the Sherman Anti-trust Law. Specifically, NCR was charged with trying to dominate the cash register business (it had about 95 percent of the market at the time), monopolizing the trade in cash registers, and attempting to maintain its market position. All charged were found guilty on the three counts; Patterson was fined and sentenced to one year in jail. While on appeal, a flood struck Dayton where the trial had taken place. Patterson and his staff helped in the recovery efforts, supplying food, blankets, tents, and medicine, and allowing use of NCR's factory grounds. The citizens of Dayton then petitioned President Woodrow Wilson to pardon Patterson and the others convicted. In 1914, the U.S. Court of Appeals overturned the original guilty verdict, and in 1915 the U.S. Supreme Court declined to hear the U.S. Department of Justice's appeal.

NCR survived the crisis and prospered. In 1922, the year Patterson died, NCR had a complex of twenty-two buildings in Dayton and 10,000 employees. Over 2 million cash registers had been sold, and annual sales totaled $30 million. The entire company was capitalized at $14.5 million. The year before, his son Frederick B. Patterson had become president of the company, while the senior Patterson remained chairman of the board. NCR continued to dominate the cash register market until the early 1970s when other companies, now also in the data processing industry, such as IBM, Burroughs Corporation,† and Sperry Rand,† introduced successful products that competed against "The Cash," as it had been known since Patterson's time. The company Patterson had established became an important member of the data processing community and by the late 1970s was enjoying annual sales of $3 billion. Patterson died on May 7, 1922.

For further information, see: Samuel Crowther, *John H. Patterson, Pioneer in Industrial Welfare* (Garden City, N.Y.: Doubleday, Page, 1923); R. W. Johnson and R. W. Lynch, *The Sales Strategy of John H. Patterson, Founder of the National Cash Register Company* (Chicago: Dartneli Corporation, 1932); Isaac F. Marcosson, *Colonel Deeds: Industrial Builder* (New York: Dodd, Mead, 1947) and his *Wherever Men Trade* (New York: Dodd, Mead, 1948); Robert Sobel, *IBM: Colossus in Transition* (New York: Times Books, 1981).

PERLIS, ALAN J. (1922–). Perlis was a key developer of ALGOL,* a programming language* intended to be a universal idiom for programmers in the late 1950s and early 1960s.

Perlis was born on April 1, 1922, in Pittsburgh, Pennsylvania, and completed his B.S. degree in chemistry at the Carnegie Institute of Technology in 1943. He spent the following year attending California Institute of Technology and in 1950 completed his Ph.D. at the Massachusetts Institute of Technology (MIT) in mathematics. He next went to work as a researcher in mathematics at the laboratory run by the Ballistic Research Lab at Aberdeen Proving Grounds. In January 1952, he accepted an appointment as a member of MIT's staff working on Project WHIRLWIND* where he wrote programs for the computer. In September, he became an assistant professor of mathematics at Purdue University. As director of its computer center (1952–1956), he used an IBM CPC* and in the spring of 1954 installed a Datatron 205. During 1955, his laboratory began work on a compiler for a language called Internal Translator (IT). It was during this period that Purdue became one of the major research centers of computer technology in the United States. In 1956, Perlis left Purdue to become director of the computer center at Carnegie Tech. He also taught mathematics while doing research on programming languages. Between 1961 and 1964, he was chairman of the Mathematics Department and, between 1965 and 1971, head of the Department of Computer Science. He became professor of computer science at Yale University in 1971 and chairman of the department in 1976.

Perlis was heavily involved in the development of ALGOL in 1957. The data processing community, sensing a need for a common universal programming language, formed a committee of vendors and representatives from universities and government agencies, to design a programming tool. Perlis was chairman of that committee, which produced ALGOL 58 and a revised version called ALGOL 60. The language, although hardly used in the United States, established new levels of discipline in the description of programming languages, identified features that would appear in subsequent languages, and established methods for the design of programming aids.

Perlis was active in the industry's professional associations, serving, for example, as the first editor in 1958 for the Association for Computing Machinery (ACM's)† journal, *Communications of the ACM* and in 1962, as its president. He was elected to the National Academy of Engineering in 1977.

For further information, see: Alan J. Perlis, ''The American Side of the Development of ALGOL,'' in Richard L. Wexelblat, ed., *History of Programming Languages* (New York: Academic Press, 1981), pp. 75–91, and for a biography, see ibid., p. 171; Jean E. Sammet, *Programming Languages: History and Fundamentals* (Englewood Cliffs, N.J.: Prentice-Hall, 1969).

PUGH, EMERSON W. (1929–). Pugh helped develop memory* for International Business Machines Corporation's (IBM's)† System 360* in the early 1960s. Subsequently, he influenced the direction of IBM's developments in the area of computer memories during the 1960s and 1970s. He also published significant materials on the history of IBM's technology.

Pugh was born on May 1, 1929, at Pasadena, California, and graduated with a B.Sc. in physics from Carnegie-Mellon University in 1951. He completed a Ph.D. in physics at the same university in 1956. He remained at Carnegie-Mellon for the next academic year, teaching physics as an assistant professor. In 1957, he joined IBM as a researcher, and between 1958 and 1961, he served as manager of the metal physics group. He was a visiting scientist at the IBM Zurich Laboratory (1961–1962) and a senior engineer within the IBM Components Division (1962–1965). At this time, he managed the development of the magnetic film memory array used in the company's largest S/360s, particularly in the Model 95 computer. In 1965–1966, he served as IBM group director of Operational Memory and, from 1966 to 1968, as director of Technical Planning within the IBM Research Division.

Pugh was first exposed to general management between 1968 and 1971 when he was a special assistant in the office of the IBM vice-president and chief scientist. Then from 1971 to 1975 he was a consultant to the IBM director of research. He returned to research more directly between 1975 and 1980 as research manager of exploratory magnetics, and during 1981 and 1982, he helped establish the IBM manufacturing research organization, a consolidation and restructuring of a key element of the company's developmental activities. In the early 1980s, he continued to study how to improve developmental research. Then in 1985 he was named manager of the IBM Technical History Project at Yorktown Heights, New York, home of much basic research within IBM. In that capacity he became responsible for research conducted on the history of IBM's early computers.

Pugh published extensively on technical subjects and on the history of IBM. He was also active in various scientific organizations. For example, in 1962 he was elected Fellow of the American Physical Society. Between 1968 and 1970, he edited the *IEEE Transactions on Magnetics*. The Institute of Electrical and Electronics Engineers, Inc. (IEEE)† named him a Fellow in 1972 and in the following year elected him president of its Magnetics Society. Pugh was elected Fellow of the American Association for the Advancement of Science in 1977. Earlier, in 1974, he had also functioned as the executive director of the National

Academy of Science for the study of automobile emissions and fuel economy, a project carried out on behalf of the Environmental Protection Agency and the U.S. Congress. He was elected vice-president of the IEEE in 1986.

For further information, see: Emerson W. Pugh, *Memories That Shaped an Industry: Decisions Leading to IBM System/360* (Cambridge, Mass.: MIT Press, 1984), with C. J. Bashe et al., *IBM's Early Computers* (Cambridge, Mass.: MIT Press, 1986), and with E. M. Pugh, *Principles of Electricity and Magnetism* (Reading, Mass.: Addison-Wesley, 1960; 2d ed., 1970).

R

RAJCHMAN, JAN ALEKSANDER (1911–). Rajchman, a computer scientist, focused on the development of memory* systems in the 1940s and 1950s while working for the Radio Corporation of America (RCA).†

Rajchman was born in London on August 10, 1911, and moved to the United States in 1935. In the same year, he completed work for his diploma in electrical engineering at the Swiss Federal Institute of Technology in Zurich and in 1938 earned a Ph.D. from that institute. Rajchman became a member of the technical staff at RCA's laboratory in Princeton, New Jersey, in 1936 and remained a researcher there until 1959 when he became the associate director of the Systems Laboratory. In 1961, RCA promoted him to director of the laboratory, and he served in that capacity until 1971. Between 1971 and 1976, he was RCA's staff vice-president of information sciences. He later consulted in the general field of computer science. Rajchman's research interests in the 1970s and 1980s involved optical devices, nonimpact printers, and display units. At RCA almost all of his work involved electrical engineering.

At RCA in the 1940s, Rajchman developed a design for magnetic core storage; similar technologies became the basis of most computer systems in the 1950s. He also led RCA into the field of computing earlier than many other competitors. In 1945, John von Neumann,** then at the Institute for Advanced Study at Princeton, began work on what would become known as the IAS Computer.* Casting about for leading-edge technologies to employ for his machine, von Neumann met with Rajchman to determine whether RCA might be of use. Initially, the IAS group sought to use Rajchman's research in the form of a new memory for their computer. Rajchman was developing a storage tube that eventually became known as the Selectron. It appeared in the RAND Corporation's first computer, known as the JOHNNIAC.* It had a 256 binary-bit structure that proved to be less reliable than was required for the IAS Computer of the late 1940s. Rajchman's tube was not as good an option as might otherwise have been the case because at about the same time the Williams tube from Great

Britain became available. It was far more reliable, leading IAS to bypass Rajchman's work.

By 1947, Rajchman was already designing a core memory that would prove to be far more reliable technologically. Similar work was underway in Great Britain, conducted by Andrew D. Booth. Rajchman's memory consisted of small doughnut-shaped ferrite substances that could be used to represent positive or negative charges—hence, on or off bits. After IAS chose other components for its memory, RCA returned to its own projects, including the development of a computer. In the early 1950s, RCA introduced a product and Rajchman worked on it.

For a brief moment in the early history of computing, Rajchman conducted important research on computer memories. His company had made it difficult for him to participate more fully in the new industry, however. For example, he urged RCA to support the development of the ENIAC* in 1944–1946, but the company apparently had other priorities. He was finally successful in getting RCA interested in the general field of data processing in the 1950s.

Rajchman was recognized by his profession for his technical contributions (made in the early days of computing) and for his many years of management at one of the important development laboratories in the United States. He received the Institute of Electrical and Electronics Engineers, Inc. (IEEE)† Liebman award in 1960, the Edison medal in 1974, and the Franklin Institute's Lerry medal in 1977.

For further information, see: Herman H. Goldstine, *The Computer from Pascal to von Neumann* (Princeton, N.J.: Princeton University Press, 1972); Jan A. Rajchman, "Early Research on Computers at RCA," in N. Metropolis et al., eds., *A History of Computing in the Twentieth Century* (New York: Academic Press, 1980), pp. 465–469, *RCA Computer Research: Some History and a Review of Current Work* (Princeton, N.J.: David Sarnoff Research Center, RCA Laboratories, 1963), "The Selectron—A Tube for Selective Electrostatic Storage," *Proceedings of a Symposium on Large-Scale Digital Calculating Machinery* (Cambridge, Mass.: Harvard University Press, 1948), pp. 133–135; Robert Sobel, *RCA* (New York: Stein & Day, 1986).

RAMO, SIMON (1913–). This American engineer was a founder of the Ramo-Wooldridge Corporation (which eventually became TRW†), serving as its chairman of the board. Like many engineers of his generation who worked in the field of defense contracting and data processing, Ramo began his career as an engineer and then spent many years managing high technology enterprises, primarily related to defense, in the United States.

Ramo was born on May 7, 1913, in Salt Lake City, Utah, where he grew up and attended public schools. He studied electrical engineering and completed his B.S. at the University of Utah in 1933. Ramo earned a Ph.D. from the California Institute of Technology in 1936 in electrical engineering and physics. Between 1936 and 1946, he worked for the General Electric Company (GE)† in Schenectady, New York, first as a research engineer and then as a manager. He had a progression of jobs which led him from section head in the engineering laboratory to head of the physics section of the electronics research laboratory. In 1946, he went to work for the Hughes Aircraft Company in Culver City, California, as director of the Electronics Department, a post he held until 1948. Between 1948 and 1950, he managed guided missile research and development for the company. He was promoted to vice-president of research, engineering, and manufacturing operations in 1950, remaining in that post until 1953. At that point Ramo joined with Dr. Dean E. Wooldridge,** also from Hughes Aircraft, to form the Ramo-Wooldridge Corporation with headquarters in Los Angeles.

The new company marketed high technology services to the military with the financial backing of Thompson Products, Incorporated, headquartered in Cleveland, Ohio. In the new firm, Ramo became the executive vice-president and director, a post he held until 1958, when Thompson Products merged with his company to form Thompson, Ramo, Wooldridge, Incorporated (which later was called TRW, Incorporated). Ramo has been TRW's chairman of the board since 1961. By the late 1970s, TRW had become one of the largest credit-rating companies in the world with over 70 million files. Its products were computer-based information.

Ramo's career alternated between engineering and management. At Hughes Aircraft he had established electronics and guided missile development, and in the process his staff built a series of systems involving airborne radar, computer and navigational equipment, and armament control devices that relied heavily on the use of computer technology. All were also for the U.S. Air Force, one of the world's largest consumers of computer-based products and a prominent supporter of the young data processing industry in the late 1940s and early 1950s. While at Hughes Aircraft, Ramo also participated in the development of the Falcon Missile, used during the Korean War. By 1953, he could correctly claim that he was managing one of the largest research operations for military applications in the United States. His work continued after he established his own company, armed with a number of defense contracts to launch the enterprise.

Because of its founders' experience in defense work, Ramo-Wooldridge won the contract to provide the overall direction for the design and engineering of the ballistic missile program for the U.S. Air Force. It soon became one of the world's largest defense projects of the 1950s. Ramo had to coordinate the activities of over 200 companies, providing guidance and management control. His firm also managed the development of the Thor, an intermediate-range ballistic missile, and later the better known Atlas, Titan, and Minuteman ICBMs. These projects in the 1950s were largely contracted to Ramo's company because of his particular skills. Between 1954 and 1958, he was also the chief scientist for the entire ballistics program sponsored by the U.S. Air Force and hence played a crucial role in encouraging the entire electronics industry to miniaturize its products. That process blended well with technical developments in computing, which by the early 1960s had led to very fast, increasingly reliable, and cost-effective computer systems. Almost every major military computer-based application sponsored by the U.S. Air Force during the 1950s and early 1960s was directly influenced by Ramo.

But like other defense contractors, Ramo and his partner recognized that they should diversify and supply services to the private sector so that they would not be completely dependent on the military. Hence, in 1954, they established Pacific Semiconductors, Incorporated, with Thompson Products holding 50 percent interest in the subsidiary. The mission assigned this organization was to manufacture transistors and diodes for the emerging computer-related component market. Further expansion and diversification within the electronics industry led to the establishment of six divisions and two laboratories by the end of 1957. Ramo, through these divisions of R-W Corporation, now had organizations to build airborne digital computers* and ground instruments for testing rockets and missiles (equipment division); an information systems division for computer equipment employed in military applications; and a communications division. Thompson-Ramo-Wooldridge Products, Inc. (another subsidiary established in 1955) designed electronic process control systems and related computers for plant automation. In 1955, Ramo was also partially responsible for creating Space Technologies Labs, Incorporated, of which he was president. It focused on providing the U.S. Air Force technical and managerial expertise in the general field of ballistic missiles. It obtained a contract to design and develop the Minuteman and Titan II, in addition to U.S. Air Force rocket boosters and spacecraft.

At the time it merged with Thompson Products, the entire corporation with which Ramo was associated had some 3,500 employees, nearly 7 million square feet of space, and $50 million in annual sales. By the standards of the early 1960s, it was a very large enterprise. After the merger, Ramo continued to devote his time to projects for the U.S. Air Force and to other projects related to the American space program. By 1963, his firm had contributed to over 90 percent of all the payloads which the United States launched into space between 1957 and 1963.

The company in its new form grew larger and remained prosperous. In 1964, for example, TRW had a payroll of 40,000 and 10 million square feet of space scattered across 100 plants and laboratories, and was generating revenues of $533 million and net after-tax income of $23 million. Ramo continued to dominate many of the company's military projects throughout the next two decades.

Ramo had other activities as well. He taught at the California Institute of Technology from 1934 to 1936, at Union College from 1940 to 1946, when he became a research associate at the California Institute of Technology, a position he has held into the 1980s. From 1964 to 1966, he was president of Bunker-Ramo Corporation, a small company that manufactured data processing equipment, including terminals.

For further information, see: TRW, *The Little Brown Hen That Could* (Cleveland, Ohio: TRW, undated [1983?]).

RAND, JAMES HENRY (1886–1968). This businessman was the founder of Remington Rand Corporation† which introduced the UNIVAC* computer, the first commercially available electronic digital computer* to be sold in quantity. His company was a leading manufacturer of office equipment and supplies, particularly of the Kardex system and typewriters.

Rand was born on November 18, 1886, in North Tonowanda, New York, and graduated from Harvard University in 1908. In 1915, he developed and sold the Kardex system of filing, a set of cards and cabinets popular in offices until the 1960s. In 1926, he created what became known as the Remington Rand Company, serving as its president until 1955. At that time his company merged with the Sperry Corporation,† and he became vice-chairman, serving until 1968. During his tenure at Remington Rand the company became a major force in the office products market during the 1920s—a claim it could maintain for its entire history.

Rand was an enterprising manager and one of the first executives in the United States to recognize the potential of computers. His company grew through a series of acquisitions and normal business expansion. He entered the computer arena by purchasing the Eckert-Mauchly Computer Corporation at the beginning of the 1950s, which was then developing the UNIVAC series of computers. Rand died in June 1968.

For further information, see: Katharine D. Fishman, *The Computer Establishment* (New York: Harper and Row, Publishers, 1981).

RANDELL, BRIAN (1936–). This computer scientist turned professor was perhaps most influential in data processing by encouraging the study of computer history. In addition to his publications on the subject, he participated in various research projects at International Business Machine Corporation (IBM)† during the 1960s. Like many who came into the industry, he studied mathematics and physics before merging those skills into what today is known as computer science.

Randell was raised in Cardiff, Wales, and attended the Imperial College of Science and Technology of the University of London, graduating in 1957 with a B.Sc. in mathematics. Between 1957 and 1964, he worked for the Atomic Power Division of the English Electric Company, Ltd., working with the DEUCE* computer. For that machine he wrote a compiler for a new language called EASICODE. By the time he left, he had become head of the Automatic Programming Section of the company and had managed or participated in the development of six programming language* compilers, including one for ALGOL* (known in his company as the Whetstone KDF9 Algol Compiler). From 1964 to 1969, Randell worked for IBM at the Thomas J. Watson Research Center at Yorktown Heights, New York, home of much of the company's basic research. Randell's first project involved the design of a new high-speed computer and soon after was moved to Los Gatos, California, as part of the Advanced Computer Systems Department. In 1966, Randell returned to Yorktown Heights as manager of System Modelling and Evaluation associated with Project IMP, a scheme to design large multiprocessing systems. In 1969, he returned to Great Britain, joining the Computing Laboratory at the University of Newcastle upon Tyne as professor of computing science; he is still there today. In England he continued to do research on operating systems and systems architecture, and launched a serious study of the history of computing. Like other professors in the field of computing, Randell was active in the International Federation for Information Processing,† particularly in the 1960s in its work with ALGOL, and served on various editorial boards: *Software Engineering* (1969), *Software Engineering Techniques* (1970), and *Acta Informatica* (1969–).

For further information, see: Brian Randell, "The COLOSSUS," in N. Metropolis et al., eds., *A History of Computing in the Twentieth Century* (New York: Academic Press, 1980), pp. 47–92, "Digital Computers: Origins," in A. Ralston and C. L. Meek, eds., *Encyclopedia of Computer Science* (New York: Petrocelli/Charter, 1976), pp. 486–490, "From Analytical Engine to Electronic Digital Computer: The Contributions of Ludgate, Torres, and Bush," *Annals of the History of Computing* 4, No. 4 (October 1982): 327–341, "Ludgate's Analytical Machine of 1909," *Computer Journal* 14, No. 3 (1971): 317–326, "On Alan Turing and the Origins of Digital Computers," in B. Meltzer and D. Mitchie, eds., *Machine Intelligence* (Edinburgh: Edinburgh University Press, 1972), 7: 3–20, *The Origins of Digital Computers* (Berlin: Springer-Verlag, 1973, revised 3d ed., 1982), and "The Origins of Digital Computers: Supplementary Bibliography," in N. Metropolis et al., eds, *A History of Computing in the Twentieth Century* (New York: Academic Press, 1980), pp. 629–653 as additional to his earlier, "An Annotated Bibliography on the Origins of Computers," *Annals of the History of Computing* 1, No. 2 (October 1979): 101–207.

REES, MINA SPIEGEL (1902–). Rees has had a distinguished career as a professor and an administrator in government and academia, and played a pivotal role in the expansion of basic research in scientific fields within the United States during the late 1940s and early 1950s. As a key administrator with the Office of Naval Research (ONR),† she supported the development of computers. In the 1960s, she led the growth of the graduate school of the City University of New York (CUNY). She was a model of the American academician in positions of management within both higher education and government.

Rees was born in Cleveland, Ohio, on August 2, 1902. She graduated from Hunter College in New York City with an A.B. in mathematics in 1923 and earned an A.M. at Columbia University in 1925 and her Ph.D. at the University of Chicago in 1931. Between 1926 and 1932, she was an instructor in mathematics at Hunter College, and then began to rise within the faculty: assistant professor (1932–1940), associate professor (1940–1950), professor and dean of the faculty (1953–1961). She was dean of the graduate school at the City

University of New York from 1961 to 1968 when she was named provost of the graduate school (1968–1969). She was president from 1969 to 1972.

While maintaining a connection with Hunter College during the 1940s, Rees also worked for the U.S. government during these years. She began her career in Washington, D.C., as a technical aide and administrative assistant to the chief of the Applied Mathematics Panel (AMP) of the National Defense Research Committee (NDRS), Warren Weaver. Both of these organizations were within the Office of Scientific Research and Development (OSRD) in the years Rees was an employee of the NDRS (1943–1946). The OSRD employed scientists to solve war-related problems. Thus, she recruited mathematicians to support the armed services during World War II.

In 1946, Rees went to work for the Office of Naval Research which underwrote many research projects in the United States under contract to professors. It also financed numerous computer-related research programs, including the Massachusetts Institute of Technology's (MIT's) WHIRLWIND.* Between 1946 and 1949, Rees ran the mathematical branch in ONR, and between 1949 and 1953, she served as director of the Mathematical Science Division where she spearheaded the U.S. Navy's major support of basic research in the United States.

Throughout the 1950s, Rees served on numerous advisory boards assisting the U.S. military community, scientific foundations, and other research-supporting institutions while also functioning as the dean of faculty at Hunter College. During the 1960s, she was a member of various national education foundation boards, including the National Science Board (1964–1970), the Council of Graduate Schools in the United States (chairman in 1970), and the Woodrow Wilson National Fellowship Foundation.

As the executive assistant to Warren Weaver, chief of the Applied Mathematics Panel within OSRD, Rees gained her first important experience in managing and supporting applied and basic research involving the government. She either knew or worked with the mathematicians employed on projects. Some of these mathematicians went on to make contributions in the general field of data processing, including Herman H. Goldstine** and John H. Curtiss.** Within the AMP, established in 1942, complex mathematical problems drew her attention. That agency eventually supported over 200 projects, including work done by John von Neumann** who at the time was an expert on the theory of games and a solid supporter of computational research projects.

Following World War II, the ONR became one of the leading government agencies supporting the work of scientists, frequently on projects concerning basic research. For example, it funded many mathematical projects, research on numerical analysis, and logistics. In the case of numerical analysis, the subject was of particular interest to Rees who believed that it could help make computational equipment effective. While at ONR Rees also took an interest in operations research, a subject born during World War II. In the late 1940s, she and others saw the need to support research on the use of quantitative methods for studying problems. That concern led to the development of various tools and techniques, which in turn became computer-based applications by the 1950s.

Linear programming and the work of George B. Dantzig,** for instance, were direct byproducts of ONR's interest in operations research. The ONR and the U.S. Navy in general saw that linear programming could improve logistics, and thus the Logistics Program was established within ONR in 1947. Two years later, the Navy created an independent agency called the Logistics Branch of the Mathematical Sciences Division. By the early 1950s, this agency's entire effort relied heavily on computers. Rees was largely responsible for launching this kind of research.

In the late 1940s, Rees encouraged research on pure and applied mathematics, statistics, and the development of the computer, with particular emphasis on numerical analysis. As a result of her emphasis on numerical analysis, she played a controversial role with Project WHIRLWIND. The development of that computer at MIT, under the direction of Jay W. Forrester,** and the financial backing it had from ONR, led the U.S. Navy to question whether the entire project was becoming too expensive. Some doubted that the machine would ever be completed and began to ask what use it might have. For her part, Rees doubted that the engineers on the project had sufficient background in mathematics to design an electronic digital computer* that would be truly a new type of computer. She thought the project had merit but little direction. Even though the U.S. Navy had doubled its financial contribution to the project in these years, she still had misgivings.

Rees, Forrester, and others both at ONR and at MIT worried, argued, and struggled over funding and management of the program throughout the late 1940s. Pressure came from the Navy Department to reduce the cost of the project. The historical records indicate that her views generally coincided with those of the Navy Department. Because of an apparent lack of progress in building a computer, combined with rising costs, she apparently came under pressure from ONR to cripple or reduce support for the project. But ONR had committed too much money and the project was too far along for the government to abandon it. Years later, Rees hinted that if the U.S. Air Force had not contributed to WHIRLWIND in its later stages, it might have died. As it turned out, WHIRLWIND was completed in the early 1950s and operated for the U.S. Air Force. It was also the largest and most expensive machine to date.

WHIRLWIND was one of several computer-based projects Rees supported. Another project involved von Neumann's computer which was under development at the Institute for Advanced Study at Princeton (called the IAS Computer*). Yet with all these projects she applied the same measure. She asked how much they would contribute to mathematics, and she took less of an interest in the engineering aspects of these programs. The establishment of the Institute for Numerical Analysis at the University of California at Los Angeles in 1947 with her support reflected this interest in mathematics.

During her years with the U.S. government, Rees met, worked with, or supported some of the early pioneers of computing. She persuaded the U.S. Navy to support many projects involving basic research in American universities in fields ranging from physics to computing, from statistics to numerical analysis. She led the government to become a major supporter of scientific research at a time when alternative sources of funding were too small. Hence, she continued the process, which started in World War II, of government support of projects.

Rees went back to Hunter College in the 1950s. In 1961, she began a new phase of her academic career, when she agreed to manage a graduate program at the newly created CUNY. During her decade-long tenure there, she established a faculty which developed doctoral programs in twenty-six fields. CUNY's first Ph.D. graduated in 1965 with a degree in English; during her tenure, 476 additional Ph.D.s were awarded. CUNY had over 2,500 students by the early 1970s. By the early 1970s, despite its youth, CUNY had already achieved a reputation for quality. Rees' administrative experience at Hunter College, followed by her maturation at ONR, taught her how to fund research projects and gave her the administrative and political skills needed to negotiate with various colleges and universities in New York, along with state and federal agencies, to launch the graduate school.

Rees received the President's Certificate of Merit in 1948 and became the first recipient of the Mathematical Association of America's Award for Distinguished Service to Mathematics (1962). Throughout the 1960s and 1970s, several academic organizations recognized her contributions to Hunter College, CUNY and the academic community at large. In recent years, she has made an additional contribution to the history of data processing: she began writing on the early history of computing, the ONR, and mathematics from the 1930s to the 1950s.

For further information, see: Kent C. Redmond and Thomas M. Smith, *Project WHIRLWIND: The History of a Pioneer Computer* (Bedford, Mass.: Digital Press, 1980); Mina S. Rees, "The Computing Program of ONR, 1946–1953," *Annals of the History of Computing* 4, No. 2 (April 1982): 102–120, *The First Ten Years of the Graduate School, The City University of New York* (New York: The Graduate School and University Center, 1972), "The Mathematical Sciences and World War II," *American Mathematical Monthly* 87, No. 8 (October 1980): 607–621.

RIDENOUR, LOUIS "MOLL" NICOT, JR. (1911–1959). As a researcher, Ridenour helped to develop such early computers as the ORDVAC* and the ILLIAC,* along with magnetic core memory* for the IBM 704,* and as an administrator he introduced computing into various government laboratories.

Ridenour was born on November 1, 1911, in Montclair, New Jersey. He completed his B.S. in physics at the University of Chicago in 1932 and his Ph.D. at the California Institute of Technology in 1936. In 1935, he received an appointment as assistant to Enrico Fermi at the Institute for Advanced Study at Princeton, but Fermi chose not to show up. Ridenour instead became a young assistant professor of physics at Princeton University and remained there until 1938 when he became an assistant professor of physics at the University of

Pennsylvania, studying the electrostatic accelerator. Then in January 1941, he joined the Radiation Laboratory at the Massachusetts Institute of Technology (MIT) which had just been established. Along with some fifty other scientists he worked on the use of microwave radar. As manager of a small team (known as Ridenour's Rangers), he directed the development of a gun-laying radar which was eventually called the SCR 584. Another system, the M–7, was developed at Bell Laboratories.† Both were the first useful antiaircraft systems to appear. Ridenour was recognized for his work by being awarded the President's Medal of Merit, the Bronze Star, and the U.S. Air Force's Exceptional Civilian Service Medal.

At the end of World War II, Ridenour was selected to edit a set of twenty-eight volumes of books on radar sponsored by the Radiation Laboratory and known as the Radiation Laboratory Technical Series. It encompassed everything known about radar as of the mid- to late 1940s. He retired from the University of Pennsylvania in 1946 (he had only been on loan to MIT), and in 1947, he became dean of the Graduate College at the University of Illinois. Between 1947 and 1950, he sponsored the establishment of the Digital Computer Laboratory at the University of Illinois. This laboratory would develop the ORDVAC and ILLIAC computers.

Ridenour once again turned his attention to the needs of the U.S. Air Force in 1949. He accepted the chairmanship of a committee to review research and development for the Air Force, and the results of this task force were presented in a series of recommendations in what became known as the Ridenour Report. Ridenour was asked to implement these recommendations as the Chief Scientist for the U.S. Air Force. One of the byproducts of his efforts was the establishment of the Air Research and Development Command (ARDC) along with MIT's Lincoln Laboratory.† These two organizations constructed the WHIRLWIND,* the largest digital computer* of the early 1950s. In 1951, Ridenour became vice-president for engineering at International Telemeter Corporation where he worked with George Brown on the early stages of magnetic core memory* for the IBM 704 computer. In 1955, he went to work for Lockheed Corporation, and, in March 1959, he was promoted to vice-president and general manager of the Electronics and Avionics Division. During his business career he also helped manage the development of the Polaris and the X–17 missiles.

A friend of Ridenour described him as a charming, intelligent, imaginative man who suffered no fools around him. Ridenour appreciated a "wordly set of values with a perhaps excessive interest in rank and wealth and symbols."

Ridenour died of a cerebral hemorrhage on May 21, 1959, in Washington, D.C. Like many of his peers in the field of data processing, he made his contributions early in life and, as it turned out, his time was short.

For further information, see: R. A. Kingery et al., eds., *Men and Ideas in Engineering: Twelve Histories from Illinois* (Urbana: University of Illinois Press, 1967); F. W. Loomis, "Louis Nicot Ridenour, Jr.," *Physics Today* 12 (September 1959): 18–21; F. Seitz and A. H. Taub, "Louis N. Ridenour, Physicist and Administrator," *Science* 131 (January 1, 1960): 20–21.

ROSEN, SAUL (1922–). Rosen was active in the creation of compilers for FORTRAN* and ALGOL,* among other programming languages.* His most productive years were the 1950s and 1960s when he and his students did considerable research on programming systems.

Rosen was born on February 8, 1922, in Port Chester, New York. He completed his B.S. at the City College of New York in 1941, his M.A. at the University of Cincinnati in 1942, and his Ph.D. at the University of Pennsylvania in 1950. He was an instructor in mathematics at the University of Delaware during the academic year 1946–1947, lecturer at the University of California at Los Angeles (UCLA) the following year, assistant professor at Drexel Institute of Technology (1949–1951), associate research engineer for Burroughs Corporation† between 1951 and 1952, and then manager of applied mathematics for the Electrodata Division between 1956 and 1958. Rosen retained his ties to academia during the early 1950s when he went into private industry. He was an assistant professor of electrical engineering at the University of Pennsylvania between 1952 and 1954, and then associate professor of mathematics at the Computer Laboratory, Wayne University (later Wayne State University), between 1954 and 1956. He was the manager of computer programming and services for Philco Corporation† from 1958 to 1960, served as a general consultant on computing and programming from 1960 to 1962, and returned to teaching as a professor of computer science at Purdue University during 1962–1966. During the academic year 1966–1967, he taught at the State University of New York (SUNY) at Stony Brook. He returned to Purdue as professor of mathematics and computer science and in 1968 became director of its University Computer Center. He remained at Purdue into the 1980s.

Between the experiences he gained in developing software* products for computer manufacturers and his own research at universities, Rosen made important contributions to the science of programming. One project he worked on while at Philco, for example, involved converting FORTRAN from an IBM 704* computer to operate on a Philco 2000. That activity may have been the first occasion when, according to Rosen, "a compiler has assumed the major burden of transition from a large scale computer of one manufacturer to an even larger scale computer of another manufacturer." He went on to develop his own fast compiler for FORTRAN. Throughout the 1960s, he continued working on compilers for a variety of languages.

Rosen's part in the development of ALGOL is of particular interest to historians. ALGOL was as much a movement as a series of compilers. In the 1950s, computer scientists in Europe and North America sought to standardize programming languages, attempting to establish a universal one that could be implemented on all kinds of computers. That movement eventually led to the creation of COBOL,* the most widely used programming language for commercial applications in the 1960s and 1970s. An earlier attempt at standardization revolved around a language called ALGOL.

The data processing industry, concerned over the lack of standards, wanted collective action. The Association for Computing Machinery (ACM)† decided to form a committee in 1957 to work on the problem and made Rosen (then an employee of Philco) a member. Standards were roughed out and were compared to those drafted by a European committee known as GAMM (Gesellschaft für angewandte Mathematik und Mechanik). The two committees met in the late 1950s to arrive at a common language which they called ALGOL. The language, almost a derivative of FORTRAN, was intended to be used in scientific computing. The language did not acquire the popularity of either FORTRAN or COBOL because it never gained the blessings of the large computer manufacturers who often had their own compilers to offer or did not want to convince their customers to convert existing programs to new ones written in ALGOL.

Nonetheless, Rosen and his colleagues had advanced the effort to standardize programming languages. A pattern for creating standards for all major languages had been established in the early 1960s, for example, when COBOL emerged and still influences almost all widely used languages in the 1980s.

For further information, see: Saul Rosen, "ALTAC, FORTRAN, and Compatibility," *Preprints, ACM 16th National Conference* (New York: ACM, 1961): 2B–2(1)–(4), "A Compiler-Building System Developed by Brooker and Morris," *Communications of the Association for Computing Machinery* 7, No. 7 (July 1964): 403–414, "Programming Systems and Languages—A Historical Survey," *Proceedings of the Spring Joint Computer Conference* 25 (1964): 1–15, as editor, *Programming Systems and Languages* (New York: McGraw-Hill, 1967); Jean E. Sammet, *Programming Languages: History and Fundamentals* (Englewood Cliffs, N.J.: Prentice-Hall, 1969); Richard L. Wexelblat, ed., *History of Programming Languages* (New York: Academic Press, 1981).

S

SAMMET, JEAN E. (1928–). Sammet is most closely associated with the design and development of COBOL,* the most widely used programming language* in the world from the late 1960s through the 1970s, primarily for commercial applications. She has also written a major history of programming languages.

Sammet graduated from Mount Holyoke College with a B.A. and from the University of Illinois with an M.A. From 1955 to 1958, she worked as a group leader in programming at Sperry Gyroscope and, from 1958 to 1961, at Sylvania Electric Products in programming. During her Sylvania years, she was involved in the initial creation of COBOL. In 1961, she joined International Business Machines Corporation (IBM)† as a programming manager, and, in 1965 she became a programming language technical manager. She was named programming technical planning manager for the Federal Systems Division in 1968; programming language technical manager in 1974; and software* technical manager in 1979. In 1981, she was named a division software technical manager, a position she still holds today.

Between 1956 and 1957, Sammet lectured on digital computer* programming at Adelphi College, and the following year she taught one of the earliest courses in FORTRAN* in the United States. While at Sylvania Electric Products, she headed up the MOBIDIC Programming project, gaining first-hand experience in the design of a computer language. She was a member of the short-range committee established for the purpose of writing the specifications for COBOL between June and December 1959. She subsequently worked on the compiler that ran with MOBIDIC. She has served as the president of the Association for Computing Machinery† (1974–1976), and, in 1977, she was made a member of the National Academy of Engineering. Earlier, in 1969, she published one of the most important histories of programming languages available, and during the 1970s she participated in a variety of historical conferences relating to the history of data processing and computing.

For further information, see: Jean E. Sammet, "The Early History of COBOL," in Richard L. Wexelblat, ed., *History of Programming Languages* (New York: Academic Press, 1981), pp. 199–277 and her *Programming Languages: History and Fundamentals* (Englewood Cliffs, N.J.: Prentice-Hall, 1969).

SCHEUTZ, PEHR GEORG (1785–1873). Better known in English-language accounts as George Scheutz, this Swedish inventor built a difference engine* similar to the one developed by Charles Babbage* and with his help. Scheutz was a man of extraordinary interests and accomplishments. He was a lawyer, the co-editor of an important Swedish newspaper, a translator into Swedish of the works of Boccaccio, Walter Scott, Shakespeare and others, and the manager of several technical and trade journals. In addition, he dabbled with mathematics and difference engines.

Scheutz's interest in calculating devices introduced him to Babbage's work. As a result of his knowledge of Babbage's difference engine, Scheutz and his son, Edvard, sought to improve on the design. By 1834 they had an operational model, built at a time when Babbage had already given up on constructing a difference engine and was devoting most of his attention to an analytical engine.* In 1851, the Swedish Academy gave the Scheutz family funding leading to the construction of a larger and more efficient difference engine in 1853. At the Exhibition in Paris during 1855, their new machine won a Gold Medal. The following year George Scheutz was knighted by his king and joined the Swedish Academy. At the same time he was recognized for his translations, particularly of Shakespeare.

Although he made a major contribution to Swedish letters by introducing British writings to his countrymen through translations, it was his two difference engines that most attracted the attention of historians. Scheutz completed his difference engine, whereas Babbage did not. Neither was a brilliant engineer, and though both were highly creative, they could almost be characterized as amateurs. Herman H. Goldstine,** who had considerable experience building computers and convincing organizations to fund their continued development, speculated that Scheutz was successful whereas Babbage was not because the Swede first built a simple prototype that worked and only after proving that a difference engine was operational did he move on to larger, more complex technology resulting in a second, also operational machine. Babbage never finished constructing the first machine and only designed the second.

In 1856, the second Swedish difference engine was purchased by an observatory in Albany, New York, which kept it until 1924 when it was sold to the Felt and Tarrant Company of Chicago. Years later it went to the Smithsonian Institution. Sometime during its career, the second Scheutz machine was used to calculate tables and to print them out. Thus, unlike Babbage's similarly designed machine, the Swedish device actually did production work.

The Swedish machines technically were variations of Babbage's. Indeed, both families communicated their findings back and forth. Once he moved on to the analytical engine, Babbage had no conflict of interest with the Swedish inventors and gave them advice. The two Swedes visited Babbage during the early 1850s, following the elder Scheutz's decade of interest in the Englishman's work. Beginning work in the early 1830s, it took nearly twenty years for the first machine to be completed, almost bankrupting the family with its costs in the process. It calculated tables, using the method of finite differences, and produced a mold to create type which in turn generated printed tables. Much of the machining for their machine was not as elegant as Babbage's, but the device functioned. At the Paris Exhibition, Babbage saw the Scheutz's machine on display, and he supported their candidacy for the Gold Medal.

Perhaps because the Swedish machine was sent off to Albany, New York, and thus away from the mainstream of European scientific activity, the engine did not generate even more interest. The Scheutz family demonstrated that the finite differences method could be mechanically successful. Ultimately, that fact was more important for history than whether or not Scheutz or Babbage should take credit for the work.

The Swedish team apparently never pursued its work on calculating machines. By 1860, George Scheutz was already in his seventies, and the son, never the prime initiator of the original project, focused his attentions on other projects. By the mid-1850s, this family's contribution to the science of computing was over.

For further information, see: R. C. Archibald, ''P. G. Scheutz, Publicist, Author, Scientific Mechanician, and Edvard Scheutz, Engineer—Biography and Bibliography,'' *Mathematical Tables and Other Aids to Computation* 2 (1947): 238–245; M. Lindgren and S. Lindquist, ''Scheutz's First Difference Engine Rediscovered,'' *Technology and Culture* 23 (1982): 207–213; M. G. Losano, *Scheutz: La Macchina alle Differenze* (Milan: Etas Libri, 1974); U. C. Merzbach, *George Scheutz and the First Printing Calculator,* Smithsonian Studies in History and Technology No. 36 (Washington, D.C.: Smithsonian Institution Press, 1977).

SCHICKARD, WILHELM (1592–1635). Schickard was a professor of Hebrew, astronomy, and mathematics at the University of Tübingen, and invented a calculator that could add, subtract, partially multiply, and divide. His was one of the first in Western Europe with an automatic carry function. This combined level of sophistication in mechanical calculators would not be matched until Gottfried Wilhelm von Leibniz** developed another important device some fifty years later.

Schickard was an extraordinarily talented man. He was a brilliant scientist who painted, a Protestant minister, and an engraver, and a serious student of both astronomy and mathematics. Through his interest in mathematics and the concerns of other mathematicians, he turned to the problem of performing mathematical calculations mechanically. There was a great need for such a device, especially for astronomy. He sought to develop a machine that would mechanize the functions of adding and subtracting employing Napier's bones (logarithms). In 1623, in a letter to his friend Johann Kepler, he reported that he was developing a machine that could carry functions to six digits. He imposed Napier's bones on cylinders that could be selected simply by the turn of a dial. Horizontal slides exposed various bones in single-digit form. Results were accumulated by turning large knobs, and the answers were viewed through small windows. Various turns of the knobs made possible viewing any number through his windows, employing these almost like a scratch pad to hold results of calculations. Thus, intermediate results being held for future calculations did not have to be jotted down on paper.

Schickard used accumulator wheels in which a tooth, through a complete rotation, would latch to another wheel, turning it and thereby causing the next highest digit to increase by one in an accumulator. Owing in large part to the strength of available materials in the seventeenth century, the machine could only carry calculations through six digits. To carry beyond six might have damaged or destroyed some of the wheels. To accommodate even larger figures than six, he proposed that the operator put a brass ring on his finger each time a carry had exceeded a six-digit carry. A small bell would signal an "overflow" condition, reminding the operator to put on another ring. Mechanically, the device could only add and subtract. Yet, by using Napier's logarithms, all multiplication and division could be reduced to additions and subtractions.

The first machine which Schickard reported to have built has been lost, perhaps another victim of the Thirty Years War which inflicted so much damage on property in the German states. A second machine, which he was building for Kepler, was destroyed in a fire while under construction. Schickard and his family died of the plague which ravished the German states in the 1630s. Conditions of war may have been the reason why his work was little known or appreciated in his day. Few others (Kepler being an exception) understood what he was doing. Indeed, models of his machine, based on scant records and one casual drawing, were not made until after World War II. However, that reconstruction has made it possible to visualize his significant contribution to mechanical calculators. Moreover, his work is a clear example of how logarithmic features can be employed on a mechanical calculator.

For further information, see: B. Von Freytag Löringhoff, "Wilhelm Schickard und seine Rechenmaschine von 1623," in M. Graef, ed., *350 Jahre Rechenmaschinen* (Munich: Carl Hanser, 1973), pp. 11–20; M. R. Williams, "From Napier to Lucas: The Use of Napier's Bones in Calculating Instruments," *Annals of the History of Computing* 5, No. 3 (July 1983): 279–296.

SCHOTT, GASPARD (1608–1666). This German scientist developed a number of methods for inscribing logarithms (Napier's bones as they were called) on mechanical devices. Although not the first to do so (Wilhelm Schickard** had many years earlier), his work represented different and still early attempts to automate these mathematical tables.

Schott studied at the University of Wurzburg, working with Athanasius Kircher (1602–1680), on mathematical problems and scientific subjects. Both were members of the Jesuit order. Probably because Sweden invaded the area in which his university was located in 1631, Schott moved to Sicily and Kircher took up residence in Rome. They worked on mathematics together and began to track the scientific efforts of other Jesuits around the world for the next twenty years. Kircher is particularly remembered for this effort. In 1651 Schott came to Rome to help him publish all the scientific data gathered from other Jesuits. By the time Schott died, they had published eleven large volumes on a broad variety of scientific issues.

Of particular interest to historians of mathematics and computing is the volume that describes mathematical methods for performing calculations common to the seventeenth century. The book catalogued algorithms, methods, and descriptions of mechanical devices to aid calculations. One device described in detail, called the *Organum Mathematicum* which used Napier's bones, may be Schott's own. This box had ten sets of tablets made from a bone-like substance for performing a variety of tasks. In this book, Schott argued that the applications were extensive: addition and subtraction via normal logarithms for arithmetic; geometric tables to enhance calculations common to surveying; tables necessary for the construction of standard fortifications; other tables to calculate dates of various Christian festivals; tables to determine parameters for sundials depending on either their inclination or direction; the ability to determine movements of the sun, especially sunrises, by season; functions necessary to calculate planetary movements (useful for horoscopes); tables for determining how much to dig or fill in construction projects; and tables to help compose music. Such reports can be regarded as the forerunners of twentieth-century promotional materials published by a computer company.

Schott's data were imposed on a series of cylinders (similar to Schickard's), each with a full set of Napier's logarithms. Data ran from end to end on the cylinders. These devices could be turned in the box so that the data could be seen through small windows at the top of the device. It also had an addition table on the top of the box. The design of the device proved impractical; data did not appear in the correct position. When users complained that the device caused mistakes, it was soon abandoned.

For further information, see: M. R. Williams, ''From Napier to Lucas: The Use of Napier's Bones,'' *Annals of the History of Computing* 5, No. 3 (July 1983): 279–296.

SHANNON, CLAUDE ELWOOD (1916–). Shannon was one of the earliest scientists working in the field of artificial intelligence (AI)* and on the theory of information. His work in these areas of study influenced the design of computers from the 1940s onward. For his important contributions he was eventually recognized with a Nobel Peace Prize, as had been other scientists working in electronics and data processing.

Shannon was born on April 30, 1916, at Gaylord, Michigan, and graduated from the University of Michigan in 1936. He continued his study of science and mathematics at the Massachusetts Institute of Technology, completing his master's degree in 1937. This thesis, using Boolean algebra, was one of the earliest documents to describe how relay and switching circuits behaved. This particular exercise led to Shannon's decades-long study of theories of information and the role of computers. In 1940, he completed his Ph.D. in mathematics, and the following year he went to work for Bell Laboratories† where he focused on the best ways to transmit information efficiently. One of his earliest tasks was to measure the efficiencies of the various methods of transmission that existed using cabling, wiring, and electricity. The problem dominated his thinking during most of the 1940s and by 1948 he had worked out methods by which one could express information efficiently in quantitative terms. Relying on the binary method of yes-no (the positive-negative of electrical impulses), he made that circumstance the basic unit of all information. In other words, either a piece of data was yes or no, or the combination of yes and no messages could be used to describe information. Today these are the 1 and 0 binary notations which are so widely used in computer science that it is difficult to imagine a time when they did not represent common ideas. These ones and zeros are important because we know them as bits—the basic unit of information in all computers analogous to atoms in biology.

During the 1940s, Shannon realized that combinations of bits could represent various pieces of information. He was able to prove that there was an upper limit or boundary on the quantity of information a specific transmission channel could carry (the result of his original assignment at Bell Labs). He illustrated that a telephone line could carry only a specific number of telephone conversations at the same time. In order to describe this quantity in specific terms, he worked on a method of measurement which, in turn, led to his use of bits—all continued outgrowths of his work for his master's thesis.

Shannon's most important description of this early research appears in his *The Mathematical Theory of Communication* (1949), one of the first treatments of the issue of information theory, a subject that became a growth industry by the early 1960s. Shannon describes a mathematical way of analyzing the quantification of information. He presents a detailed discussion of how data can be changed as a system "drops" bits, distorts them, and so on, while illustrating how such elements in communications as redundancy and noise can be quantified and thus be measured. Thus was born what today is known as information theory. The significance of this field of study can not be overemphasized since work on this subject has contributed to the design of circuits up to the present time. It also influenced the architecture and technologies embedded in computers and telecommunications equipment. By the 1980s, his ideas were even being applied to the study of biology, psychology, phonetics, semantics, and literature.

In the 1950s, Shannon encouraged others to work in the field of artificial intelligence and information theory. During the summer of 1956, an important conference was held at Dartmouth College in which the giants of the then small community of people interested in artificial intelligence gathered for their first major collaborative session. Present were Shannon, John McCarthy** the host, Marvin L. Minsky**, and Nathaniel Rochester, then manager of research on information for International Business Machines Corporation (IBM).† The conference was important because it aroused interest in the subject and led to an exchange of ideas that later influenced a number of researchers. Shannon had already established himself in the field before this event, first with his book in 1949 and next with a paper published in *Scientific American* in 1950. The article described a chess-playing machine, and he suggested that machines could be made to do more than they currently did. Rather than limit them to numerical calculations, these devices could execute instructions that caused them to work "symbolically with elements representing words, propositions or other conceptual entities." He said that computers did not have to be just calculating devices—a new idea at the time. The article also examined how a machine had to face specific problems when playing chess, and described the magnitude and dynamics of such an effort.

In 1953, in a paper entitled "Computers and Automata," Shannon posed a list of questions that scientists are still trying to answer. They included such questions as could machines be organized into hierarchies, similar to the human brain, with learning progressing up through these levels? Could digital computers* be designed and programmed so that eventually 99 percent of all their instructions would be created by the machines themselves? Could a computer diagnose its own problems and then repair them? This last question was answered by the 1980s with computers that identified their own problems in many cases and in some could either go around them or actually resolve the crisis.

The year 1952 was an important one for those working in the field of artificial intelligence, for both Minsky and McCarthy came to work for Shannon at Bell Labs that year. The three launched the conference at Dartmouth, and Shannon published the book on automata as part of that effort. In 1956, this shy, quiet scientist left Bell Labs to join the faculty of MIT where he still teaches today. He built intelligent mice machines which, as a result of the experience of running through mazes, learned how to get through these mazes better as well as a chess-playing device. Yet Shannon never devoted himself exclusively to artificial intelligence as his other colleagues did; instead, he preferred the study of information theory. He has continued to work on a precise measurement of the amount of information that can be carried by a particular machine and its efficiency. His best work was behind him, however, in the 1940s and 1950s when he influenced the whole course of the computer's future.

For further information, see: Pamela McCorduck, *Machines Who Think: A Personal Inquiry into the History and Prospects of Artificial Intelligence* (New York: W. H. Freeman and Co., 1979); Claude Shannon, ''A Chess-Playing Machine,'' *Scientific American* 182 (February 1950), ''Programming a Digital Computer for Playing Chess,'' *Philosophy Magazine* 41 (March 1950): 356–365, and with John McCarthy, eds. *Automata Studies. Annals of Mathematical Studies* 34 (Princeton, N.J.: Princeton University Press, 1956), and with W. Weaver, *The Mathematical Theory of Communication* (Urbana: University of Illinois Press, 1949); Joseph Weizenbaum, *Computer Power and Human Reason* (New York: W. H. Freeman and Co., 1976).

SHAW, J. CLIFF (1922–). Shaw worked in the general field of artificial intelligence* during the 1950s and 1960s and developed list processing languages. These languages made it possible to establish the next instruction in a given sequence on the basis of previous sequences; thus, computers could determine the best way to allocate the use of their storage. Recursive computing, as it was known, increased the overall utilization and ease of use of computers in the 1950s and 1960s when the availability of memory* in a computer remained expensive and limited.

Shaw was born and raised in California, the son of a paint store owner. During World War II, he flew military aircraft as a navigator, and after the war, he worked as an actuary for an insurance firm in Los Angeles, taking advantage of his considerable talent in mathematics. In 1950, he came to work for the RAND Corporation, the home of much research in computing and the owner of an important, early electronic digital computer* called the JOHNNIAC.* In his initial years at RAND, Shaw worked on a radar simulator with another scientist who would become important in the annals of artificial intelligence, Allen Newell.** Shaw primarily handled administrative matters but became interested in automating the company's processes while improving its calculators. In 1954, he and Newell began discussing how to build a chess machine. By 1956, along with Newell and Herbert (Herb) A. Simon,** he had built the Logic Theory Machine, which performed ''thinking'' activities. That program performed list processing on the JOHNNIAC. Shaw contributed to the project with programming. This project was the first significant proof that artificial intelligence could be called a science.

Like others interested in artificial intelligence in the 1950s and 1960s, Shaw was interested in chess machines because he believed such devices did, in his words, ''seem to have penetrated the core of human intellectual endeavor.'' Chess machines had fascinated scientists for centuries and no more so than in the 1950s and 1960s. The three—Newell, Simon, and Shaw—focused additional attention on chess-playing into the 1960s.

Earlier, in 1957, they developed a program called the General Problem Solver whose design was based on the notion that human intelligence had some general principles of operation that were applied to all manner of problems. They tried to emulate this set of generalities with software.* GPS, as it was called, became one of the first general-purpose problem-solving packages available in the United States. Subsequent versions and variants of these and other simulators appeared throughout the 1960s, along with a variety of list processing languages. In sum, Shaw helped create the new field of artificial intelligence during the 1950s.

For further information, see: Pamela McCorduck, *Machines Who Think* (New York: W. H. Freeman and Co., 1979); Jean E. Sammet, *Programming Languages: History and Fundamentals* (Englewood Cliffs, N.J.: Prentice-Hall, 1969).

SHOCKLEY, WILLIAM BRADFORD (1910–). One of the giants of twentieth-century electronics, Shockley was one of the fathers of the transistor and a major force in the development of the semiconductor industry which manufacturered the chip,* the basic building block of modern computers. His work with both transistors—which were used to build computers in the 1950s and early 1960s—and chips profoundly encouraged the rapid growth of computers in the second half of this century.

Shockley was born in London on February 13, 1910, and in 1932 completed his undergraduate studies in physics at the California Institute of Technology. In 1936, he was awarded a Ph.D. by the Massachusetts Institute of Technology (MIT). While a graduate student at MIT he was a teaching fellow and after graduation he joined the technical staff of Bell Laboratories,† staying until 1942. After World War II, he returned to Bell, and, in 1954, he became director of its transistor physics research facility. It was at Bell Labs in the late 1940s that transistors were first developed, and thus the position he held was a critical one. Wishing to do more research on integrated circuits, he left Bell to form the Shockley Semiconductor Laboratory which, in 1958, he renamed the Shockley Transistor Corporation. The firm underwent other name changes in the 1960s. From 1958 to 1963, he also lectured at Stanford University where he was already recognized as a giant in electronics. From 1963 to 1975, he held the Alexander M. Poniatoff professorship of engineering science and applied science at Stanford; subsequently, he was professor emeritus. From 1965 to 1975, he served on an advisory council for Bell Labs, and earlier, during 1954–1955, he was director of research for the Weapons Systems Evaluation Group in the U.S. Department of Defense. He was no stranger to the military establishment, for he had served as a technical advisor in the War Department during World War II. From 1942 to 1944, he had also been director of research within the Anti-submarine Warfare Operations Research Group of the U.S. Navy.

For his work on transistors, Shockley and his co-workers John Bardeen and Walter H. Brattain shared the Nobel Prize in Physics in 1956. Other awards followed, including induction into the Inventor's Hall of Fame in 1974.

Shockley's more than ninety patents were a testimonial to his contributions to the invention of the junction transistor and other semiconductor-type of devices.

Shockley had the necessary background and intelligence to make a significant contribution in the field of electronics. He came out of MIT with a solid academic background and interest in solid-state physics. When he joined Bell Labs, he spent some time in the vacuum tube department, working for Clinton Davidson, himself a Nobel laureate. Thus, like Walter Brattain, he had both the education and experience with current technology to appreciate its limitations and the need for improvements in miniaturization and reliability. The real hope in the 1940s lay with semiconductors, especially silicon and germanium. By "doping" these semiconductors with other elements, scientists at Bell Labs realized that the electrical properties of silicon and germanium could be influenced. Beginning in 1945, Shockley headed up a research team dedicated to the further study of both silicon and germanium. This research represented some of the earliest work on what would ultimately become the computer chip. John Bardeen also joined the staff, bringing with him a theoretical background in solid-state physics. Shockley managed, while Brattain played the role of the nuts-and-bolts experimenter. Late in 1947, the three attached thin wires very closely together on one side of a piece of germanium. When they ran an electrical current through this device, electrical signals were amplified. On December 23, they demonstrated to other Bell scientists how to amplify the human voice with this semiconductor of electricity.

This crude device was the start of what would very quickly become a nonmoving electronic apparatus which these scientists then described as a "transfer resistance" unit. A colleague nicknamed it a "transistor." Bardeen and Brattain did most of the work on this early device. In fact, their names appeared on the application for patent granted Bell Labs for the transistor. Shockley continued to improve this transistor, and in 1950, he created a germanium junction transistor. It was Shockley's more reliable device, not the earlier one, that became the basis of Bell's transistor products of the 1950s. By 1956, when the three men shared the Nobel Prize for having discovered the effect of transistors, twenty-six companies had been licensed by American Telephone and Telegraph† to produce the device.

Shockley dominated the thinking about semiconductors at the start of the decade by having written the first and, for many years, the definitive work on the subject. In 1954, when he left Bell Labs, he continued his research on semiconductors back in California, which was not only his home state but also the source of much engineering talent that would be needed to launch new ventures in the field of semiconductors. Setting up the Shockley Semiconductor Laboratories at Palo Alto, Shockley created a staff in Mountain View, near Stanford University, that conducted considerable research on semiconductors. He encountered little difficulty in hiring high-quality researchers since he was already recognized as a "father of the transistor." Shockley envisioned making even smaller transistors for which a market existed. Some of his scientists, however, found it difficult to work for his company and in 1957 left to form Fairchild Semiconductor.† While he worked with germanium and silicon, these scientists preferred to concentrate entirely on silicon as the more efficient semiconducting material. In the subsequent five to eight years, Fairchild spawned the semiconductor industry and, in effect, Silicon Valley. While these successes were occurring, Shockley was no longer in the forefront of technological evolution and focused most of his attention on research and teaching at Stanford from 1965 through the 1970s.

For further information, see: Dirk Hanson, *The New Alchemists* (Boston: Little, Brown and Co., 1982); National Geographic Society, *Those Inventive Americans* (Washington, D.C.: National Geographic Society, 1971); T. R. Reid, *The Chip: How Two Americans Invented the Microchip and Launched a Revolution* (New York: Simon and Schuster, 1984); William Shockley, *Electrons and Holes in Semiconductors, with Applications to Transistor Electronics* (New York: Van Nostrand, 1950); see also his lectures along with those of the others who shared in his Nobel prize, *Nobel Lectures—Physics, 1942–62* (New York: Elsevier, 1964); George L. Trigg, *Landmark Experiments in 20th-Century Physics* (New York: Crane, Russak, 1975).

SIMON, HERBERT ALEXANDER (1916–). One of the major figures in the world of artificial intelligence (AI)* and cognitive psychology, Simon has also written on economics, political science, organization, management, and public administration.

Simon was born in Milwaukee, Wisconsin, in 1916, studied political science, and completed his Ph.D. at the University of Chicago in 1943. He taught first at the University of California at Berkeley and then at the Illinois Institute of Technology. In 1949, he joined the faculty of the Carnegie Institute of Technology (now known as the Carnegie Mellon University) to help establish the Graduate School of Industrial Administration. This particular business school based much of its teaching on scientific knowledge in such fields as economics, psychology, and operations research, representing a departure from other programs of the late 1940s. Simon had already published his first book and thus established a line of interest consistent with the mission of this new graduate program. Simon has remained at Carnegie down to the present, and beginning in the 1970s, also teaches in the Psychology Department as the Richard King Mellon Professor of Psychology and Computer Science.

This American scholar represents the eclectic side of research in the field of artificial intelligence. In 1954, he published a textbook on public administration and, in 1960, another book on servomechanism analysis of factory production and control. His first book, which appeared in 1947, centered on how the administrator functioned in formal organizations and concluded that humans were molded and constrained by institutional, informational, and even computational limits which he called "bounded rationality." This book led to models of economic decision-making, focusing on the idea that people try to satisfy a situation rather than to optimize circumstances (the latter being the current feeling of the 1930s and 1940s). In his early work, he also argued that computers and people could be understood in those terms, particularly their intelligence. That first book played a major role in his winning the Nobel Prize in Economics in 1978.

Simon worked with John Cliff Shaw** and Allen Newell** during the 1950s on various AI projects. In 1956, they produced some of the earliest heuristic programs *(Logic Theorist)* and additional material two years later *(General Problem Solver)*. In 1957, they presented the first list processing language (called IPLs at the time) which made up an important class of early programming languages* to emerge out of the 1950s. They continued to work in the general field of AI into the 1980s. Simon was deeply interested in modeling human thought (known in the late 1950s and throughout the 1960s as "simulation of thought processes"). Simon and his associates wrote software* to study the subject and published additional books and articles. His programs analyzed problem-solving, the act of memorization, inducting, behavior in semantical domains, and learning in general. Throughout the 1970s, Simon published important works on human problem-solving.

In addition to the Nobel Prize, in 1975 Simon received the Association for Computing Machinery (ACM)† Turing award, one of the most important awards in the data processing industry. The American Psychological Association had honored him earlier in 1969 with its Award for Distinguished Scientific Contribution. He was also inducted into the National Academy of Science. By the early 1980s, he had published a dozen books and some 500 articles. Simon has continued to research and write.

For further information, see: Pamela McCorduck, *Machines Who Think* (New York: W. H. Freeman and Co., 1979); Allen Newell, "Hert A. Simon," in Anthony Ralston and Edwin D. Reilly, Jr., *Encyclopedia of Computer Science and Engineering* (New York: Van Nostrand Reinhold, 1983), pp. 1324–1325; Herbert A. Simon, *Administrative Behavior* (New York: Macmillan Co., 1947), *Models of Thought* (New Haven, Conn.: Yale University Press, 1979); *The Sciences of the Artificial* (Cambridge, Mass.: MIT Press, 1969; 2d ed., 1981), and with Allen Newell, *Human Problem Solving* (Englewood Cliffs, N.J.: Prentice-Hall, 1972), with L. Siklóssy, *Representation and Meaning* (Englewood Cliffs, N.J.: Prentice-Hall, 1972), with et al., *Public Administration* (New York: Alfred A. Knopf, 1954), and *Planning Production, Inventories, and Work Force* (Englewood Cliffs, N.J.: Prentice-Hall, 1960).

STIBITZ, GEORGE ROBERT (1914–). Stibitz worked for the Bell Telephone Laboratories† and designed relay computers in the 1930s. His first machine was one of the initial binary quasi-programmable devices ever built, and by the late 1940s its descendants were very sophisticated and reliable machines.

Stibitz was born in York, Pennsylvania, on April 20, 1904, but was raised in Dayton, Ohio, where his father taught theology. As a child, he became interested in science and had an aptitude for engineering. While in high school, he took a class taught by Charles Kettering who invented the ignition systems first used in automobiles. After high school, Stibitz attended Denison University at Granville, Ohio, and then studied at Union College in Schenectady, New York, where he completed his M.S. in 1927. He immediately went to work for General Electric (GE)† which had a research laboratory in the same town. Soon after, Stibitz decided to complete his graduate studies and matriculated at Cornell University, completing a Ph.D. in mathematical physics in 1930. Upon graduation he went to work for Bell Labs.

The problem he and Bell Labs struggled with was quite simple in concept. The telephone company needed to perform many mathematical calculations and the manipulation of numbers to design electromagnetic relays used in telephonic networks. Much of the work involved calculating complex numbers. For example, these involved square roots of minus one which called for a person to track two separate results with only the aid of a desktop mechanical calculator. One total represented the results of a real part of a number, and the other for an imaginary part of a number represented imaginary parts of a telephone network. The inability to compute the numbers fast enough slowed down the development of new equipment in the early 1930s. Stibitz therefore set out to find a faster way to get the job done.

In the fall of 1937, working with other mathematicians and engineers at Bell Labs on the design of relay switching equipment, Stibitz began to play with old relays at home. He was able to assemble them so that they could add two binary digits using lights, one of which glowed for the digit 1 and remained dark to represent 0. Stibitz's "breadboard" circuit worked. Since he constructed it on the kitchen table at his home, his colleagues nicknamed the device the "Model K." His gadget was a collection of two relays, strips from a metal can, two light bulbs from flashlights, two batteries, and with everything nailed down to a piece of wood. He showed the device to colleagues, and by late 1938 Bell Labs had given him permission to start building a machine that could perform calculations.

By early 1939, Stibitz and his colleagues at Bell had designed a computer and began its construction in April. The machine first ran on January 8, 1940, and remained in operation at Bell Labs until 1949. Called the Complex Number Calculator, it could add, subtract, multiply, and divide complex numbers. It operated like a large capacity desktop calculator because it was not programmable in the sense that computers would be within a decade. This machine was also the first to handle remote job entry. On September 11, 1940, while attending a session of the American Mathematical Society at Dartmouth College, Stibitz transmitted his computer work to be done over a specially rigged telephone line and received the results back in the same manner. By 1970, that form of processing had become a regular event within the data processing community.

Stibitz left Bell Labs in 1941 and during World War II served on the National Defense Research Committee (NDRC) where he conducted additional research on computation and war-related projects. Following World War II, he moved to Vermont to work as a private consultant. He became a professor of physiology at Dartmouth Medical School in Hanover, New Hampshire, in 1964 and conducted research on computers and medicine into the 1980s. After his retirement, he donated his private papers to Dartmouth. Stibitz received the Harry Goode Memorial award in 1965, given by the American Federation of Information Processing Societies (AFIPS).†

For further information, see: Jeremy Bernstein, *Three Degrees Above Zero. Bell Labs in the Information Age* (New York: Charles Scribner's Sons, 1984); Paul E. Ceruzzi, *Reckoners: The Prehistory of the Digital Computer, From Relays to the Stored Program Concept, 1935–1945* (Westport, Conn.: Greenwood Press, 1983); Bernard D. Holbrook and W. Stanley Brown, *A History of Computing Research at Bell Laboratories (1937–1975)* (Murray Hill, N.J.: Bell Telephone Laboratories, 1982); E. Loveday, "George Stibitz and the Bell Labs Relay Computer," *Datamation* (September 1977): 80–85; George R. Stibitz, "Early Computers," in N. Metropolis et al., eds., *A History of Computing in the Twentieth Century* (New York: Academic Press, 1980), pp. 479–483; Henry S. Tropp, *An Inventory of the Papers of George Robert Stibitz Concerning the Invention and Development of the Digital Computer* (Dartmouth, N.H.: Dartmouth College, 1973).

STRACHEY, CHRISTOPHER (1916–1975). This British scientist made important contributions to the design of computers and to programming during the 1950s and 1960s and to denotational semantics in the early 1970s. In the 1950s, he was active in the National Research and Development Corporation (NRDC) and, in the 1960s and 1970s, at Cambridge and Oxford universities.

Strachey was born on November 16, 1916, into a well-known British family and attended King's College, Cambridge, during the 1930s. During World War II, he did research on radar and immediately afterwards taught. He began work with computers in 1951 when he started programming processors at both the National Physical Laboratory and at the University of Manchester, both of which were already leading centers for data processing activities in Great Britain. Between 1952 and 1959 he worked for NRDC where he continued to write programs and do logical design for computers. He became a private consultant in computer design in 1959 but joined the University Mathematical Laboratory at Cambridge in 1962. He had done some work with this laboratory as far back as 1959. His primary contribution during the 1959–1965 was in helping to develop a programming language* called CPL. In 1965, he established the Programming Research Group at Oxford University where he stayed until 1975. While there, he continued work on the theory of programming and denotational semantics.

Despite his numerous changes in jobs, Strachey managed to work consistently in several distinct fields. Immediately after joining NRDC on June 3, 1952, he continued to program the Mark I* at the University of Manchester. Later that year he presented one of his earliest papers on data processing concerning programming the Manchester computer. He helped to write programs at the

University of Toronto soon after for the St. Lawrence Seaway Project. He next helped to design the Elliott 401 British computers and in 1954–1955 the Pegasus computer made by the Ferranti Company.† Strachey primarily did the design of programming for them. He maintained close ties to each of the thirty-six Pegasus computers built between the mid-1950s and 1962. In 1957, he became involved in a British project to design a large computer along the lines and size of the American STRETCH* and LARC* machines. Strachey also encouraged research into multiprogramming techniques mainly at the University of Manchester but also elsewhere in Great Britain. He defined many of the characteristics that time-sharing systems would need during the late 1950s and filed for a patent in the field in February 1959. He received considerable credit for originating the concept.

The years from 1959 to 1965 were taken up with private consulting and with research at Cambridge University. Strachey was involved in a variety of projects concerning the design of systems, the use of programming languages, and application development. He participated in the logic design for the Elliott 502 (1960) and in a computer for Decca Radar (1961). In these years the question of using a programming language called ALGOL* occupied the attention of many computer scientists. Strachey thought that the language had many deficiencies and, along with another major computer scientist Maurice V. Wilkes,** described how to improve it. By the early 1960s, he was also criticizing his British colleagues for allowing developments in computer science to race ahead of Great Britain's in the United States.

In June 1962, Strachey formally became part of Wilkes' team at the University Mathematical Laboratory at Cambridge where he worked on programming languages, most notably on the Cambridge Programming Language (CPL). It was an ALGOL 60–based language designed in the early 1960s for use essentially on the Titan, a computer in use at Cambridge by 1964. He had also become very interested in describing the semantics of CPL while it was still being designed. However, CPL was not completed, nor was a compiler developed in this period.

Strachey spent a portion of 1965–1966 at the Massachusetts Institute of Technology (MIT) and came to Oxford University in April 1966. He wrote on programming languages during the next ten years while teaching on the theory of programming. He did additional work on CPL and, by the early 1970s, was heavily involved in conducting research on mathematical semantics. In 1971, his work in programming was recognized with his designation as a Distinguished Fellow of the British Computing Society. While working on a book-length manuscript entitled "A Theory of Programming Language Semantics," he contracted infectious hepatitis and died on May 18, 1975. At the time of his death, many in his profession recognized his contributions to the logics design of British computers and to programming languages in general. His long-term significance vis-à-vis denotational semantics is still being assessed.

For further information, see: J. Alton et al., *Catalogue of the Papers of Christopher Strachey (1916–1975)* (Oxford: Contemporary Scientific Archives Centre, 1980); D. W. Barron, "Christopher Strachey: A Personal Reminiscence," *Computer Bulletin, Series 2* (1975): 5, 8–9; Martin Campbell-Kelly, "Christopher Strachey, 1916–1975, A Biographical Note," *Annals of the History of Computing* 7, No. 1 (January 1985): 19–42; S. H. L. Clarke, "The Elliott 400 Series and Before," *Radio and Electronic Engineer* 45, No. 8 (1975): 415–421; Christopher Strachey, "Systems Analysis and Programming," *Scientific American* 25, No. 3 (1966): 112–124, and with R. E. Milne, *A Theory of Programming Language Semantics* (London: Chapman and Hall, 1976), and with M. V. Wilkes, "Some Proposals for Improving the Efficiency of Algol 60," *Communications, ACM,* 4, No. 11 (1961): 488–491. For an interpretation of his contribution, see D. S. Scott, "An Appreciation of Christopher Strachey and His Work," in J. E. Stoy, ed., *Denotational Semantics: The Scott-Strachey Approach to Programming Language Theory* (Cambridge, Mass.: MIT Press, 1977), pp. xiv–xxx.

SVOBODA, ANTONIN (1907–1980). Svoboda worked in the area of analog computers,* particularly for fire-control systems for the military, fault tolerance in the design of digital computers,* and the general field of computer-related mathematics. He was instrumental in the development of computer research in his native Czechoslovakia and in the design and construction of the EPOS computer.

Svoboda was born on October 14, 1907, in Prague and completed his undergraduate studies in 1931 in electrical engineering at the Czech Institute of Technology. At Charles University in Prague, he pursued graduate work in experimental physics, completing his Ph.D. in 1936. During these early years, he published a book on the scientific approach to bidding in bridge *(New Theory of Bridge)* and played piano with the Prague Wind Quintet. In the fall of 1936 he was drafted into the army where he worked on antiaircraft fire-control devices. In 1938 he was mustered out of the service and returned to teaching at his undergraduate college. When the Germans invaded Czechoslovakia, he fled to Paris where, with his country's encouragement, he gave the Allies much fire-control equipment and information on its use. While in Paris, he built his first analog computer for use with this equipment. After the fall of Paris, he went to the United States.

Svoboda continued work on his antiaircraft equipment and in 1943 joined the Massachusetts Institute of Technology's (MIT's) Radiation Laboratory's staff which was then actively developing a number of military-related devices. While there, he also worked on radar equipment while refining his own devices and published a book on computing based on work done at MIT, *Computing Mechanisms and Linkages* (1946).

At the end of the war, Svoboda returned to his own country where he hoped to establish a computer industry. Recognizing that digital computers were more useful than either analog devices or electromechanical calculators, he sought to use his exposure to various computing projects in the United States and in Western

Europe to expand his nation's impact on the new field. He established a laboratory to develop equipment and entered into an agreement with what eventually became known as the National Enterprise Aritma which manufactured keypunch and punched card equipment in Prague. He built several devices, some of which were programmable. The best recognized of his projects was a calculating punch (1949–1950) which was a relay computer that relied on card input/output and was programmed by using a plugboard. It soon became the centerpiece of Aritma's equipment and systems. Several hundred were constructed during the 1950s.

In 1950, Svoboda joined the Central Institute of Mathematics where he created a laboratory which in time became the Research Institute of Mathematical Machines associated with the Czechoslovakian Academy of Sciences. It became the center of computer research, offering the Ph.D. in computer science. Since 1952, it has also published an annual entitled *Information Processing Machines*. By 1964 his institute had 900 employees or students, making it one of the largest computer research facilities in Eastern Europe. During the 1950s and 1960s, its researchers worked on the SAPO and EPOS computers, along with numerical analysis and software.*

SAPO was the first large automatic digital computer constructed in Czechoslovakia. Built between 1950 and 1956, it was a relay computer equipped with magnetic drum memory.* It had five-address instructions and relied on a floating-point architecture of 32-bit binary arithmetic. The uniqueness of this machine lay in its fault-tolerant aspects, that is, it had considerable design characteristics to insure that it would not break down. For example, it had three central processing units (CPUs) that concurrently worked on the same problem and compared each other's work; two of the units voted on the correct answer for each calculation. This design concept is fairly common today.

Svodoba's staff also worked on the EPOS between 1958 and 1963. This machine relied on vacuum tubes, germanium diode logics, delay-line registers, and ferrite core memory, components that were more reliable than those employed in SAPO. The EPOS even looked like a modern computer, whereas SAPO looked like a wall-length metal box out of an old science fiction movie. EPOS was a decimal machine, and was not based on hexadecimal notation as would be the case with many of the world's computers. Like SAPO it had fault-tolerant features. It relied heavily on transistor technology, and it had the capability of concurrently managing multiple tasks at various stages of computation. Thus, it could be printing out data from one problem while calculating on another.

In the early 1960s, the Czech government forbade Svoboda from appearing at international conferences. When the political environment at home made research increasingly difficult, together with family and friends, he returned to the United States in 1965. The following year he joined the faculty of the University of California at Los Angeles where he taught logic design, computer architecture, and other technical courses related to the construction of computers, and continued research on logic design for computers. He retired in 1977 and died on May 18, 1980.

For further information, see: J. G. Obonsky, "Eloge: Antonin Svoboda, 1907–1980," *Annals of the History of Computing* 2, No. 4 (October 1980): 284–292; Antonin Svoboda, *Computing Mechanisms and Linkages* (New York: McGraw-Hill, 1948), *Advanced Logical Circuit Design Techniques* (New York: Garland STPM Press, 1979) and "From Mechanical Linkages to Electronic Computers: Recollections from Czechoslovakia," in N. Metropolis et al., eds., *A History of Computing in the Twentieth Century* (New York: Academic Press, 1980), pp. 579–586.

T

TARSKI, ALFRED (1901–1983). Tarski, a mathematics professor at the University of California at Berkeley, is considered one of the greatest logicians of the Western world and one of the major mathematicians of the twentieth century. He wrote over 300 papers and seven books on mathematics, truth, logic, and the mathematics of semantics, and his work on Euclidean geometry in the 1930s led to the birth of theoretical computer science, which is concerned with what types of problems such technology can resolve.

Tarski was born and raised in Warsaw, Poland, where he completed his Ph.D. in 1924 at the University of Warsaw. He immigrated to the United States at the start of World War II and joined the faculty at the University of California in 1942, becoming full professor in 1946. His research focus was on theories of models, theory of sets, and algebra. He died on October 27, 1983.

For further information, see: J. W. Addison, "Eloge: Alfred Tarski, 1901–1983," *Annals of the History of Computing* 6, No. 4 (October 1984): 335–336.

TORRES Y QUEVEDO, LEONARDO (1852–1936). Torres y Quevedo proposed an electromechanical means in 1893 for applying many of the ideas originally elucidated by Charles Babbage** in the 1820s and 1830s. His work contributed to the early history of digital computational devices, going far beyond the available technology of his day which consisted mainly of card input/output devices of the Hollerith type or mechanical cash registers and adding machines. He also developed two chess-playing machines.

Torres y Quevedo was born in the province of Santander, Spain. He was trained as a civil engineer, although he spent the majority of his life inventing machines that performed functions automatically. During his lifetime, his work received attention in both France and Spain. He directed a research laboratory, was a member of the French Academy of Science, and was president of the

Academy of Science of Madrid. His early work involved the construction of a mechanical analog device to perform mathematical calculations. In 1906, he completed work on a model boat that was radio controlled, and he demonstrated the boat in Bilbao's harbor before the king of Spain. The boat was commanded by radio to adjust rudder and speed. Torres y Quevedo later applied this same technology to a torpedo. He also designed a semirigid airship which both sides used during World War I in considerable number. In 1916, an Aero Car of his design was installed at Niagara Falls and is still there today.

Torres y Quevedo also devoted considerable attention to devices relating to the history of data processing and computing. In 1911, he demonstrated an automated chess-playing machine for the first time. This was the first of two chess-playing machines he designed. Historians claim that his 1911 machine was the first chess-playing device that was completely automatic. The Spanish machine electrically sensed the position of the pieces and employed a mechanical arm to move them. Torres y Quevedo developed his second device in 1922, this time moving the pieces by the use of magnets placed under the board.

In his devices, Torres y Quevedo took advantage of the electromechanical capabilities of newly available technologies. The radio was yet another example of his use of recent scientific developments. Enhancements in the field of electricity and its applications allowed Torres y Quevedo to call attention to the usefulness of this source of energy in the resolution of mathematical operations. The issues he struggled with and for which he suggested the application of electricity had been investigated by European inventors for the previous 300 years. He concluded that devices run by electricity could do more than their predecessors, and he suggested that electrically driven machines could be instructed to "do certain things which depend on certain conditions." In addition, he argued that using preestablished arbitrary rules to govern the behavior of machines would make the use of electricity and its properties ideal for calculations. His comments in part anticipated the definition of a programmable machine.

In 1913, Torres y Quevedo published a paper which historians would later consider a milestone in the history of computing. "Essais sur l'Automatique" (Essays on Automatics) allowed him to fully develop many of his ideas and experiences with computational devices. He acknowledged Charles Babbage and described the Englishman's work, placing it within the context of recent developments in calculators. The Spaniard also developed his argument that electromechanical methods be applied to the construction of analytical devices. He described the design of a calculator capable of automatically calculating the value of $a\,x\,(y - z)^2$ for a sequence of sets of values of the variables involved. He envisioned a device that could store numbers, automatically conducting arithmetic procedures using tables stored in the machine and comparing two quantities. He suggested that it could be controlled by a read-only program. Torres y Quevedo also described how to use conditional branching at the same time.

This paper may even contain a primitive description of floating-point arithmetic function. Over the years the document has become recognized as an important contribution to the literature of universal automated calculating devices employing electronics. Through its detailed descriptions, it suggested the possibility of a digital computer* with program control.

Torres y Quevedo next went to work on another calculating device and this time actually constructed a working model. He demonstrated the machine in France in 1920, proving that an electromechanical analytical machine could be built and operate correctly. Neither this machine nor at least two others that he made were fully and completely electromechanical calculators but the portions that were illustrated the possibility of such computational units. His primary concern was to show that machines could become "automatics." The device he demonstrated in 1920 put to rest any doubt that his ideas were not feasible.

This machine, called the "electromechanical arithmometer," consisted of an arithmetic unit attached via cable to a typewriter. Commands would be fed to the calculator by typing them on the console and subsequently being printed out by the calculator. Torres y Quevedo never fully completed the machine, refusing to turn it into a commercial venture. All the same, he has a prominent place in the history of modern Spanish science. A laboratory has been named after him, and some of his machines are on exhibit and still operational in Madrid.

Torres y Quevedo died in 1936, at the start of the Spanish civil war. Serious scientific research on computational devices would not resume in Spain until the late 1950s.

For further information, see: Leonardo Torres Quevedo (Madrid: Colegio de Ingenieros de Caminos, Canales y Puertos, 1978); Brian Randell, "From Analytical Engine to Electronic Digital Computer: The Contributions of Ludgate, Torres, and Bush," *Annals of the History of Computing* 4 No. 4 (October 1982): 327–341; Alcalde L. Rodríguez, *Biografía de D. Leonardo Torres Quevedo* (Santander: C.S.I.C., 1974); José García Santesmases, *Obra e inventos de Torres Quevedo* (Madrid: Instituto de España, 1980); and for samplings of his writings in English, see Brian Randell, ed., *The Origins of Digital Computers* (Berlin: Springer-Verlag, 1982) pp. 89–120.

TURING, ALAN MATHISON (1912–1954). This British mathematician contributed to our understanding of the relationship between the process of thinking and electronic computing. Considered a pioneer in the development of computers and in defining their role, he also became a saint for those who later studied artificial intelligence.* At the age of twenty-five, he described what later became known as the Turing machine, an abstract theoretical device that gave firm definition to the characteristic of a mathematical algorithm while defining the "intelligent" role that a computational device might play.

Turing was born on June 23, 1912, in London, the second son of a British family that had earlier been connected with the Indian Colonial Service. Turing was raised in Great Britain, and, in 1931, he matriculated at King's College, Cambridge University. He so distinguished himself in mathematics that upon completion of his studies in 1935, he was made a Fellow of King's College. Between 1936 and 1938, he worked and studied at Princeton University, exploring the ramifications of his Turing machine. During World War II, he lived in Great Britain and was employed at the Department of Communications where he worked on decoding German ciphers. Following the war, he participated in the design of electronic computers and articulated his thoughts concerning the possibility of computers replicating the thinking process of human minds. Turing was convinced that by the year 2,000 that would be possible. Like many of his generation working in the general areas of computing and information theory in the late 1940s, he also studied the mathematical basis for creating nonsymmetric living systems out of symmetric egg cells.

Turing enjoyed an active social life and played the violin which he taught himself. During World War II, when plagued by severe hay fever, Turing, always the eccentric, solved his problem by wearing a government-issued gas mask while riding his bicycle to work. According to legend, one day his bicycle chain slipped off the hub gear when he rode it back and forth to work. Rather than simply have it repaired, he noted that the mishap occurred after a certain number of rotations of his pedals; he counted them, dismounting just before the chain came off. Later, Turing mounted a mechanical counter on the pedals so that he would not have to focus his attention on counting the number of rotations. He subsequently did a mathematical analysis which indicated that the problem existed after x number of chain links had passed over the gear mechanism. A deformed chain link caught with a slightly damaged gear tooth that caused the chain to pop off. Turing's experience with this problem suggests his creative approach to problems. This amiable, though impractical, man constantly revealed his clever, curious mind.

One of Turing's most important contributions to the subject of computer science was his paper introducing the Turing machine, published in 1937, called "On Computable Numbers with an Application to the *Entscheidungsproblem*." In the paper, Turing proved that certain types of problems could not be solved by fixed or definite processes; this was a major assertion in the field of mathematical logic. He observed that a "definite process" was that which could be done by a machine, the one he described in the paper. It was, in short, the abstract computation device which we know as the Turing machine.

Turing asked his readers to envision a device with a tape drive, the tape being partitioned into squares and each square having either a finite number of symbols or remaining blank. His device would theoretically scan one square at a time, moving the tape forward or backward one square at a time. His machine could erase symbols or print them, and nothing else. He then illustrated how his machine, with only two commands, could do a variety of problems, provided they were expressed as zeros and ones, in other words, in binary code. Turing argued that in this manner one could detail the steps required to accomplish a particular task which could then be done by his machine. In theory then, the Turing machine executed any task asked of it and served as a digital computer* before one had been invented. Any procedure in mathematics or list of instructions required to solve a particular problem could be executed when set up the way he wanted. This was especially the case, Turing said, for algorithms which up to then had not been defined precisely enough for resolution. When viewed within the framework he created, however, certain classes of problems could not be solved using algorithms. Turing borrowed his ideas from Charles Babbage** (who designed difference and analytical engines* in the early 1800s) and from George Boole** (who developed logic using binary codes). The computer he conceived was similar to those built in the 1940s and 1950s.

While at Princeton, Turing worked with Alonzo Church, a logician, and met John von Neumann** who, in the mid-1940s, wrote a seminal paper on the architecture of computers which is still influencing their design in the 1980s. During World War II, while working for the British government, Turing gained experience with computational gear at Bletchley Park,† an estate just north of London. There, a group of mathematicians, scientists, expert chess players, and others worked with a German cipher device called Enigma, developing methods for decoding German military and civilian transmissions throughout the war. They cracked German codes quickly, and as a result, British and American leaders learned almost every German military plan ever made. Although most of the work done at Bletchley Park remains shrouded in government secrecy, this effort was one of the most important in World War II.

Turing's contribution to the project required the use of a computer called COLOSSUS,* a complex of servomechanical parts and brass. This computer was an automatic operating electronic device used to decipher German messages. Ultra, as the project was known, required Turing's talents as a mathematician and logician to help design and build variations of the machine throughout the war. In 1942, he came to the United States, probably met with von Neumann who was then deeply interested in what became known as electronic digital computers,* and then returned to continue producing electromechanical devices to break codes generated by the German Enigma class machines (also known as the *Geheimschreiber* devices).

Following World War II, Turing went to work for the National Physical Laboratory (NPL) where he continued to help design computers. The first of these was called the Automatic Computing Engine (ACE),* built during the second half of the 1940s and completed in the late 1950s. During the year-long sabbatical he took from King's College in 1947, he wrote a paper, published in 1969, called "Intelligent Machinery." This essay speculated on subjects which were of interest to such proponents of artificial intelligence (the term did not come into vogue until the second half of the 1950s) as Norbert Wiener,** who visited him in early 1947. In exploring how equipment could show behavior which one might characterize as intelligent, Turing said that his purpose was to use the human brain "as a guiding principle." He outlined the case against the possibility of intelligent devices, such as the notion that humans would never admit to the possibility of an entity as intelligent as they. Turing described other arguments, including the one that no machine could solve all the problems that a human could, and then proceeded to refute each objection.

Turing next argued that there already existed devices which imitated parts of humans, such as television cameras acting as "eyes" and microphones as "ears." Servomechanical devices (later called robots) also replicated functions inherent in such human limbs as legs and arms. He observed that "electrical circuits which are used in electronic computing machinery seem to have the essential property of nerves." In other words, electricity, like nerves, could transmit and receive information. Similar concepts were already being explored by Wiener, the father of the term *cybernetics*. Turing envisioned the construction of a brain which could play games (chess, bridge, poker, tic-tac-toe, for example), learn languages and translate them, handle cryptography, and perform the functions of mathematics. Drawing an analogy with the baby's brain as a clean slate to be written upon, he observed that a machine might have routines, general knowledge, and so on, added to it in order to make it possible for that device to perform certain assigned tasks, much as we train a baby. But the important idea was that all problems of an intellectual sort could be described to the computer as "find a number *n* such that and so forth." The ideas described in this paper reflected many of the research concerns of workers later concerned with artificial intelligence. The essay was not published until 1969, by which time its ideas had become common fare among AI researchers.

The timely availability of another paper (it appeared in October 1950), entitled "Computing Machinery and Intelligence," explored the issue of whether or not a computer could think. The way Turing chose to answer the question resulted in the famous Turing's Test. Questions would be asked by an interrogator to a person in one room and a computer in another. Both communicated theoretically back to the questioner only by teletype. Turing argued that if the interrogator—after a reasonable amount of dialogue—could not tell whether he was communicating with a machine or a person, then the machine could think.

Although many of his contemporaries took Turing to task for his ideas, the Turing Test survived because it was at least a reasonable and, for a time, the only methodology offered to define thinking in a machine. He expanded on this idea until the end of his life.

In the early 1950s, Turing began to work on yet another project: the Manchester Automatic Digital Machine (MADAM) at the University of Manchester. Impatient with the slowness with which the ACE was being built (it was not completed until 1958), he developed an interest in biology and tried to find mathematical ways to examine biological problems. During these years, he continued to argue that the difference between what machines and humans did was one less of substance than of degree. Thus, he was an early subscriber to the idea that a theory of information existed, one that was not limited to machine or human. This concept came into its own in the 1950s.

Always controversial, even his death on June 7, 1954, proved intriguing. The medical examiners finally concluded that he had died of self-administered potassium cyanide while in a moment of mental imbalance. His mother, however, suggested that Turing enjoyed tinkering with household chemicals to investigate whether additional substances could be developed from them; she theorized that he had been careless, possibly accidentally using a spoon coated with the poison. Some of his friends argued that he killed himself because he was homosexual and could not face the embarrassment of his sexual orientation perhaps being uncovered. In the world of the 1950s, such a revelation would have been devastating to his life and career.

Decades after his death, Turing's ideas that computers might replace men and women received increasing attention from scientists and philosophers who looked back to his work as vital, first steps in their own research. By the early 1980s, his concepts suggested a new definition of people as processors of information rather than as animals. Nature was more and more being viewed as "information to be processed"; truly Turing's ideas were compatible with the information age. With the wide use of computers by then, the analogy of the man/computer had become a reality, an idea some honestly called "Turing's man." Turing was a pioneer in his concepts of the intellectual role of computers, he suggested lines of investigation that have become known as artificial intelligence, and he built electronic computers which greatly advanced the science of data processing in Great Britain.

For further information, see: J. David Bolter, *Turing's Man: Western Culture in the Computer Age* (Chapel Hill: University of North Carolina Press, 1984); B. E. Carpenter and R. W. Doran, eds., *A. M. Turing's ACE Report of 1946 and Other Papers* (Cambridge, Mass.: MIT Press, 1986); A. Hodges, *Alan Turing: The Enigma* (New York: Simon and Schuster, 1983); Pamela McCorduck, *Machines Who Think* (New York: W. H. Freeman and Co., 1979); Alan M. Turing, "Computing Machinery and Intelligence," *Mind* 59 (1950): 433–460, "Intelligent Machinery," in B. Meltzer and D. Michie, eds., *Machine Intelligence* (Edinburgh: Edinburgh University Press, 1969), 5: 3–23, "On Computable Numbers, With an Application to the *Entscheidungsproblem,*" *Proceedings, London Mathematical Society,* Series 2, 42 (1937): 230–265 and with et al., *Proposals for Development in the Mathematics Division of an Automatic Computing Engine (A.C.E.),* Report E. 882. Executive Committee, National Physical Laboratory (Teddington, Middlesex, U.K.: National Physical Laboratory, 1945), reprinted 1972; S. Turing, *Alan M. Turing* (Cambridge: Heffer, 1959); Gordon Welchman, *The Hut Six Story: Breaking the Enigma Codes* (New York: McGraw-Hill Book Co., 1982).

U

UTMAN, RICHARD (1926–). Utman is representative of a new, little known profession within the data processing community: that of the manager of a computer facility. He worked with many military and civilian applications in the United States during the 1950s and 1960s and with some of the most widely used computers of these times.

Utman was born and raised in California, primarily in the Los Angeles region. He studied physics and mathematics at Pomona College in Claremont, completing his B.A. there before working on an M.L.S. in information science at Rutgers University in New Brunswick, New Jersey. After completing his studies in the late 1940s, he went to work for the U.S. Navy-Aviation Supply Office in Philadelphia where he established and ran one of the earliest business data processing centers in the United States, having it fully operational by 1950. Between then and 1955, he used the UNIVAC I,* the IBM 701,* the 702–705,* and the 650* computers. In 1955, he accepted a position at the Stanford Research Institute where he ran its Southern California EDP and OR Group in Economic Research and Development facilities. In 1956, he left to join the ElectroData Division of the Burroughs Corporation.† His mission at the newly established division was to establish and manage departments in market development and product planning. Between 1956 and 1959, he participated in the development of that company's 205 and 220 systems. He then served as a field manager for the U.S. Army Tactical Data Systems Project at Fort Huachuca, Arizona, on behalf of the Ramo-Wooldridge Corporation (which later became known as TRW†). Utman used a 709/7090 to simulate the activities of a Raytheon MOBIDIC computer along with other small computer-like devices before they were bought by the military.

Utman soon left to become the first director of Systems Programming within the UNIVAC Division of Sperry Rand† in 1960, remaining for two years. While there, he managed the design and development of the systems software* which ran on the UNIVAC III, the 490, and the widely used 1107 computers. In 1963, he once again moved, this time to become director of standards for the Business Equipment Manufacturers Association (BEMA). In this critical period in the evolution of software (1963–1964), Utman was an active member of the American Standards Association X3 Standards Committee on Computers and Information Systems Administration while it was establishing guidelines for the newly developing or modified ASCII, FORTRAN,* and ALGOL.* Between 1964 and 1968, Utman was an independent consultant, developing information systems for the J. C. Penney Company, Union Carbide Corporation, D'Arcy Advertising Company, and Computer Usage Corporation. Acting as a data processing consultant in a period when there were few such individuals, he also did work for First National City Bank of New York, Information Management Inc., and Applied Logic Corporation. These last three projects involved operating systems,* use of programming languages,* time-sharing systems, and automatic machine-language translation applications.

As administrative officer for Princeton University (1967–1972), Utman was responsibile for developing and operating information systems for the management and administration of the university. He also developed what became a very controversial library management system which, according to many press reports in the mid-1970s, did not work well enough to keep (apparently not his fault, however). After 1972, he became a faculty member at Princeton and today is a library systems analyst.

Over the years Utman published numerous articles, primarily on business data processing, library applications, and systems programming. He has also enjoyed studying the history of the deserts and mountains of the Southwestern United States.

For further information, see: Franklin M. Fisher et al., *IBM and the U.S. Data Processing Industry: An Economic History* (New York: Praeger Publishers, 1983).

V

VAN DER POEL, WILLIAM LOUIS (1926–). This Dutch computer scientist's life provides a window into the voluminous, but poorly documented, developments within data processing in the Netherlands. Van der Poel was specifically involved in the construction of early Dutch electronic computers in the 1950s.

Van der Poel was born in The Hague and because of World War II attended only one year of classes at a local technical college. After the war, he matriculated at the Delft University of Technology to study physics (1945), completing his degree in engineering physics in February 1950. While a student, he was exposed to computer science as a research assistant to N. G. de Bruyn and A.C.S. van Heel, helping them to design and build a relay computer to do optical ray tracing. Although van der Poel did not participate in the completion of the project, he played a critical role in its design. The machine, TESTUDO, as it was called, operated from 1952 to 1964 and was a slow operating computer. After completing his studies, van der Poel went to work for the Dr. Neher Laboratory which belonged to the Netherlands Postal and Telecommunications Services. While there he worked on the design and construction of what became the first electronic computer built in Holland, called the PTERA. It was a magnetic drum computer similar in design to other European devices; it operated from 1953 to 1958. He also worked on the design of another device which was faster and more complex. Building on knowledge gained with the first machine, the staff built what became the prototype of ZEBRA, a later Dutch computer. The prototype, named ZERO, used functional bit microcoding. Based on experience with these machines van der Poel wrote his doctoral dissertation for the University of Amsterdam in 1956 on the design of computers. Much of the data available on ZEBRA's design appeared in that thesis. In the late 1950s, fifty-five ZEBRAs were constructed, nine of which were used in the Netherlands.

In 1956 van der Poel became chief engineer and manager of the Mathematics

Department at the Dr. Neher Laboratory. In addition to the work done on ZEBRA, used primarily by the Dutch telephone service, he had begun working with a blind and deaf mathematician, Gerrit van der May, since 1952, on aids for such people. The results included a mechanical conversation machine and an improved Braille telephone. An earlier device which he had worked on to provide telephone capabilities with Braille had earned him the Visser-Neerlandia Prize. The improved machine was completed and demonstrated in 1957. Van der Poel went on to build a Braille teleprinter, which attached to a normal teleprinter hooked into a telex network. He also worked with his friend to construct a tape-driven Braille printer. These devices were made in multiple copies at the laboratory for use by others in Holland.

In 1961, van der Poel became head of the Switching Laboratory within the Dr. Neher Laboratory. In the early 1960s, he devoted his energles to the design of machines to perform character and mark recognition, primarily for use by the Dutch Postal Giro System. The following year after his promotion, he was also named professor at Delft University of Technology to teach computer science. He left the laboratory in 1967 in order to take up full-time teaching responsibilities at the university. His research interests during the 1970s and into the 1980s shifted to programming languages* and to the design and implementation of compilers.

Beginning in the 1960s, van der Poel helped to develop an ALGOL compiler for ZEBRA, five LISP* systems, and a variety of SNOBOL* compilers. He served as chairman of the International Federation for Information Processing (IFIP)† W.G. 2.1 Committee, which was responsible for nurturing ALGOL since its formation in 1962. He left the committee in 1969. He also served as treasurer and president of the Dutch Computer Society (Nederlands RekenMachine Genootschap, otherwise known as NRMG). As part of his role as a lecturer at large for the scientific council of the Nederlands Studiecentrum voor Informatica, van der Poel designed a simulated language called Seemingly Easy Reckoning Automation (SERA). This hypothetical language, designed for educational purposes, was implemented in simulated form and continues to be used in the Netherlands as part of the state examinations for students on programming.

In recognition of his contributions to Dutch data processing, van der Poel was appointed a member of the Netherlands Royal Academy of Science on June 1, 1971. This academy (Koninklijke Nederlandse Akademie van Wetenschappen) strongly supported the work of computer scientists in the country. During the 1980s, he also represented his country in IFIP.

For further information, see: William Louis van der Poel, "Comment on the Composition of Semantics in ALGOL 68," *Communications of the ACM* 15, No. 8 (August 1972): 772, and "SERA, A Hypothetical Machine for Teaching Purposes," in *Proceedings of the First IFIP World Conference on Computer Education, Amsterdam, The Netherlands* (1970), passim.

VAN WIJNGAARDEN, ARIE (1933–). Van Wijngaarden, a professor of physics, represented his country and Europe in discussions leading to the development of ALGOL,* one of the first universal programming languages.* The standards it set and the cooperation among international computer scientists it fostered were milestones in the early history of software.*

Van Wijngaarden was born on April 8, 1933, in Holland; he was educated in his country but took his Ph.D. in physics from McMaster University in Canada in 1962. He held various teaching assignments in Holland throughout the 1960s and, beginning in 1973, served as professor of physics at the University of Windsor, Ontario, Canada. In the 1950s, when both American and European computer scientists sought to standardize computing languages, he became involved. The movement led to discussions on both sides of the Atlantic on developing a language called ALGOL. In January 1960, a major conference was held in Paris to work out details for the new language. In eighteen days of intense work, representatives of European and American data processing communities hammered out the elements of the language. Their output was a report defining ALGOL 60. Van Wijngaarden had actively pushed for reforms of an earlier version of the language called ALGOL 58 and thus made specific recommendations involving syntax, use of complex numbers, vectors, and matrices. He drafted portions of the ALGOL 60 report, an important document in the history of programming.

The language appeared during the 1960s in various forms, all closely related. Although compilers were developed by several manufacturers, ALGOL did not take off as the universal programming language. FORTRAN* remained the language of choice for engineers and scientists, while COBOL* served the same purpose for commercial applications. Nevertheless, van Wijngaarden continued to support ALGOL. His work and that of his fellow scientists involved with ALGOL created a pattern of cooperation in setting standards for programming languages. In the 1960s and 1970s, for example, standards were established and updated periodically for most major languages, such as COBOL, through the formal sponsorship of industrywide organizations.

For further information, see: Peter Naur,'' The European Side of the Last Phase of the Development of ALGOL 60,'' in Richard L. Wexelblat, ed., *History of Programming Languages* (New York: Academic Press, 1981), pp. 92–171, and with Alan J. Perlis, ''The American Side of the Development of ALGOL,'' ibid., pp. 79–91; Jean E. Sammet, *Programming Languages: History and Fundamentals* (Englewood Cliffs, N.J.: Prentice-Hall, 1969).

VEREA, RAMÓN (1838–1899). In 1878, Verea, a Spaniard living in New York, built a direct multiplication machine. This device was a mechanical aid to multiplication operating on the general principle common with such devices as the ''Millionaire''* that ''one turn of the crank for each figure in the multiplier'' produced a result. He used cylinders containing holes of different diameters into which needles could be inserted to varying depths. Projecting parts served the same function as tongues, for example, in the calculating plates of Léon Bollée's** mechanical device of the same period. It was more efficient then older designs that had called for multiple additions to reach an answer. Verea never capitalized on his invention although he applied for a U.S. patent on it; thus, it remained an historical curiosity. The *New York Herald,* however, quoted him as saying that he "did not make the machine to either sell its patent or put it into use, but simply to show that it was possible and that a Spaniard can invent as well as an American." Whether or not that was an excuse for failing to market his invention, he had formal drawings of the machine which would have been suitable for patent registration properly witnessed and documented, and he constructed one as required by the patent office.

For further information, see: Charles Eames and Ray Eames, *A Computer Perspective* (Cambridge, Mass.: Harvard University Press, 1973).

VON KEMPELEN, WOLFGANG (1734–1804). This *enfant terrible* was the maker of a fraudulent automaton. During the eighteenth century, automatic machines were the popular playthings of many European courts as mechanical ducks and fancy clocks appeared in every country. Von Kempelen built a chess-playing machine which over the centuries became the most famous of many frauds. His device involved an android dressed in Turkish-like clothes sitting at a chess table. The table was supposed to contain the mechanism to make the machine work. In fact, it housed a little man who actually played the game. One unfounded story that later circulated was that the little man was a legless Pole wanted by the Russian police. Von Kempelen took his "machine" all over Europe, playing chess against other people in courts and fairs,—and always winning, of course. The story is told that in 1809, several years after von Kempelen died, his machine was still making the rounds and, depending on which version of the story is accepted was pitted against either Napoleon or some of his generals. But the machine won. It was brought over to the United States where it went from state to state. Known by then as the Turk, it drew considerable attention, until 1854 when it burned in a fire in Philadelphia.

This device, albeit an illusion of mechanized intelligence, was so well known that it inspired many who sincerely sought to build such a device. Researchers who eventually became involved in artificial intelligence* were also fascinated by the thought of a chess-playing machine and, particularly during the 1950s and 1960s, attempted to construct one, based on their feeling that intelligence, as used by humans, was best illustrated and employed in the game of chess. The idea was that constructing a device that could play the game would help clarify how the mind works while replicating intelligent behavior mechanically—the dream of the original builders of automata. It was also von Kempelen's idea, his fraud notwithstanding.

For further information, see: J. F. Racknitz, *Uber den schachspieler des herrn von Kempelen und dessen nachbildung* (Dresden: n.p., 1798).

VON NEUMANN, JOHN (1903–1957). Von Neumann, a dominant figure in the history of twentieth-century mathematics, is best remembered for his work in mathematics, his research on computers and related technologies, and his encouragement of others to work on both. He was also associated with Los Alamos and the invention of the atomic bomb and published over 125 articles and books. No history of data processing can be complete without considering his influence on computers of the 1940s and 1950s. His work in mathematics contributed to the basis of modern computer technologies, as did his early efforts with high-speed digital computers.* Von Neumann described the theoretical concepts of the stored program in contemporary computer technology, and his work in this area reflected research by other scientists, particularly at the University of Pennsylvania.

Von Neumann was a child prodigy, born into a Jewish banking family in Budapest, Hungary, on December 3, 1903. When only six years old, he could divide eight-digit numbers without using paper and pencil. Following the defeat of the Austro-Hungarian Empire in World War I, he attended universities in Germany and Switzerland, studying physics, mathematics, and chemistry. He completed his formal studies at the University of Budapest where, in 1926, at the age of twenty-two, he earned a Ph.D. in mathematics. In the following year, he joined the University of Berlin as a *Privatdozent*. By 1930 he had published a number of papers concerning set theories, algebra, and quantum mechanics. He joined the faculty at Princeton University as a visiting professor in 1930, staying until 1933 when he took a position at the newly formed Institute for Advanced Study (IAS) where he remained until his death. By 1940, he had earned a worldwide reputation as an outstanding and prolific mathematician with seventy-five publications to his record. His papers were in various fields of mathematics, and he encouraged others to study the subject while legitimizing it as an area for modern scientific investigation.

World War II opened a new chapter in von Neumann's life. He became a consultant to various military organizations in the United States, in which other scientists were attempting to apply mathematics to war-related projects; he advised the Ballistics Research Laboratory, the U.S. Navy Bureau of Ordnance, and the Los Alamos Scientific Laboratory; and he became involved in windtunnel research, nonlinear systems of equations, and atomic energy. He spent increasing amounts of time with applied mathematics. Von Neumann grasped technical problems quickly, drew up creative solutions, and had the necessary administrative talent to make workable recommendations. He was also a persuasive person, which helped him obtain financial support for his pet projects. Von Neumann enjoyed the prestige which his success brought him. He simultaneously influenced U.S. scientific policies and enhanced the field of mathematics.

Throughout World War II, although von Neumann became heavily involved with administration, he continued to conduct research. For example, in 1944, he presented the results of considerable work on quantum mechanics that proved

Erwin Schrödinger's (1887–1961) wave mechanics and Werner K. Heisenberg's (1901–1976) matrix mechanics were mathematically equivalent, once again illustrating the possible use of mathematics.

In addition to his interest in mathematics and quantum mechanics, von Neumann made critical contributions to the development of the modern computer. During the course of his work at Los Alamos, he had already discovered that manual methods of calculation were slow and cumbersome, particularly when calculating implosions requiring solutions for nonlinear systems of equations. Although some work on computers had already started during the 1930s and picked up sharply in the early 1940s, only with America's entry into World War II did von Neumann become more interested in digital computers.* Indeed, he thought such devices could be of considerable use to scientists, specifically to mathematicians.

Von Neumann's interest in computers deviated sharply from that of his peers in that he expressed great concern for applied mathematics and the use of calculators. In the early 1940s, mathematicians in academia had little interest in calculators, their primary focus being theoretical. Von Neumann's biographer, Nancy Stern, argues that his interest legitimized the study of applied mathematics and computational equipment, particularly for the mathematics community and scientists in general during World War II. Apparently, his interest in computers was initially shown in his correspondence in January 1944 in which he inquired about work being done by various groups in applied mathematics. Von Neumann next communicated with Howard H. Aiken** who was building the Mark I* at Harvard, the first electromechanical relay computer; with George R. Stibitz*** at Bell Laboratories† who had considerable experience with electromechanical relay equipment; and with Jan Schilt who was doing work for International Business Machines Corporation (IBM)† at Columbia University's Watson Scientific Computing Laboratory.

Von Neumann learned about the project at the Moore School of Electrical Engineering† to build the ENIAC* in August 1944. He met Herman H. Goldstine**, who was then associated with the project as a military liaison, and they discussed it. Von Neumann subsequently made many visits to the Moore School where he made recommendations concerning the ENIAC and encouraged the scientists at Los Alamos, then working on the atomic bomb, to test the new computer at their facilities. Von Neumann's concern for the work at the Moore School brought attention to the project fom a variety of U.S. government agencies which until this time had ignored the attempt to build a computer there. Soon after his visits began, the school received a government contract to begin work on a follow-on device, called the EDVAC.* While his precise role in the issuance of this important contract is not known, historians believe that von Neumann's interest in the EDVAC may have brought negotiations that had been in progress for some time to a positive conclusion.

As a result of von Neumann's regular visits to the ENIAC project, the staff working on the computer regularized their administrative procedures. Minutes were now taken of staff meetings, discussions and research were documented, and time tables were established for the completion of work. Von Neumann made suggestions on technical issues concerning EDVAC's design when not advising at Los Alamos. He greatly influenced the staff at the Moore School, many of whom were young scientists. With the end of the war, his relationship with the school also came to an end. Two schools of thought emerged regarding the role and function of mathematics and computers. One group, led by John W. Mauchly** and J. Presper Eckert,** was interested in the commercial applicability of computers, and the second, von Neumann among them, was more committed to pure scientific research.

Tensions came to a climax as Eckert and Mauchly applied for a patent on the ENIAC. The president of the University of Pennsylvania, Harold Pender, wanted the patent assigned to the university. Eckert and Mauchly, having made no written agreement that their work belonged to the university, protested and hired legal counsel. Pender, wanting to avoid a long battle, dropped the university's claim to the patent. Meanwhile, the question of who owned the rights to EDVAC became a critical one. In March 1946, the Moore School required all of its scientists to sign patent release forms on any work done there. Eckert and Mauchly refused and so resigned from the Moore School and formed the Eckert-Mauchly Computer Corporation to build computers. Their departure, along with that of other scientists, signalled the end of an important era in the development of computers at the University of Pennsylvania, but it did not put an end to the controversy over ENIAC and EDVAC involving von Neumann.

In 1945, von Neumann became embroiled in a controversy with Eckert and Mauchly over who developed the concept of the stored program in computers. The argument today may seem like a moot one since all computers are based on it; yet at the time it was an important issue. The stored program concept simply meant that a computer could take whole programs and data into its memory, which could then be executed (that is run or performed) and the results released to people in the form of reports or data on an output device, such as cards or disk and tape files. The idea that programs should be moved in and out of the computer in order to perform transactions, and under the control of some master control program—which today we call the System Control Program, or SCP—was revolutionary. Until then, the few computational devices in existence did not function this way. To program a computer, people had to turn switches on and off on the device; this was a slow and inefficient process. Once written, a stored program could be used again and again and, in later years, be moved in and out of computers in fractions of a second. The idea of putting programs into a computer thus represented a radically new approach to the design of a computer's architecture and promised incredible gains in productivity, not to mention speed of processing and automation.

The controversy regarding the idea of stored program computers arose from design policies established at the Moore School. In order to have the ENIAC ready quickly for the war-related projects, such as automatic programming capabilities, computer's designers decided not to develop any new procedures which might add months of added work. Yet the idea of modifications was discussed during the design of ENIAC and came up again during the early design phases for EDVAC in early 1944. At that time, Eckert raised the possibility of constructing the machine based on the concept of a stored program and prepared a short paper on the subject for internal use by the staff at the Moore School. Discussions were held on the matter even before von Neumann became involved with the Moore School.

Then, in June 1945, von Neumann wrote "First Draft of a Report on EDVAC", a paper that described the concept of a stored program and its possible application to the design of a computer. This draft paper circulated among a number of scientists working at the Moore School as well as in other academic circles, making it one of the first reports on the design of digital computers* to be widely read within the scientific community. Von Neumann discussed logic control and the idea of a stored program as it might relate to a new computer. The concepts discussed in the report became the foundation of much computer design work in the United States for the next fifteen years. Von Neumann, whose name was the only one to appear on the cover of the report, was given the credit for the stored program idea.

A lively debate developed over the true authorship of the idea; even today a consensus has not been reached. Herman Goldstine credited him for developing the concept, despite Eckert's earlier one-page report on the idea. Arthur W. Banks, a scientist who worked with ENIAC and EDVAC, concurred with Goldstine. A scientist at Los Alamos, Nicholas C. Metropolis, however, maintained that, while von Neumann contributed to the concept, he did not develop it alone; rather, the idea emerged out of preliminary design specifications for the EDVAC drawn up by a group of scientists at the Moore School. Harry Huskey, then at the Moore School, shared this view. In fact, controversy continued for decades. When Eckert and Mauchly applied for patents on EDVAC, they could do so because von Neumann's document had circulated widely for over a year prior to patent application, and under the law the stored program concept was in the public domain. Von Neumann's biographer notes that his failure to make any serious attempt to deny sole authorship probably fueled the suspicion that he had attempted to take credit for work done by others. At the time the controversy first developed, von Neumann had already split with Eckert and Mauchly over their desire to commercialize computer research.

Von Neumann's original motivation for writing the paper was not sinister: he simply wanted to formalize ideas that had already been discussed at the Moore School, much as any academic would want to write down the results of research and thinking in a particular field. The paper was highly theoretical and conceptual, not technical. He intended it to be applicable to any digital computer and not just to EDVAC.

Von Neumann divided the concept of a computer's architecture into five components: arithmetic function, central control, the idea of memory,* and the roles of input and output tasks. He focused attention on the theoretical aspects of computers, whereas Eckert and Mauchly, engineers by training, was concerned chiefly with the actual design of machines. Von Neumann relied on his background in mathematics to establish a logical structure for his concepts. He even drew analogies between the computer and the central nervous system, predating by several years work done by Norbert Wiener** of the Massachusetts Institute of Technology (MIT) on cybernetics, comparing theories of information systems to biological and neurological realities.

By providing a conceptual overview in a paper that was widely circulated, von Neumann stimulated discussion and influenced the design of computers using the stored-program concept and programming. Even today, computers contain input and output devices or functions, memory, control programs, and application programs and data that come in and out of computers under the command of software* and hardware. Von Neumann's greatest contributions to the development of the computer, therefore, lay as much in his collation of ideas as in the dissimination of recent work done by others, all laced with the theoretical underpinnings of an academic mathematician. His "First Draft of a Report on EDVAC" may thus be seen as part of a much larger role he had carved out for himself.

In addition to his relations with the Moore School and the controversy over the stored program, von Neumann influenced other computer-related activities. The most important of these involved computing projects at the Institute of Advanced Studies at Princeton. He wanted a computer at the IAS as early as World War II and campaigned for the necessary funds and people. In 1945, the IAS agreed to launch a computer project, probably out of fear that von Neumann might leave if they did not do so. Von Neumann wanted this project in order to illustrate the benefits of a computer as a research tool, as well as to prove the social utility of many forms of mathematics. With funding from IAS and various U.S. government agencies, von Neumann proceeded to acquire a staff and build a computer. Goldstine joined his team, as did others from the Moore School. During 1946 and 1947, considerable basic design work was done. While the Radio Corporation of America (RCA)† worked on the memory portion of the new machine, the team developed circuits and then an arithmetic logic unit. After years of intensive work, by the spring of 1951 the IAS computer* was operational, running programs and doing calculations for the Los Alamos Laboratory. Von Neumann encouraged the use of this machine to do metereological studies based on the idea that weather systems could be expressed mathematically and thus be tracked and predicted through computations.

Unlike earlier computers, the IAS computer could be partially programmed directly in machine-readable code. Although primitive by the standards of devices produced only a decade later (for example, it had no floating-point registers), it was semi-programmable. It was used for various mathematical computational projects throughout the 1950s and, at the end of the decade, it was retired, a contribution to the Smithsonian Institution.

By the mid-1950s, von Neumann had turned his attention to the Atomic Energy Commission. He also continued to publish papers in various fields of mathematics until his death on February 8, 1957.

For further information, see: S. Bochner, "John von Neumann," *National Academy of Sciences Biographical Memoirs* 32 (1957): 438–457; S. J. Heims, *John von Neumann and Norbert Wiener: From Mathematics to the Technologies of Life and Death* (Cambridge, Mass.: MIT Press, 1980); Nancy Stern, "John von Neumann's Influence on Electronic Digital Computing, 1944–1946," *Annals of the History of Computing* 2, No. 4 (October 1980): 349–362; A. H. Taub, ed., *John von Neumann: Collected Works,* 6 vols. (Oxford: Pergamon Press, 1963); S. Ulam, "John von Neumann, 1903–1957," *Bulletin, American Mathematical Society* 64 (1957): 1–49, reprinted in *Annals of the History of Computing* 4, No. 2 (April 1982): 157–181; John von Neumann, *The Computer and the Brain* (New Haven, Conn.: Yale University Press, 1958), *First Draft of a Report on the EDVAC,* Contract No. W–670–ORD–492, Moore School of Electrical Engineering (Philadelphia: University of Pennsylvania, 1945), and *Theory of Self-Reproducing Automata* (Urbana: University of Illinois Press, 1962), and the convenient *Papers of John von Neumann on Computing and Computer Theory,* edited by William Asproy and Arthur Burks (Cambridge, Mass.: MIT Press, 1987).

W

WANG, AN (1920–). Wang, an engineering executive, founded Wang Labs, Inc., one of the leading manufacturers of word processing equipment in the United States, primarily during the 1970s. He was also an early researcher in the general field of memories for computers.

Wang was born in Shanghai, China, on February 7, 1920. He came to the United States in 1945 and was naturalized as a U.S. citizen in 1955. He earned a B.S. from Chiao Tung University in 1940, and an M.S. (1946) and a Ph.D. (1948), both from Harvard University. Wang worked as a teacher at Chiao Tung University (1941–1945) and was an employee of the Chinese Government Supply Agency in Ottawa, Ontario, Canada (1946–1947). Between 1948 and 1951, he was a research fellow at Harvard. In 1951, Wang established his own company, Wang Labs, located in Cambridge, Massachusetts, operating there until 1955. In that last year he became chairman of the board and chief executive officer of the new Wang Labs, Inc., headquartered in Lowell, Massachusetts (a position he held until the 1980s).

This scientist was trained on computers at the Harvard Computer Laboratory, the source of over a dozen important research scientists in the early years of computing. While associated with that laboratory, he developed what became known as magnetic core memory* for computers. Various individuals were working on the general problem of magnetic core memories in the 1940s as a way of expanding the amount of data storage on a computer in the early years of computing. Prominent among these people were Jay W. Forrester** at the Massachusetts Institute of Technology (MIT) who was the father of the WHIRLWIND* computer, Jan A. Rajchman** at Radio Corporation of America (RCA),† and even a worker at the Los Angeles Department of Public Works who sold his rights to International Business Machines Corporation (IBM)† in 1956. Wang's efforts were made while he was at Harvard working for Howard H. Aiken,** builder of the Mark* series of computers in the 1940s. Wang made enough progress on his own memory to apply for patent rights in 1949. RCA first filed for patent rights, deferred to IBM after IBM acquired additional rights to those claimed by MIT, and added those accumulated in 1956. Thus, Wang was precluded from fully capitalizing on a technology that became relatively common by the late 1950s.

The creation of Wang Laboratories ultimately became Wang's most time-consuming effort. Initially financed with $15,000 in 1951, by the mid-1980s it was worth $620 million. An Wang personally owned $200 million in shares, his wife an additional $120 million, and his children a total of $300 million. The machines he built were high-function, competitive devices which became major factors in the world of word processing in the mid-1970s. Earlier, his products had ranged across a wide variety of computer-related components and devices. In November 1986 his son Frederick A. Wang was named president of the company, with An Wang remaining as chairman and chief executive officer.

Wang's work was recognized by the Institute of Electrical and Electronic Engineers,† which made him a fellow. He was also inducted into the American Academy of Arts and Sciences and received nearly a dozen honorary degrees from such universities as Suffolk, Syracuse, and Tufts.

For further information, see: Stephen T. McClellan, *The Coming Computer Industry Shakeout* (New York: John Wiley and Sons, 1984); Jack B. Rochester and John Gantz, *The Naked Computer* (New York: William Morrow and Co., 1983); An Wang, with Eugene Linden, *Lessons: An Autobiography* (Reading, Mass.: Addison-Wesley, 1986).

WARE, WILLIS HOWARD (1920–). Ware worked on the construction of the IAS Computer* in the late 1940s and early 1950s. He subsequently took charge of research on computers at the RAND Corporation where he built the JOHNNIAC* and became a leading expert on Soviet computers.

Ware was born on August 31, 1920, in Atlantic City, New Jersey. He completed his B.S. in electrical engineering at the University of Pennsylvania, home of the Moore School of Electrical Engineering.† At the time of his graduation, this school was about to enter its golden era of computer building that led to the construction of the ENIAC,* the world's first electronic digital computer,* followed by the EDVAC.* Ware earned an M.S. in electrical engineering from the Massachusetts Institute of Technology (MIT) in 1942, also an early center for research on computers, particularly analog computers.* He completed his Ph.D. in electrical engineering at Princeton University in 1951. Ware was a research engineer at Hazeltine Electronics Corporation in Little Neck, New York, from 1942 to 1946, after which he joined John von Neumann's** staff at the Institute for Advanced Study at Princeton to build a digital computer; he remained until 1951. Between 1951 and 1952 Ware worked for the North American Aviation Corporation which had become active in the use of computers. His last move came in 1952 when he accepted a position at the RAND Corporation in the field of computing. He remained there for the duration of his career. He was also an adjunct professor at the University of California at Los Angeles (UCLA) between 1955 and 1968.

Ware was also a leader in several professional organizations. He became the first chairman of the American Federation of Information Processing Societies (AFIPS)† (1961–1962), a member of the Privacy Protection Study Commission (1975–1977) and its vice-chairman. Since 1959, he has also served on a variety of government-sponsored advisory panels focusing on issues related to data processing, computers, and their use. Ware authored a two-volume study on the design of computers which was published in 1963.

At the Institute for Advanced Study, Ware helped to design the IAS Computer and organize the laboratory that constructed it. Ware left IAS just before its computer was completed to join North American Aviation but quickly moved to RAND in 1952 for the immediate purpose of building a computer similar to that under construction at IAS. Called the JOHNNIAC after von Neumann, it replicated most of the features of the IAS machine. The construction of IAS's system and others by various individuals, including Ware's, made the IAS Computer important. The JOHNNIAC completed its acceptance tests in March 1954 and was subsequently used for government-related research and in various scientific projects. His project encouraged others to construct computers that were installed at Los Alamos, Argonne, and Oak Ridge, all national laboratories supported by the government. The IAS machine at Los Alamos was called the MANIAC,* at Argonne the AVIDAC, and at Oak Ridge the ORACLE. Even the ILLIAC II at the University of Illinois was of a similar type.

Ware was recognized for his contributions to the art of designing computers and for his role in the construction of several early digital processors. In 1975, the Data Processing Management Association (DPMA)† named Ware Computer Sciences Man of the Year. He received the U.S. Air Force's Exceptional Civilian Service medal in 1979. Ware was also made a Fellow of the Institute of Electrical and Electronics Engineers (IEEE).†

For further information, see: Herman H. Goldstine, *The Computer from Pascal to Von Neumann* (Princeton, n.g.: Princeton University Press, 1972); Michael R. Williams, *A History of Computing Technology* (Englewood Cliffs, N.J.: Prentice-Hall, 1985).

WATSON, THOMAS JOHN (1874–1956). One of the great American business executives of the first half of the twentieth century, Watson is best known as the founder of the International Business Machines Corporation (IBM).† Before creating IBM he was a manager and executive at the National Cash Register Company (NCR).† Throughout the first half of this century, Watson played a key role in the office products marketplace as head of IBM. He was also active in charitable affairs. He avoided involvement in politics, believing that IBM required his full attention. In addition to his role as a businessman, he created one of the most famous mottos in business, IBM's one-word expression, "Think."

This natural salesman was born on February 17, 1874, at Cambell in Steuben County, New York. This rural area of the Finger Lakes district in upper New York was home for Watson where his father ran a lumber business and raised his family on a farm. Watson's formal education took place at the Addison, New York, Academy and later at the Elmira, New York, School of Commerce. In 1892, he took a job as a salesman selling pianos and sewing machines and an occasional organ off the back of a yellow wagon near Painted Post, New York. That experience, along with several other jobs, provided him with some exposure to the art of selling but never as much as he would gain as a salesman for NCR. In 1896, the year he joined the cash register firm, he began a fifteen-year career at "the Cash" that would see him rise through the ranks to executive management with a brilliant career. Ultimately, he became general sales manager at corporate headquarters in Dayton, Ohio.

In his years, Watson learned sales and management, and developed distinct ideas about business which he would one day apply to his own firm. Watson learned the value of having well-lighted factories and pleasant working conditions. He recognized that physical fitness meant mental fitness. Although not himself an athlete, Watson was an ardent believer in the importance of good health among his employees.

Watson worshipped sales; he was once quoted as saying that he liked collecting salesmen. Accordingly, he gave them more formal training in sales, marketing, and his products than could be found in other companies. Watson also assigned them territories that other representatives of the same company could not encroach upon and assigned them quotas which, when fulfilled, gave these salesmen recognition and status. At IBM salesmen who made their annual sales quotas became members of the 100 Percent Club. Other executives at NCR taught Watson that making salesmen adhere to a conservative dress code and forbidding them to drink alcoholic beverages were sound business practices. He was intelligent and willing to learn and sufficiently ambitious to apply his lessons effectively. These years at NCR molded a disciplined salesman into a creative manager who subscribed to the belief that salesmen were the most important community within a company. He learned the wisdom of what was then a revolutionary and enlightened notion, that people were his greatest asset. In those days and later at IBM, he argued that manufacturing divisions and laboratories had the responsibility of developing products for his salesmen to market. He learned that a tactician's mind was required to turn a sale. Continuous quality service to customers was also an important element in a successful marketing operation. While many of these ideas seem obvious today, they were not at the turn of the century and thus set executives from NCR apart from their peers in most American and European companies.

Even at NCR, a well-run company with outstanding management, Watson stood out. Within three years after he became a salesman, he had become one of the company's best recognized peddlers, selling NCR's cash registers in upper New York State. In 1899, he left Buffalo for Rochester to become branch manager of the local NCR office. In the rough and tumble world of turn-of-the-century cash register sales—and it was a very competitive business—he faced considerable competition in Rochester. But he prospered, and his ability to deal roughly with competition impressed NCR's upper management. In 1903, he was asked to head up an effort to put second-hand register sales firms out of business. To do so he established a company, funded quietly by NCR, to sell second-hand registers. Watson's Cash Register and Second Hand Exchange, located in New York City, did so well that he established other outlets in Philadelphia and Chicago. In 1907, the undercover operation went public when NCR announced that Watson would manage the company's second-hand market. He was now stationed at corporate headquarters in Dayton, Ohio. By 1910, Watson had become sales manager for the company.

That same year Watson's good fortunes changed. The American Cash Register Company, an arch-rival of his firm, sued NCR for violating the Sherman Antitrust Act and in 1912, the government joined in the suit with American Cash. Coming at the peak of antitrust litigation in the United States, the suit against NCR led to a conviction on February 13, 1913, of the company and a number of defendants, including Watson who was fined $5,000 and sentenced to one year in jail. While appealing the sentence, which he never served, in March a severe flood hit Dayton, causing considerable damage, leaving some 90,000 people homeless. NCR helped to revive the community, cleaned up the mess, and provided housing. These efforts helped to rehabilitate NCR's management, including Watson, in the public mind. In 1915, an appeals court found the original convictions faulty, but by this time Watson had left NCR. He had been fired in April 1914 following a quarrel over whether or not to sign a consent decree on behalf of the firm with the U.S. government concerning many of the practices that had led to the original litigation. He refused to sign because he felt he had never violated the law. His firing ended a successful, if controversial, career at one of the great American companies of the day. Watson was later to comment that he left NCR having vowed to form a company greater than "the Cash" as NCR was known in the industry.

This extremely well-trained NCR executive was soon offered a job with a company that had recently been formed (1911) out of a number of small firms called the Computing-Tabulating-Recording Corporation, or simply C-T-R. The young firm claimed assets of $17.5 million and was a conglomerate of card record equipment, meat scale manufacturing, and time-recording devices. Watson found this company ideal for his purposes, for he would be given a free hand to operate it as he sought fit, even though its business condition was weak. He joined the firm as general manager while the cloud of conviction still hung over him and was given an annual salary of $25,000 and stock, not an inconsiderable deal in 1914. Only a decade later this firm would become IBM.

Watson began work by focusing on cards and card-tabulating equipment. He imposed many of the management techniques common at NCR on all portions of C-T-R, including the One Hundred Percent Club, education for salesmen, and a more modern product line, and he instilled a philosophy that later emerged at IBM as "Respect for the Individual." Salesmen were elevated to an elite class within the firm, but they were also expected to succeed. And the company did well. Sales went from $4.2 million in 1914 to $8.3 million in 1917. Card production went up, new sorters and tabulators were brought out, and the company's arch-rival—the Powers Tabulating Company†—lost sales. In 1924, with all possible rivals within C-T-R vanquished, Watson was now chief executive officer. In February of that year, with a strong vision of things to come, Watson changed the name of C-T-R to International Business Machine, or simply IBM. A great American venture had been born.

The 1920s was a period of prosperity for the office equipment market. At IBM Watson continued to introduce new card punch machines while building a strong marketing organization. The new name for the company presaged his image of a major firm. By the end of 1929, one out of every five card punch devices in the U.S. market were IBM's, and by the middle of the 1930s, NCR had seen better days. By the end of 1938, IBM's revenues had grown to $34.7 million, and the company was a giant in the information processing industry.

The years before World War II represented a formative period in the company's life during which Watson nurtured a culture that bore the imprint of his management style and personal philosophy. Watson visited branch offices and plants, and he talked about hard work, honesty, and customer service. His homilies from this period eventually filled an 800-page book. He assumed a paternalistic management style toward workers in the plants and demanded absolute obedience from his managers. Watson extolled the virtues of his salesmen and created folk heroes at each of the annual One Hundred Percent Club conventions. Think signs appeared all over the company, along with flip chart stands. Conservatively dressed salesmen, always in white shirts and usually in blue suits, lived relatively uncontroversial lives in their communities while becoming a familiar sight in the offices of government agencies and America's companies. Watson also publicized the company at every opportunity. In 1939, at the New York World's Fair, he set up an exhibit on IBM and had the officials at the fair declare an IBM Day. The exhibit alone was far more extravagant than the size of the company would indicate. But his vision of a firm was always larger than reality.

The 1920s and 1930s witnessed the development and acceptance within the company of IBM's Basic Beliefs. They called for respect for the individual, best customer service always, and excellence from all. To the hard-working Watson, they were articles of faith and sound business practices. By World War II, probably no IBMer was unaware of these beliefs. Decades later in the 1970s when American businessmen asked why Japanese companies were doing so well, the Oriental businessmen pointed to IBM's precepts as time-proven methods for obtaining success. The three concepts were as strongly advocated at IBM in the 1980s as they had been in the 1930s.

Watson continued to broaden and strengthen the company throughout the Great

Depression. On the eve of World War II, he had achieved the image he wanted, and IBM was strong and successful. *The New York Times* labeled Watson "an industrial giant" and *Time* acknowledged that he was an "astute businessman." On the negative side, Watson was also called a despot who hated unions. Yet he was a booster, supported the government's economic programs, voted Democratic and liked President Franklin D. Roosevelt, and believed the New Deal was a square deal for the nation. In 1939, Watson was sixty-five years old, in good health, and was enthusiastically managing a corporation already reputed to have one of the best sales organizations in the world.

During the 1920s and 1930s, Watson became interested in art, supported such cultural organizations as symphonies and museums, and served on the board of Columbia University. His company contributed millions of dollars and a great deal of equipment to Columbia for the study of astronomy and card punch applications. In the early 1940s, showing some interest in computational devices, Watson backed research at Harvard University leading to the construction of Professor Howard H. Aiken's** Mark I* computer in 1944, along with additional research and development within IBM itself on electronic calculators. The unveiling of the computer at Harvard infuriated Watson who believed IBM did not receive as much credit for the project as he had thought appropriate. Most of the engineering work on the machine had been done at IBM's facility at Endicott, New York, under the direction of Professor Aiken and IBM engineering managers. Thereafter, Watson did not support any additional research at Harvard on computational devices.

In the years immediately following World War II, IBM leaders vigorously debated whether or not the company should enter the business of making computers. IBM had been built up on card punch equipment and related office products. Some computational products had been developed, such as electromechanical calculators, particularly as a result of joint projects with Columbia University. However, no work had been done on computers. Watson favored some tentative steps into the computer arena and authorized the construction of what would become IBM's first major product in that area: the Selective Sequence Electronic Calculator (SSEC).* It was easy to justify this small step since the SSEC would be a replacement product for electromechanical calculators which IBM had been selling for years and the demand for which was well understood. Historians have concluded that in the late 1940s Watson moved slower into the emerging computer field than he had into other areas of opportunity in the office equipment market. This slowness was uncharacteristic of Watson.

In his defense, some historians have pointed out that the market for computers was still undefined, his customers had yet to express strong interest in such devices, competition from other firms was not visible, and IBM's current product line was selling well. Some have also argued that Watson, already seventy-one in 1945, was no longer inclined to make changes as quickly or as aggressively as he had in the past. Yet he was still in good health, with all his mental faculties intact, and wanted the company to prosper and grow. Without doubt, however, he was obviously reluctant to make major changes. In the mid-1940s, he was heard to say that "the job is to protect a business after you built it up."

Some groups within the company supported IBM's involvement with computers. In that camp was Watson's son, Thomas Watson, Jr.,** the man who would inherit responsibility for the firm. Watson, Sr., began grooming his son in the late 1940s to take over the company. Both Watsons supported the company's continued research in electronic calculating machines. Meanwhile, Watson, Jr., noticed that other firms had begun to build computers, the most important of which was the UNIVAC* brought out by Remington Rand* in 1951. Even then, debate still raged in many companies regarding the size of the market for computers. Forecasting worldwide demand of fewer than twenty machines was not unusual. Such low figures made it difficult for some firms to invest heavily in a narrow market. The business case for jumping into the computer field thus had as much to do with IBM's slowness in moving in as any concerns Watson, Sr., might have had about change and new directions.

It was Watson, Jr., who, as a result of experience with calculators in IBM and the success of the UNIVAC, convinced his father to sponsor a full-blown effort to get into the computer market. In 1952, Watson, Jr., became president of IBM while his father remained chairman of the board. Watson, Sr., however, increasingly let his son make the basic decisions within IBM, offering advice and support when asked for it. Watson, Jr., committed IBM to the new computer industry. He was prepared to move away from the traditional office products arena which his father had done so well with and to dedicate the full resources of the company to a new era. Watson, Sr., backed his son, who showed enormous energy in pushing IBM into new product development and marketing in the early 1950s. The traditional green walls of IBM gave way to a new image of glass and steel as the decade marched on. Watson, Jr., could accurately take credit for moving IBM into the world of computers. Revenues continued to grow, supporting and encouraging changes. Between 1952 (with annual revenues of $333 million) and the end of 1955, sales doubled.

By May 1955, although still chairman of the board at least nominally, Watson, Jr.'s two sons now fully managed the firm. Tom Watson Jr., ran the company as a whole as president, while his brother, Dick Watson, ran World Trade—the overseas portion of IBM. In June 1956, Watson, Sr., had a heart attack and died on June 19. Obituaries across the nation repeatedly hailed him as a giant of American industry, recalled his singlemindedness of purpose in building up IBM and his simple views on business values, and acknowledged his penchant for moralism. He had built a great company out of pieces of small enterprises, taking an eclectic group of marketing and manufacturing concerns and welding them into a high-tech, fully integrated company with gross revenues in the year of his death in excess of $700 million. Watson, an intelligent, self-made man, whose severe countenance hardly changed from one decade to another, had slipped into history as the father of one of the great institutions in the history of data processing.

For further information, see: T. G. Belden and M. R. Belden, *The Lengthening Shadow: The Life of Thomas J. Watson* (Boston: Little, Brown and Co., 1962); W. Rodgers, *Think: A Biography of the Watsons and IBM* (New York: Stein and Day, 1969); Robert Sobel, *IBM. Colossus in Transition* (New York: Times Books, 1981); "Thomas John Watson (1874–1956)," *Think* 23, No. 7 (July, August, September 1956); T. J. Watson, *As a Man Thinks* (New York: IBM Corporation, 1936) and his *Human Relations* (New York: IBM Corporation, 1949), and the massive compendium cited in the entry, *Men-Minutes-Money* (New York: IBM Corporation, 1934).

WATSON, THOMAS JOHN, JR. (1914–). This American businessman succeeded his father, Thomas J. Watson, Sr.**, as president and chairman of the board of International Business Machines Corporation (IBM).† His most important contribution to the data processing industry came when he forced IBM into manufacturing computers. During his tenure, IBM produced many computer systems, the most famous of which were the S/360* and the S/370* introduced in the 1960s and 1970s, respectively. IBM became the largest vendor of data processing equipment and software* in the world during his years as chief executive officer, and he led the company through a period of growth that made it one of the largest in American industry.

Watson, Jr., was born on January 8, 1914, in Dayton, Ohio, while his father was the sales manager for the National Cash Register Company (NCR).† The younger Watson was the oldest of four children. One of his siblings, Arthur (better known as Dick), was five years his junior and also worked for IBM as an executive. His father intended that Tom would someday take over IBM and so throughout his childhood made the boy visit factories belonging to the company, and attend one Hundred Percent Club meetings, and exposed him to many facets of the firm. Unlike his father, Tom Watson, Jr., enjoyed sports, became an excellent skier and yachtsman, and later flew airplanes. He graduated from Brown University in 1937 and immediately joined the company as a sales trainee. Watson was next a salesman in New York until 1940 when he was called to active duty by the U.S. Army (he was already in the New York National Guard). Watson flew airplanes as an officer in the U.S. Army Air Corps throughout World War II. In 1945, he was mustered out as a lieutenant colonel, and a more serious young man returned to IBM in January 1946.

In the late 1940s his father groomed him for the presidency by exposing him to many of the key issues confronting top management at the firm. During this period Watson, Jr., came to the conclusion that IBM should get into the business of making and selling computers. Others in the firm, along with his father, had reservations, remembering that the company had been built on the card-tabulating business—an enterprise that was still doing well. Watson, Jr., however, believed that in time computers would represent a much larger market than unit record equipment. Therefore, he pushed hard to have IBM introduce electronic calculators in the late 1940s, playing a key role in the company's initial steps taken in the computer market.

In 1952, Tom Watson, Jr., was named president of IBM but his father retained the title of chairman of the board. By this time, the younger Watson was moving IBM into the full commitment to computers needed to make it a leading force in that market while taking over the day-to-day operations from his father. New products appeared: the 650* and the 700 series which placed IBM squarely in the computer arena while opening a new chapter in the history of the company.

During his early years as president, IBM's revenues continued to grow, doubling between 1952 and 1955. Line executives gained more power as increasing numbers of decisions were decentralized away from corporate headquarters in New York. The process was encouraged by Watson, Jr. The stiff collars worn by Watson, Sr., gave way to the soft collars worn by his son. Equipment in the mid-1950s was redesigned to look more modern as the black Queen Anne legs on tabulating equipment gave way to grey molded forms. New buildings were constructed and relied heavily on glass and steel, changing IBM's image from an old-fashioned brick and mortar manufacturing company into a sleek, contemporary one. In June 1956, Watson, Sr., died, leaving the reins of power completely in the hands of Watson, Jr. Dick Watson had been made head of IBM's overseas operations (called World Trade) in 1949 and continued in that capacity.

Watson remained president of IBM until 1961 when he was elected chairman of the board, a post he held until 1971. Between 1971 and 1979 he was chairman of the all-important executive committee, the most powerful policy group within the company. During the years he ruled IBM, the firm grew from annual sales of $333 million in 1952 to $23 billion in 1979. Another important event came in 1956 when Watson, Jr., agreed to sign a consent decree with the U.S. government restricting the company's practices in the card-tabulating equipment market, thereby ending the shadow of antitrust litigation. Watson agreed to the decree because he believed the real future lay with computers which were not being regulated by such an agreement. In the late 1950s, he pushed for new products in the computer arena, a marketplace that by the mid-1950s he had come to dominate. His last vacuum tube computer, the 709,* was marketed in the late 1950s on the eve of a technological revolution for the company and the data processing industry as a whole.

In 1961, Watson made decisions that were as dramatic and risky as any his father had ever made and that, in retrospect, were the most important of his life or in the history of IBM. In 1961 and 1962, he and his executives committed the company to build what would eventually be called the S/360 computers. These machines revolutionized the entire data processing industry and caused the kind of breakthrough that made both IBM and the industry explode with growth and change rarely seen in American history. The introduction of the

S/360 family of computers has been hailed as the most successful product introduction in the history of American business. At the center of the decision-making process was Watson, Jr.

In 1960, the company recognized that by 1964 its current line of computers would no longer be competitive, and therefore work had to begin now on the next generation. In 1961, three-quarters of all revenues at IBM were coming from the sale of computer equipment, making the concern over obsolete products a serious one. Watson created a high-level task force called the SPREAD* Committee to make recommendations. In December 1961, this group suggested a new line of computers be developed, all of which would be compatible with each other, along with a new set of peripherals, operating system,* and other software at price/performance ratios better than in the past. These managers also urged that these computer systems be based on such new technologies as transistors and integrated circuits, leaving behind the vacuum tube era. Watson and the other key decision-makers agreed with these recommendations, and, in 1962, IBM began a crash effort to create the new product line. Over $100 million a year went into the project, the company borrowed money, plants were totally renovated with more automation and new tools, people were retrained, and the company transformed. Watson and his aides recognized that failure to produce this new product line would eventually mean the demise of IBM since other competitors would take the technological and, hence, competitive lead.

On April 7, 1964, IBM announced the System 360 family. This was the largest single set of products announced by any vendor in the industry to date. The initial set of products included five computers, the OS (operating system), and a complete new line of peripherals: disk, tape, card input/output equipment, and the 1403 N1—the most popular printer of the 1960s and 1970s. The company spent the next five years expanding all of its plants even further, while bringing out additional members of the family to meet a market demand that far exceeded all of IBM's previous forecasts.

The data processing industry recognized at once that a quantum leap forward in technology and price/performance had taken place. IBM enjoyed the benefits of this recognition, doubling its revenues within the next five years. The company changed forever, evolving rapidly from a medium-size firm to one of the largest in the world. By 1968, IBM's S/360 computers accounted for over 15 percent of all installed processors in an industry that during the 1960s had grown at a rate of over 30 percent. Watson had done nothing less than totally reshape the data processing industry. Even today, historians frequently think of the history of the industry as either pre-or post-360. After the S/360, it was no longer the era of atomic energy: it was the information age. IBM had neatly reinforced what economists such as Kenneth Galbraith or social critics such as Alvin Toffler would variously label a postindustrial or service-oriented economy. What Henry Ford had done for automobiles and industrial economy, Tom Watson, Jr., had done for data processing and the American economy of the late twentieth century.

Success with the S/360 brought Tom Watson and IBM problems in the 1970s in the form of antitrust litigation. Throughout the 1970s, various competitors sued IBM, although all ultimately lost their cases. The most serious case involved an antitrust suit filed by the U.S. government on the last working day of the Johnson administration. The suit consumed much of top management's attention throughout the decade, hundreds of millions of dollars, and a large crop of corporate lawyers. In the end, during the first term of the Reagan administration, the government dropped the case and it became history.

By the mid-1970s, power at IBM was shifting as the large organization grew, and it moved away from Watson to various portions of the company. This process was in part encouraged by Watson and in part the result of the business becoming so large and complex. In 1973, Frank T. Cary** took over as chairman of the board and John R. Opel** was named president. In July 1979, President Jimmy Carter appointed Watson ambassador to the Soviet Union, a country Watson had flown in and out of during World War II while working with the Lend/Lease Program. By taking this diplomatic job, he had to resign from the board of directors at IBM. He also served as ambassador to the U.S.S.R. during the Reagan administration and subsequently retired from politics. Watson retained his concern for IBM and influence even in the mid-1980s, nearly a half century after joining the firm. But by the late 1970s, the Watson era was over, and the company was in other hands.

For further information, see: Franklin M. Fisher et al., *Folded, Spindled, and Mutilated: Economic Analysis and U.S. v. IBM* (Cambridge, Mass.: MIT Press, 1983) and his *IBM and the U.S. Data Processing Industry: An Economic History* (New York: Praeger Publishers, 1983); William Rogers, *Think: A Biography of the Watsons and IBM* (New York: Stein and Day, 1969); Robert Sobel, *IBM: Colossus in Transition* (New York: Times Books, 1981).

WEIZENBAUM, JOSEPH (1923–). This American scientist was one of the most important figures in the world of artificial intelligence (AI)* and in later life its greatest critic. He developed ELIZA, an early and important programming language* which incorporated characteristics of AI. His views also helped temper the enthusiasm of many, particularly outside the academic community, for the potential of AI. ELIZA was important in the 1960s, as Weizenbaum's critical and often hostile views on AI were in the 1970s.

Weizenbaum was born in Berlin, Germany, on January 8, 1923; during the 1930s his family immigrated to the United States, where he subsequently became a U.S. citizen. He graduated from Wayne University (later Wayne State University) in 1948 with a major in engineering and earned an M.S. from the same institution in 1950. He held a number of positions in the 1950s in the general area of systems engineering and computer science, but the most important of these were with General Electric (GE)† at its Computer Development Laboratory (1956–1963). During his years at GE, he became active in the new field of AI, an area of research that had come into existence at the same time he was at GE.

In 1963, Weizenbaum left GE to join the faculty of the Massachusetts Institute of Technology (MIT), long the home of much research on data processing and already a major center for the study of artificial intelligence. He began as a visiting associate professor of engineering, next was an associate professor (1964–1970), and thereafter full professor of computer science and engineering. During the 1970s, he also served as the editor of the *International Journal of Man-Machine Studies,* was a member of the advisory committee on automated personal data systems established by the Secretary of the U.S. Department of Health, Education and Welfare, beginning in 1972, and was a Fellow at the Center for Advanced Study of Behavioral Science (1972–1973). In the following academic year, he was the Vinton Hayes Senior Research Fellow at Harvard University.

Weizenbaum's claim to importance lay in his work in the field of AI between the late 1950s and the mid-1970s. While at GE, he had been assigned to a project that called for the integration of hardware and software* on behalf of the Bank of America. The project, called ERMA, was one of the first and largest efforts to automate banking functions with computers. As part of that effort, Weizenbaum had developed a new programming language* that had qualities later associated with AI. That project led him to become interested in the fundamental characteristics of language, a key issue in the field of AI.

When Weizenbaum joined the faculty at MIT, his university provided him with a terminal at home on which he soon programmed an application that could answer questions. These included, for example, "Is this April?" and "Is today Wednesday?" That success led him to explore the possibilities of mind-machine relations, probing the level of complexity one could achieve with such devices. He became more interested in patterns of words and sentences, leading to a set of programs called ELIZA, named after Miss Doolittle, the heroine in "My Fair Lady." ELIZA simulated conversation between a Rogerian psychoanalyst and his patient. The computer played the part of the analyst or, to quote Weizenbaum, the "caricature" of the analyst. He explored the idea that certain comments could elicit specific responses based on the context in which the original statements were made. For example, these might be likened to responses typically made at a party by someone in reaction to another person's statements, regardless of whether the original comment being reacted to was fully understood.

ELIZA allowed Weizenbaum and others to dabble in the psychoanalytical process of conversation. It simulated a patient-doctor conversation to the point where it appeared very real, as if human intelligence and sensitivities were functioning in the "mind" of the computer instead of being simulated responses generated in reaction to predefined (context) situations. In 1965, he published his first paper on the language. There soon emerged a group of scientists who thought the program was therapeutic since patients using it would expose their feelings more readily to a computer and this program than to a human psychoanalyst. Weizenbaum did not feel ELIZA merited that kind of endorsement. Hence, his experience with ELIZA led him to question the validity and value of some lines of research in the general field of AI by the end of the 1960s.

Nonetheless, ELIZA was an important contribution. It had natural English responses to natural English statements received. When it failed to comprehend a comment made by a human, it would even react with "I see," or "Go on," much like people did, giving additional credence to the software's "intelligence." In comparison to similar languages of the period, therefore, it appeared to be very sophisticated and obviously effective.

Weizenbaum's efforts to define the relationship of mind and machine led to his publication of several items concerning the value of AI. The most important of these was his book of early 1976, *Computer Power and Human Reason*. It instantly became one of the most important publications in the field of AI; it was also an attack on the subject, one that received considerable attention. The fact that it was authored by an important scientist in the field (because he developed ELIZA) and he was from MIT (a bastion of much research in AI) brought the publication much attention. Equally significant, he was one of the first scientists to raise moral questions concerning the role of artificial intelligence and thus of computers.

The main theme of the book centered on the concept that computers should be blocked from three specific domains, regardless of whether or not they were technically capable of doing specific tasks: where computers might attack life itself; where effects were irreversible and side effects not fully appreciated; and where a computer substitutes for a human function that requires respect for individuals, understanding, and love. Weizenbaum also argued that much of the work being done in AI was technique rather than science. He stated that much of the effort in AI was also being done by programmers, not scientists. He was uncomfortable with the notion that a machine could serve as a metaphor describing people. He feared that that form of description would discount such human traits as love, dignity, and trust, surrendering them to artificial forms (such as machines) that might turn on humans (worst case) or, at a minimum, did not deserve such attributes.

The intellectual atrophy that allowed people to rely primarily on science to explain ideas worried him since that approach would diminish the role of the human spirit. He did not subscribe to the idea that computers could do everything, at least potentially. Weizenbaum thought there were limits. He criticized his colleagues for overstating the case for AI, and in turn, they attacked him, as well as his ideas. Some, for example, thought he could no longer do scientific research and hence had felt compelled to write the book. Nonetheless, this publication caused much debate in the world of data processing. He had already become the *enfant terrible* of his own profession by the early 1970s, so the book was not news to colleagues. Outside academia, it made him a minor celebrity and an effective antidote to the enthusiasm that had existed for AI. As it turned out, the 1970s witnessed less progress in the general field of AI than the 1950s had. It was only in the 1980s that progress comparable to what had come in AI's earliest years again became evident in the form of specific "expert systems."

For further information, see: Pamela McCorduck, *Machines Who Think* (New York: W. H. Freeman, 1979) and her *Universal Machine: Confessions of a Technological Optimist* (New York: McGraw-Hill Book Co., 1985); Joseph Weizenbaum, *Computer Powers and Human Reason: From Judgment to Calculation* (New York: W. H. Freeman, 1976) and his "ELIZA—A Computer Program For the Study of Natural Language Communication Between Man and Machine," *Communications of the Association for Computing Machinery* 9, No. 1 (January 1965): 36–45.

WHEELER, DAVID JOHN (1927–). Wheeler developed computers at Cambridge University during the 1950s.

Wheeler was born on February 9, 1927, and graduated from Trinity College before beginning work on a Ph.D. under Maurice V. Wilkes** at the University Mathematical Laboratory in the late 1940s. In 1951, he completed his graduate work with a thesis entitled "Automatic Computing with EDSAC".* Between 1951 and 1958, he served as a research fellow at Trinity College. Between 1951 and 1953, he was also an assistant professor at the University of Illinois, years when the university was developing its own program of computer-related studies.

Between 1956 and 1966, Wheeler was assistant director of research at the Mathematical Laboratory, home of EDSAC. From 1966 to 1977, he held the position of Reader in Computer Science. From 1977 on, he held a chair in computer science. During the past three decades he has done research on digital computers* and played a key role in building one of the earliest British stored-program computers. In 1951, along with Wilkes and Stanley Gill,** he wrote a book called *The Preparation of Programs for an Electronic Digital Computer, with Special Reference to the EDSAC and the Use of a Library of Subroutines,* which was the first textbook published on programming.

Wheeler has taught at the University of California at Berkeley and at the University of Sydney in Australia. He has advised Bell Laboratories† on computing projects and in 1981 was elected to the prestigious Royal Society.

For further information, see: the introduction by Martin Campbell-Kelly to the reprint of Wheeler's book, *The Preparation of Programs for an Electronic Digital Computer* (Los Angeles: Tomash Publishers, 1982, originally Cambridge, Mass.: Addison-Wesley Press, 1951).

WIBERG, MARTIN (1826–1905). This Swedish inventor built an enhanced version of the difference engine* constructed by Pehr G. Scheutz** of Stockholm. Scheutz had developed a machine that could perform mathematical functions, generate astronomical tables, and print them in the mid-1800s. In 1874, Wiberg completed construction of a more advanced device for the same purpose. Wiberg's machine received considerable attention when completed and was the subject of a detailed study by the Académie des Sciences de Paris. Much of the work on difference engines was intended to produce mathematical tables to help in the study of astronomy. By the late nineteenth century, scientists had to rely heavily on mathematics, creating a new branch of science called numerical astronomy. The work of Charles Babbage,** and then Scheutz and Wiberg, supported this application, work that continues today.

For further information, see: R. C. Archibald, ''Martin Wiberg, His Tables and His Difference Engine,'' *Mathematical Tables and Other Aids to Computation* 2 (1947): 371–373 and his *Mathematical Table Makers* (New York: Yeshiva University, 1948); M. Wiberg, *Tables de Logarithmes Calculées et Imprimées au moyen de la Machine à Calculer* (Stockholm: n.p., 1876).

WIENER, NORBERT (1894–1964). Wiener developed the term *cybernetics* and described the concept more emphatically than anyone else in the early stages of modern computer science. This brilliant mathematician stimulated important research on automata both in the physical sciences and in biology and medicine. He is considered to be one of the great mathematicians of the twentieth century.

Wiener was born on November 24, 1894, in Columbia, Missouri. His father, a well-known Russian-Jewish linguist who immigrated to the United States, stressed education to his son. Wiener, a brilliant prodigy, began to read at the age of three, entered Tufts University at the age of eleven, and earned his A.B. degree in 1909. His father taught languages at Harvard and perhaps for that reason Wiener pursued his graduate studies there, completing his Ph.D. in 1913 just before his nineteenth birthday. His doctoral dissertation was on mathematical logic—then an emerging field within mathematics. Harvard University awarded Wiener the Sheldon Fellowship (1913–1915) which allowed him to continue his studies in mathematics. During this two-year period, he worked with such great mathematicians as Alred North Whitehead, Bertrand Russell, G. H. Hardy, David Hilbert, Edmund Landau, and J. E. Littlewood in Europe. During World War I, he worked at the Aberdeen Proving Ground designing range tables for artillery. In 1919, he joined the faculty at the Massachusetts Institute of Technology (MIT) as an instructor and remained on the faculty until his death in 1964.

The decades between World War I and World War II were years of teaching and research in mathematics for Wiener. In 1926 he studied in Copenhagen and Gottingen as a Guggenheim Fellow. He also served as a visiting professor at Cambridge University in 1931–1932 and at the University of Peping, China, in 1935–1936. His research during these years centered on feedback theories and on servomechanisms. Bertrand Russell wanted him to study the theory of electrons and matter in general, while Harold L. Hazen** at MIT encouraged him to focus on physics as well. Wiener also became interested in neurophysiological research through a friend, Arturo Rosenblueth, and then engaged in research on this subject. The exposure to mathematics, physics (particularly electronics), and medicine would later shape his ideas concerning cybernetics.

World War II drew the attention of many faculty members at MIT to war-related projects, some of which involved computers. Wiener focused on antiaircraft defense systems from the perspective of a mathematician. This work proved essential to his contributions to computing because he had to work out the calculations necessary for artillery on the ground to be aimed correctly at an airplane in motion, taking into account speed, impact of wind, directions of wind and plane, and the velocity and weight of the shells being fired.

Following World War II, Wiener brought together his various experiences with mathematics, physics, medicine, and ballistics in his most important work on computing: *Cybernetics, or Control and Communication in the Animal and Machine* (1948). The title embodies the shortest and clearest definition of the term. The word itself comes from the Greek phrase for steersman. Wiener argued that the human nervous system and computers could be described in essentially the same way, or, put more elegantly, the study of cybernetics involves the mathematics of optimizing and the study of communications and control in physiological systems and machines. He envisioned such work as being interdisciplinary, covering activity in the study of neural networks, learning theories, computers, theories of communication, servomechanisms, and automatic control systems, both animal and machine-like. This book and his subsequent work in the field established the need for philosophical relations between more traditional mathematics and mechanistic theories. He described some of those relationships so effectively that his work stimulated research in the scientific community on how the brain works and consequently on artificial intelligence.* Taken one step further, his cybernetics has influenced the nature of the computer's thought processes since the 1950s. The very word *cybernetics* in the Romance languages is another term for computer science. The concept of artificial intelligence which, by the 1970s, was a major issue in the design of future systems in computer systems started with Wiener. This also helps explain why MIT has focused on artificial intelligence so strongly for more than two decades.

In addition to this highly sophisticated and technical contribution to the study of the simulation of thinking, Wiener became concerned about the impact of automata (particularly machines) on humankind and spent the rest of his career and life talking and writing about the subject. His second most important book dealt with the impact of his ideas: *The Human Use of Human Beings; Cybernetics and Society* (1950). In this book he called attention to the future impact of computers on humankind and warned of the dangers associated with the abuse of such technology.

Although Wiener is best remembered for cybernetics, his work in the field of mathematics is equally impressive. As early as 1933, his peers recognized his research: that year he was awarded the Bocher Prize of the American Mathematical Society. During the interwar period he had studied and written about the Fourier integral, Brownian motion, time series, even relativity and quantum theory and other specialized aspects of mathematics that would later prove important fields of study: postulational theory, vector and differential spaces, and potential theory, including mathematical foundations. This brilliant mathematician died on March 18, 1964, while in Stockholm.

For further information, see: Norbert Wiener, *Cybernetics, or Control and Communication in the Animal and Machine* (Cambridge, Mass.: MIT Press, 1948), *The Human Use of Human Beings; Cybernetics and Society* (Boston: Houghton Mifflin, 1950), his memoirs: *Ex-Prodigy* (Cambridge, Mass.: MIT Press, 1953) and *I Am a Mathematician* (Cambridge, Mass.: MIT Press, 1956), *God and Golem, Inc.* (Cambridge, Mass.: MIT Press, 1966).

WILKES, MAURICE VINCENT (1913–). Wilkes built the EDSAC,* one of Great Britain's first stored-program computers of the 1940s. He also trained a generation of computer scientists in England.

Wilkes was born in 1913 and did his university training at Cambridge, studying mathematics and physics. Like many physicists during World War II, he worked on war-related projects and made important contributions to the development of radar. In 1945, he took charge of the Mathematical Laboratory at Cambridge University. Later renamed the Computer Laboratory, it was there that the EDSAC was constructed following a class he attended at the University of Pennsylvania during the summer of 1946 where he was exposed to the ENIAC* and the EDVAC.* Both of these American computers provided him with design examples which he used in constructing the EDSAC. He completed initial construction of the machine in May 1949, although his staff continued to refine and improve its structure during the early 1950s.

Wilkes quickly became a well-known specialist on computers, publishing articles and books on the subject. In 1956, he became a fellow of the Royal Society and then first president of the British Computer Society, serving in that capacity until 1960. He represented his country in various computer organizations during the 1960s, including the American Federation of Information Processing Societies (AFIPS).†

Wilkes' scientific contributions lay primarily in the construction of EDSAC. As a consequence of that project, his design team developed a library of subroutines used by the machine. These were frequently used commands which users employed in their programs rather than rewrite them each time a new program was created. This represented a major productivity boost for users of computers. Today the use of subroutines is normal, but Wilkes' use of them was a first. The use of macros in computer languages emerged from his laboratory. In the 1940s, however, he used the term *synthetic orders* to describe these functions. A second innovation was microprogramming, which became a design feature of EDSAC II in the mid-1950s. This allowed frequently used machine-level commands to be housed in the computer and represented a major refinement of his original idea of synthetic orders. These commands did not have to be loaded into the computer by a user each time a particular program was run; rather, these new instructions were resident in the machine all the time. That concept is now a reality in every computer. Third, he helped to develop a language called Wisp. Between 1965 and 1970, his laboratory worked on time-sharing systems. Professor Wilkes has published five books.

For further information, see: S. Gill, "Maurice V. Wilkes," in A. Ralston and C. L. Meek, eds., *Encyclopedia of Computer Science* (New York: Petrocelli/Charter, 1976): 1558–1559; M. V. Wilkes, *Automatic Digital Computers* (New York: John Wiley and Sons, 1956), "The Design of a Practical High-Speed Computing Machine," *Proceedings of the Royal Society,* Series A, No. 195 (1948): 274–279, "Early Computer Development at Cambridge: The EDSAC," *Radio Electrical Engineering* 45, No. 7 (1975): 332–335, *Memoirs of a Computer Pioneer* (Cambridge, Mass.: MIT Press, 1985).

WILLIAMS, FREDERIC CALLAND (1911–). Williams designed and built the Williams tube, one of the most important of the early memory* systems employed on early British and American computers. His most important work with memories came immediately after World War II with Tom Kilburn at the University of Manchester.

Williams, born and raised in Great Britain, worked at Bletchley Park† during World War II; there scientists, mathematicians, engineers, and champion chess players decoded the German ENIGMA* code, using computer-like devices. After the war, he left the Telecommunications Research Establishment (TRE) and joined the staff of the Manchester University. There, within the Electrical Engineering Department, he developed various designs for electrostatic memory devices. On June 2, 1948, he tested his first memory device. The machine constructed to exercise it was the first electronic device built to execute a stored program.

Williams designed a cathode ray tube (CRT) which could have impressed on its phosphate surface an electrical charge representing data. A series of these CRTs were then lashed to a computer. Each thus carried data, and the more memory one wanted the more CRTs were needed. Williams' original system handled thirty-two words of memory. The Williams tube allowed a user random access to data. Other memory devices of the late 1940s, including mercury delay tubes, still forced a user to access information sequentially. In addition, the tube was more reliable, with fewer errors and down-time. The machine used in 1948, called the Manchester Mark I* or MADM, along with the new memory, marked an important milestone in the early history of electronic digital computers* because it influenced the design of other British machines. The memory system he developed was also employed on the U.S. IAS Computer* and in WHIRLWIND* at the Massachusetts Institute of Technology. For his contributions he was knighted.

For further information, see: Simon Lavington, *Early British Computers* (Bedford, Mass.: Digital Press, 1980); Maurice Wilkes, *Memoirs of a Computer Pioneer* (Cambridge, Mass.: MIT Press, 1985); F. C. Williams, "Early Computers at Manchester

University,'' Radio Electrical Engineering, 45, No. 7 (1975): 327–331; Michael R. Williams, *A History of Computing Technology* (Englewood Cliffs, N.J.: Prentice-Hall, 1985).

WOOLDRIDGE, DEAN EVERETT (1913–). Wooldridge was a co-founder, with Simon Ramo,** of the Ramo-Wooldridge Company, which in time became TRW, Incorporated,† both major suppliers of computer-based defense systems in the United States during the 1950s and 1960s. Because of the U.S. Air Force's strong support for computer systems, and the managerial expertise of his company in implementing such technology, Wooldridge became a developer of computer applications in the early years of the data processing industry.

Wooldridge was born on May 30, 1913, in Chickasha, Oklahoma, where he attended public schools and graduated from high school at the age of fourteen. He studied physics at the University of Oklahoma and graduated in 1932. Wooldridge remained at the university to complete his master's degree in 1933. In 1936, he earned a Ph.D. in physics from the California Institute of Technology. He then went to work for Bell Laboratories† to conduct research on electronics. By the end of the decade, he was managing Bell's physical electronics research. During World War II he led the project to create the first airborne fire-control system. He also managed the study which in time led to the development of the Nike guided missile for the U.S. Army.

In 1946, Wooldridge became director of research and development at Hughes Aircraft Corporation in California. In 1950, the firm named him a director and, in 1952, vice-president of research and development. The following year he and Ramo resigned from Hughes Aircraft and established their own firm to market defense systems to the military; they quickly won three contracts from the U.S. Air Force. It appeared that Wooldridge more than Ramo was the driving force in the establishment of the Ramo-Wooldridge Company. Yet both had excellent credentials and reputations within the defense community. In 1954, they participated in the Air Force Strategic Missiles Evaluation Commission which studied the possibility of constructing a missile to carry the hydrogen bomb. The commission recommended the construction of intercontinental ballistic missiles, known as ICBMs, with civilian management of the project. In August 1954, the U.S. Air Force established the Western Development Command, later known as the Ballistic Missile Division. Ramo-Wooldridge won the contract to manage the development of the ICBM for this organization. The firm later acquired similar contracts to manage the Atlas, the Titan I and II, and the Minuteman.

During the 1950s, the company diversified into various civilian markets in order to reduce its dependence on the U.S. Air Force, establishing divisions and subsidiaries to do the work. Pacific Semiconductors, for example, came into existence in 1954 to build components for computers. Within three years the R-W Corporation owned six major divisions and two laboratories. Thompson-Ramo-Wooldridge Products, Incorporated, which had been created in 1955, began working on process control systems for manufacturing plants and was closely followed by the establishment of Space Technologies Labs, Incorporated. Space Technologies contained the firm's entire ICBM effort and later worked on other projects for the U.S. Air Force involving rocket boosters and spacecraft.

Wooldridge served as president of the parent company during this period of enormous growth. He was instrumental in the reorganization of the company of 1957, which led to the complete merger with Thompson Products, the company that originally had capitalized the Ramo-Wooldridge Company. The new entity, called Thompson-Ramo-Wooldridge, Incorporated, later became known as TRW, Incorporated. Wooldridge served as president of the firm until 1961. The next year Wooldridge resigned and went back to the California Institute of Technology as a research associate. He had no further dealings with TRW.

For further information, see: TRW, *The Little Brown Hen That Could* (Cleveland, Ohio: TRW, undated [1983?]).

WOZNIAK, STEPHEN GARY (1950–). Wozniak was the technical brains behind the Apple Computer. He designed and built Apple I and II along with other microcomputers sold by Apple Computer, Inc.† His microcomputers differed from earlier machines in their ease of use by nontechnical people (including nursery school students) and relatively low price. His devices ushered in the personal computer as an important segment of the data processing industry in the late 1970s.

Wozniak was born on August 11, 1950, in San Jose, California. From an early age he expressed interest in electrical and engineering subjects, and, like many inventors of computers before him, became an avid ham radio operator as a child. While in high school in Sunnyvale, California, he designed and built microcomputers. In the fall of 1968, he matriculated at the University of Colorado where he displayed less interest in classes than in furthering his knowledge of electronics. That year he met Steven P. Jobs,** the future founder of Apple Computer, Inc. Wozniak spent his second year at De Anza College and then transferred to the University of California at Berkeley where he could study electrical engineering. In 1971, he and Jobs entered into their first business venture: making "blue boxes" that allowed people to make free long-distance telephone calls. Before he could complete his studies at Berkeley, at the age of twenty-three, he went to work for Hewlett-Packard† where he worked on the design of new calculators. During the mid-1970s, he and Jobs built microcomputers for the Homebrew Computer Club† and other computer outlets, learning about that technology—the basis for their future Apple Computer Company.

Jobs encouraged Wozniak to resign from H-P to devote full attention to the design of a new micro which they named Apple. (Jobs had been working at an apple orchard at the time.) Jobs wanted to form a company to market the devices and Wozniak to design the machines. In the late 1970s, their firm now established, they designed and built a series of Apple II machines: IIe, III, and IIc, and in the early 1980s, Lisa and Macintosh. The technical genius behind each of these projects was Wozniak. Jobs and Wozniak were millionaires within several years. All of Wozniak's Apples were designed as small, lightweight, easy-to-use devices which made them marketable. Wozniak's machines ushered in a new wave of computer literacy in the United States. Hundreds of thousands were sold by the mid-1980s, creating a multibillion subset of the data processing market. As of 1986, "Woz" as he is frequently called, was still designing microcomputers for Apple.

For further information, see: Paul Freisberger and Michael Swaine, *Fire in the Valley: The Making of the Personal Computer* (Berkeley, Calif.: Osborne/McGraw-Hill, 1984); Doug Garr, *Woz: The Prodigal Son of Silicon Valley* (New York: Avon, 1984); Michael Moritz, *The Little Kingdom: The Private Story of Apple Computer* (New York: William Morrow, 1984).

Z

ZEMANEK, HEINZ (1920–). Zemanek is a prolific writer of programming languages* and information theory and is active in European data processing organizations. He also managed one of the most important data processing laboratories in Europe for International Business Machines Corporation (IBM).†

Zemanek was born in 1920 in Vienna, Austria, where he was raised. He studied communication theory at the Technical University of Vienna until 1944. After World War II he completed his graduate work at the Technical University of Stuttgart, and between 1947 and 1961, he taught at the Technical University of Vienna. Between 1948 and 1949, he lived in Paris, studying at the Sorbonne, the École Normale Supériore, and at the PTT Laboratories, all under the auspices of a French scholarship. In 1950, he finished writing his Ph.D. thesis on time multiplexing in telegraphy. In 1964, he was appointed professor at the Technical University of Vienna. Zemanek's research interests from 1952 to 1962 centered on cybernetics and artificial intelligence.* As early as 1954, even before the term *AI* was used, he co-chaired a group interested in cybernetics at the European Forum Alpbach. Between 1955 and 1959, he led a team that built a computer called "Mailüfterl."

Zemanek's first exposure to IBM came as a consultant to the company in 1960 and 1961. He joined the firm in 1961 as manager of the IBM Science Group in Vienna. Since 1964 he has served as the director of the IBM Laboratory in that city in various other management capacities. He has also played an active role in industrywide organizations. Between 1968 and 1971, he was International Federation for Information Processing's (IFIP)† vice-president and chairman of its Technical Committee 2 on Programming Languages. In 1971, he became IFIP's president, serving a three-year term.

Zemanek has published extensively on information theory, computer voice-output, the theory of programming languages, the Vienna method of defining methods of programming languages, and other topics for over 200 publications (articles and books).

For further information, see: Heinz Zemanek, "Die algorithmische Formelsprache ALGOL," *Elektronische Rechnanlagen* 1 (1959): 72–79, 140–143, "Formalization—History, Present and Future," in *Programming Methodology: Lecture Notes in Computer Science* 23 (New York: Springer-Verlag, 1975), pp. 477–501, "Semiotics and Programming Languages," *Communications of the ACM* 9, No. 3 (March 1966): 139–143.

ZUSE, KONRAD (1910–). Zuse built an automatic calculating machine in the late 1930s, a programmable device which may have been the first that worked. He went on to build a series of computers which he named Z during the 1940s and 1950s. They represent the most important German contribution to the early development of the modern computer.

Zuse was born in Berlin on June 22, 1910, and was raised in East Prussia. In 1927, he began his studies in engineering at the Technical College at Berlin-Charlottenburg. His concern in these years was engineering, not mathematics or computers, but, like many engineers in these years who went on to invent computers in the 1940s, Zuse was annoyed at the tedium associated with manually calculating solutions or with using slow and primitive electromechanical calculators, not to mention slide rules.* None of these devices ever allowed the user to calculate more than six equations of six unknown values. Yet much of his work and that of contemporaries required more sophisticated calculations. With a growing body of theoretical work suggesting that mathematics could be used to enhance knowledge of particular fields, including physics, engineering, weather analysis, and chemistry, the limitations were frustrating. His inability to do the kind of work asked of engineers created the incentive he needed to invent what would become an early computer.

While a student in Berlin, Zuse worked out a theoretical design for a calculator that was similar to that designed by Charles Babbage** in the 1820s and 1830s. Zuse was not aware of the earlier work until 1939, long after he had worked out many of the details of his own engine. He graduated from college in 1935 and found employment as a stress analyst at the Henschel Aircraft Company where, on weekends and on free evenings, he began to build a device to automate calculations. According to the story now famous in the annals of data processing history, he built his machine in his parents' living room. (A surviving photograph showing the minute size of that room only leaves one to wonder how anyone could even have moved around in it.)

Zuse's design for a computer called for the machine to be an automatic calculator capable of carrying digits from column to column (an unresolved problem in the mid-1930s in decimal-based calculators) and able to do multiplication. His first major decisions were to reject purely mechanical means and to rely, where possible, on electricity, and to employ a binary numbering system in which data could be represented as being either on or off, much as a switch was used to control the flow of electricity or, now, data. All characterizations of data would thus be either one or zero. Zuse's objective was to keep the machine as simple as possible to design and build. The use of binary systems was a profound change from the past. (Today all computer systems use binary mathematics). In theory, it was not a new idea; Gottfried Wilhelm von Leibniz** had discussed the issue as early as 1680. Zuse had learned on his own that a machine based on two elements instead of ten (necessary for a decimal system) would be easier to work with.

By the fall of 1936, Zuse had worked out enough details to begin building a machine. It included a memory* unit to house numbers that were being used in a calculation. Slots on a metal dish stored numbers, similar to Babbage's idea. Thus, the position of a pin in a dish's slot represented a single digit. One side of the dish represented one and the other zero; hence, a binary system existed. A second plate against the first, sensing where the pins were, could thus "read" the numbers, while a third (placed perpendicular to the first two) caused the reading and writing functions to take place. This first memory could house sixteen binary digits, which equaled between four and five decimal digits. Most exciting of all, the memory device worked as intended.

The next component of the system was the arithmetic unit; that is, that portion of the computer that had to conduct calculations. In a binary system, multiplication (e.g., $1 \times 1 = 1$) is simple. In this case other products would simply be zero. Thus, the complex problem of how to multiply in decimals, which had so stumped manufacturers of calculators, did not really exist in binary machines. What was more difficult was the expression of all calculations in binary form so that the machine could understand them.

Zuse had to work out a set of notations to describe the functions of the arithmetic unit in terms relevant to a binary device. For this he relied on mathematical notation and the symbolism of relays cast into a yes/no mode. Everything had to be designed so that answers were either yes or no, on or off—in effect, using Propositional Calculus to describe his transactions. In crude layman's terms, Propositional Calculus was logic in mathematical notation. Logic, as a branch of mathematics, was the subject of considerable study during the 1920s (especially by David Hilbert, W. Ackermann, Alfred North Whitehead, and Bertrand Russell). In the end, all of Zuse's problems resulted in statements consisting of *and, or,* and *not*. By Christmas of 1938, he had built an arithmetic unit to attach to his memory and it worked, somewhat. After he had built several more devices, Zuse named this first one the Z1 and subsequent machines Z2, Z3, and so on until he gave up inventing Z series machines in the 1950s. The naming of generations of machines or follow-on products by name (Z) and numbers (1, 2, 3, etc.) was a common practice during the 1950s.

Zuse decided to feed programs to his machine by way of used movie film which could be perforated and thereby "read." It was inexpensive. With the help of friends, especially of Helmut Schreyer who had more knowledge and experience than Zuse with electricity, he examined the possibility of using electrical components, such as vacuum tubes, instead of nearly all mechanical pieces, or plates. Zuse relied on telephone relays in his arithmetic and control units, fearing that tubes were still too unreliable. With the introduction of electrical components, the Z2 was born.

It appeared, however, that this particular machine was quickly surpassed by rapid modifications. In December 1941, Zuse completed an enhanced version of the machine, now called the Z3. It was a fully operational program-controlled computer. Historians considered it the first such device in the world. World War II now hampered additional work. The Z3 was destroyed by bombing in 1945 but remains an important machine in the history of computing. Documents, pictures, and comments provided by its inventor help historians appreciate its significance. It had two cabinets of relays, a keyboard display arrangement, programs were read into the system on film tape, and its memory functioned well. Relays provided the technology for both memory and the arithmetic unit. They had on/off switches which had to be timed properly to coordinate correctly the flow of data back and forth. Statistically, the machine was small. It had 2,600 relays, of which 1,800 were dedicated to memory, another 600 to the arithmetic unit, and the remaining 200 to input/output device control. The system never performed normal day-to-day work, remaining instead an experimental machine whose potential was frustrated by the circumstances of war more than by the inclinations of its inventor. Furthermore, the Z3 never had as great an impact on the development of the modern computer as did American machines (e.g., the ENIAC*) because so few people knew of the Z3's existence. Nonetheless, it was programmable, had floating-point features, used a binary form, and it worked.

During World War II, Zuse managed to construct a Z4 that relied more extensively on relays than the Z3. He began work on it in 1942 when the Z3 proved his design would work well. Differences were few but important. Because Zuse wanted more capacity, he expanded the word length from twenty-two to thirty-two binary digits. The Z4's memory was larger, too, which he wanted to be 1,024 numbers but only managed to expand to 512, which was still eight times that of the Z3. Because of Allied bombings, Zuse had to keep the machinery moving from one location to another, which slowed the work. His equipment eventually ended up in the Bavarian Alps, and in 1950 the Z4 began to perform routine work for the Federal Technical Institute at Zurich. At that point his ideas drew attention, inspired considerable work in Europe, and brought appreciation to his prewar findings.

As a result of his success with this machine Zuse established a company to make and market computers. The Z5 was one of his company's earliest products. Various follow-on models were constructed. The last Z series built was in the mid-1950s. These relay machines were not the only devices he made, however; during the 1950s he also manufactured electronic computers. By the early 1960s, Zuse had difficulty raising sufficient capital to support research and development. The net result was that his concern was taken over by the Siemens Company, a major European computing firm throughout the 1960s and 1970s and, during the 1980s, an important vendor of European computer equipment.

After Siemens took possession of his company, Zuse returned to his theoretical research on computing while writing on his earlier work. His experience was proof that most inventions are evolutionary and that they are invariably developed concurrently by various people. While he worked alone, others at the Massachusetts Institute of Technology, Oxford University, Manchester University, the University of Pennsylvania, and even in Australia were also designing and building machines.

For further information, see: Paul E. Ceruzzi, *Reckoners: The Prehistory of the Digital Computer, From Relays to the Stored Program Concept, 1935–1945* (Westport, Conn.: Greenwood Press, 1983); H. Zemanek, "Konrad Zuse," in A. Ralston and C. L. Meek, eds., *Encyclopedia of Computer Science* (New York: Petrocelli/Charter, 1976): 1463–1464; Konrad Zuse, "German Computer Activities," in *Computers and Their Future* (Llandudno: Richard Williams and Partners, 1970): 6/3–6/17, "Installation of the German Computer Z4 in Zurich 1950," *Annals of the History of Computing* 2, No. 3 (July 1980): 239–241, "Some Thoughts on the History of Computing in Germany," in N. Metropolis et al., eds., *A History of Computing in the Twentieth Century* (New York: Academic Press, 1980), pp. 611–627, "The Working Program-controlled Computer of 1941," *Honeywell Computer Journal* 6, No. 2 (1972): 49–58.

Appendix A: Individuals Listed by Date of Birth

The persons listed below all are subjects of entries in this dictionary.

Ramon Lull (1235–1315)
John Napier (1550–1617)
Henry Briggs (1561–1630)
Wilhelm Schickard (1592–1635)
René Grillet (1600s)
Gaspard Schott (1608–1666)
Blaise Pascal (1623–1666)
Samuel Morland (1625–1695)
Gottfried Wilhelm von Leibniz (1646–1716)
Pierre Jacquet-Droz (1700s)
Wolfgang von Kempelen (1734–1804)
Joseph-Marie Jacquard (1752–1834)
Baron Jean-Baptiste-Joseph Fourier (1768–1830)
Pehr Georg Scheutz (1785–1873)
Charles Babbage (1791–1871)
George Boole (1815–1864)
Countess of Lovelace, Augusta Ada (1816–1852)
Martin Wiberg (1826–1905)
William Stanley Jevons (1835–1882)
Henry Adams (1838–1918)
Ramón Verea (1838–1899)
John Shaw Billings (1839–1913)
John Henry Patterson (1844–1922)
John K. Gore (1845–1910)

Willgodt Theophil Odhner (1845–1905)
Joseph Boyer (1848–1930)
George Barnard Grant (1849–1917)
Charles Ranlett Flint (1850–1934)
Ernst Georg Fischer (1852–1935)
Leonardo Torres y Quevedo (1852–1936)
Allan Marquand (1853–1924)
George Winthrop Fairchild (1854–1924)
William Seward Burroughs (1855–1898)
Lyman Frank Baum (1856–1919)
Alfred Blake Dick (1856–1934)
Carl George Lange Barth (1860–1939)
Herman Hollerith (1860–1929)
Vilhelm Bjerknes (1862–1951)
Maurice d'Ocagne (1862–1938)
Dorr Eugene Felt (1862–1930)
Annibale Pastore (1868–1936)
Léon Bollée (1870–1913)
Edward Andrew Deeds (1874–1960)
Thomas John Watson (1874–1956)
James Wares Bryce (1880–1949)
Percy E. Ludgate (1883–1922)
Clark Hull (1884–1952)
James Henry Rand (1886–1968)
Theodore Henry Brown (1888–1973)
Clair D. Lake (1888–1958)
Vannevar Bush (1890–1974)
William Frederick Friedman (1891–1969)
Leslie John Comrie (1893–1950)
Alfred Blake Dick, Jr. (1894–1954)
Norbert Wiener (1894–1964)
Leslie Richard Groves (1896–1970)
Douglas Rayner Hartree (1897–1958)
Ernest Galen Andrews (1898–1980)
Boris Artybasheff (1899–1965)
Howard Hathaway Aiken (1900–1973)
Donald Alexander Flanders (1900–1958)
Harold Locke Hazen (1901–1980)
Alfred Tarski (1901–1983)
Wallace John Eckert (1902–1971)
Mina Spiegel Rees (1902–)

John Vincent Atanasoff (1903–)
John von Neumann (1903–1957)
John Grist Brainerd (1904–)
Alston Scott Householder (1904–)
Derrick Henry Lehmer (1905–)
Grace Brewster Murray Hopper (1906–)
Reynold B. Johnson (1906–)
Gordon S. Brown (1907–)
John William Mauchly (1907–1980)
Antonin Svoboda (1907–1980)
James Franklin Forster (1908–1972)
John Hamilton Curtiss (1909–1977)
William Bradford Shockley (1910–)
Konrad Zuse (1910–)
Richard Goodman (1911–1966)
Cuthbert C. Hurd (1911–)
John Aleksander Rajchman (1911–)
Louis ''Moll'' Nicot Ridenour, Jr. (1911–1959)
Frederic Calland Williams (1911–)
Dave Packard (1912–)
Alan Mathison Turing (1912–1954)
Julian Bigelow (1913–)
Herman Heine Goldstine (1913–)
Simon Ramo (1913–)
Maurice Vincent Wilkes (1913–)
Dean Everett Wooldridge (1913–)
George Bernard Dantzig (1914–)
Walter W. Jacobs (1914–1982)
George Robert Stibitz (1914–)
Thomas John Watson, Jr. (1914–)
Arthur Walter Burks (1915–)
Börje Langefors (1915–)
Nicholas C. Metropolis (1915–)
Harry Douglas Huskey (1916–)
Claude Elwood Shannon (1916–)
Herbert Alexander Simon (1916–)
Christopher Strachey (1916–1975)
Jule Gregory Charney (1917–1981)
Dov Chevion (1917–1983)
Frank August Engel, Jr. (1917–)
Robert Mario Fano (1917–)

Ralph Ernest Meagher (1917–)
Andrew Donald Booth (1918–)
Jay Wright Forrester (1918–)
John R. Pasta (1918–1981)
John Presper Eckert, Jr. (1919–)
Niels Ivar Bech (1920–1975)
Robert William Bemer (1920–)
Frank Taylor Cary (1920–)
An Wang (1920–)
Willis Howard Ware (1920–)
Heinz Zemanek (1920–)
Isaac Levin Auerbach (1921–)
Robert Rivers Everett (1921–)
Gene Myron Amdahl (1922–)
Alan J. Perlis (1922–)
Saul Rosen (1922–)
J. Cliff Shaw (1922–)
John Weber Carr (1923–)
Victor Mikhaylovich Glushkov (1923–1982)
Jack St. Clair Kilby (1923–)
Herman Lukoff (1923–1979)
Joseph Weizenbaum (1923–)
John Backus (1924–)
Julien Green (1924–)
John R. Opel (1925–)
William Michael Blumenthal (1926–)
Fernando José Corbató (1926–)
John Diebold (1926–)
Stanley Gill (1926–1975)
Kristen Nygaard (1926–)
Kenneth Harry Olsen (1926–)
Richard Utman (1926–)
William Louis Van Den Poel (1926–)
Robert (Bob) Overton Evans (1927–)
Charles Katz (1927–)
John McCarthy (1927–)
Marvin Lee Minsky (1927–)
Allen Newell (1927–)
Robert Norton Noyce (1927–)
David John Wheeler (1927–)
Bernard Aaron Galler (1928–)

Peter Naur (1928–)
Jean E. Sammet (1928–)
Gordon E. Moore (1929–)
Emerson W. Pugh (1929–)
Frederick Phillips Brooks, Jr. (1931–)
Ole-Johan Dahl (1931–)
Andrei Petrovich Ershov (1931–)
Arie Van Wijngaarden (1933–)
Chester Gordon Bell (1934–)
Ralph E. Griswold (1934–)
Brian Randell (1936–)
Philip Don Estridge (1938–1985)
Donald Ervin Knuth (1938–)
Adam Osborne (1939–)
Gary A. Kildall (1942–)
Stephen Gary Wozniak (1950–)
William H. Gates (1955–)
Steven Paul Jobs (1955–)

Appendix B: Individuals Listed by Profession

This appendix only lists the names of persons whose biographies have been included in this volume. Additional biographies of a more limited nature appear within entries in the first volume (*Organizations*) of the three-volume *Historical Dictionary of Data Processing*.

The first section lists inventors and scientists who developed or contributed to the precursors of modern electronic data processing. In general regard to the post-1930 period, several categories have been devised for the convenience of the reader. Persons were assigned to particular categories on the basis of when and in which area they made their most important contributions. This is not to say, for example, that contributions made in hardware were not followed by achievements in another period and in, perhaps, business or government.

INVENTORS AND SCIENTISTS (PRE-1930)

Charles Babbage (1791–1871)
Carl George Lange Barth (1860–1939)
Vilhelm Bjerknes (1862–1951)
Léon Bollée (1870–1913)
George Boole (1815–1864)
Henry Briggs (1561–1630)
Theodore Henry Brown (1888–1973)
James Wares Bryce (1880–1949)
Dorr Eugene Felt (1862–1930)
Ernst Georg Fischer (1852–1935)
Baron Jean-Baptiste-Joseph Fourier (1768–1830)
John K. Gore (1845–1910)
George Barnard Grant (1849–1917)
René Grillet (1600s)
Herman Hollerith (1860–1929)
Clark Hull (1884–1952)

Joseph-Marie Jacquard (1752–1834)
Pierre Jacquet-Droz (1700s)
William Stanley Jevons (1835–1882)
Reynold B. Johnson (1906–)
Clair D. Lake (1888–1958)
Gottfried Wilhelm von Leibniz (1646–1716)
Countess of Lovelace, Augusta Ada (1816–1852)
Percy E. Ludgate (1883–1922)
Ramon Lull (1235–1315)
Allan Marquand (1853–1924)
Samuel Morland (1625–1695)
John Napier (1550–1617)
Willgodt Theophil Odhner (1845–1905)
Blaise Pascal (1623–1666)
Annibale Pastore (1868–1936)
Pehr Georg Scheutz (1785–1873)
Wilhelm Schickard (1592–1635)
Gaspard Schott (1608–1666)
Leonardo Torres y Quevedo (1852–1936)
Ramón Verea (1838–1899)
Wolfgang von Kempelen (1734–1804)
Martin Wiberg (1826–1905)

HARDWARE SCIENTISTS AND ENGINEERS (POST-1930)

Howard Hathaway Aiken (1900–1973)
Gene Myron Amdahl (1922–)
Ernest Galen Andrews (1898–1980)
John Vincent Atanasoff (1903–)
Niels Ivar Bech (1920–1975)
Chester Gordon Bell (1934–)
Julian Bigelow (1913–)
Andrew Donald Booth (1918–)
John Grist Brainerd (1904–)
Frederick Phillips Brooks, Jr. (1931–)
Gordon S. Brown (1907–)
Arthur Walter Burks (1915–)
Vannevar Bush (1890–1974)
John Presper Eckert, Jr. (1919–)
Robert (Bob) Overton Evans (1927–)
Robert Rivers Everett (1921–)
Jay Wright Forrester (1918–)
Stanley Gill (1926–1975)
Victor Mikhaylovich Glushkov (1923–1982)
Herman Heine Goldstine (1913–)
Douglas Rayner Hartree (1897–1958)
Harold Locke Hazen (1901–1980)
Cuthbert C. Hurd (1911–)

Harry Douglas Huskey (1916–)
Charles Katz (1927–)
Jack St. Clair Kilby (1923–)
Börje Langefors (1915–)
Herman Lukoff (1923–1979)
John William Mauchly (1907–1980)
Ralph Ernest Meagher (1917–)
Nicholas C. Metropolis (1915–)
Gordon E. Moore (1929–)
Peter Naur (1928–)
Robert Norton Noyce (1927–)
Emerson W. Pugh (1929–)
John Aleksander Rajchman (1911–)
Louis ''Moll'' Nicot Ridenour, Jr. (1911–1959)
William Bradford Shockley (1910–)
George Robert Stibitz (1914–)
Antonin Svoboda (1907–1980)
Alfred Tarski (1901–1983)
Alan Mathison Turing (1912–1954)
William Louis Van Der Poel (1926–)
John von Neumann (1903–1957)
Willis Howard Ware (1920–)
David John Wheeler (1927–)
Maurice Vincent Wilkes (1913–)
Frederic Calland Williams (1911–)
Konrad Zuse (1910–)

SOFTWARE SCIENTISTS

John Backus (1924–)
Robert William Bemer (1920–)
John Weber Carr (1923–)
Fernando José Corbató (1926–)
Ole-Johan Dahl (1931–)
George Bernard Dantzig (1914–)
Andrei Petrovich Ershov (1931–)
Robert Mario Fano (1917–)
Bernard Aaron Galler (1928–)
Julien Green (1924–)
Ralph E. Griswold (1934–)
Grace Brewster Murray Hopper (1906–)
Donald Ervin Knuth (1938–)
Allen Newell (1927–)
Kristen Nygaard (1926–)
Alan J. Perlis (1922–)
Saul Rosen (1922–)
Jean E. Sammet (1928–)
Christopher Strachey (1916–1975)

Arie Van Wijngaarden (1933–)
Heinz Zemanek (1920–)

BUSINESS LEADERS

Isaac Levin Auerbach (1921–)
William Michael Blumenthal (1926–)
Joseph Boyer (1848–1930)
William Seward Burroughs (1855–1898)
Frank Taylor Cary (1920–)
Edward Andrew Deeds (1874–1960)
Alfred Blake Dick (1856–1934)
Alfred Blake Dick, Jr. (1894–1954)
John Diebold (1926–)
George Winthrop Fairchild (1854–1924)
Charles Ranlett Flint (1850–1934)
James Franklin Forster (1908–1972)
Leslie Richard Groves (1896–1970)
Gary A. Kildall (1942–)
Kenneth Harry Olsen (1926–)
John R. Opel (1925–)
Dave Packard (1912–)
John Henry Patterson (1844–1922)
Simon Ramo (1913–)
James Henry Rand (1886–1968)
An Wang (1920–)
Thomas John Watson (1874–1956)
Thomas John Watson, Jr. (1914–)
Dean Everett Wooldridge (1913–)

SCIENTISTS IN ARTIFICIAL INTELLIGENCE

John McCarthy (1927–)
Marvin Lee Minsky (1927–)
Claude Elwood Shannon (1916–)
J. Cliff Shaw (1922–)
Herbert Alexander Simon (1916–)
Joseph Weizenbaum (1923–)
Norbert Wiener (1894–1964)

USERS OF COMPUTERS

Jule Gregory Charney (1917–1981)
Leslie John Comrie (1893–1950)
Wallace John Eckert (1902–1971)
Frank August Engel, Jr. (1917–)
Walter W. Jacobs (1914–1982)
Derrick Henry Lehmer (1905–)
Richard Utman (1926–)

GOVERNMENT LEADERS AND SCIENTISTS

John Shaw Billings (1839–1913)
Dov Chevion (1917–1983)
John Hamilton Curtiss (1909–1977)
Donald Alexander Flanders (1900–1958)
William Frederick Friedman (1891–1969)
John R. Pasta (1918–1981)
Mina Spiegel Rees (1902–)

WRITERS AND PUBLICISTS

Henry Adams (1838–1918)
Boris Artybasheff (1899–1965)
Lyman Frank Baum (1856–1919)
Maurice d'Ocagne (1862–1938)
Richard Goodman (1911–1966)
Alston Scott Householder (1904–)
Brian Randell (1936–)

DEVELOPERS OF MICROCOMPUTERS AND THEIR SOFTWARE

Philip Don Estridge (1938–1985)
William H. Gates (1955–)
Steven Paul Jobs (1955–)
Adam Osborne (1939–)
Stephen Gary Wozniak (1950–)

Index

Boldface numbers indicate the location of the main entries in the text.

About the Author

JAMES W. CORTADA is Senior Marketing Programs Administrator for the IBM Corporation. He is the author of numerous books on the history and management of data processing, including *EDP Costs and Charges*, *Managing DP Hardware*, and *An Annotated Bibliography on the History of Data Processing* (Greenwood Press, 1983), as well as two companion volumes to the *Historical Dictionary of Data Processing: Biographies* which cover organizations and technology in the history of data processing. Dr. Cortada has also published numerous articles in a variety of journals.